FISHERIES & AQUACULTURE ECONOMICS MANAGEMENT

FISHERIES & AQUACULTURE ECONOMICS MANAGEMENT

Editor

Bojan Negrescu

scitus academics

Fisheries & Aquaculture Economics Management
Edited by **Bojan Negrescu**

This Book is Printed on Acid-Free Paper. ♾

ISBN: 9781681179537
Library of Congress Control Number: 2017949097

616, Corporate Way, Suite 2, 4766,
Valley Cottage, NY 10989

www.scitusacademics.com

Notice

Printed in United States of America.

Preface

The global fish production industry is highly varied with respect to species, production technology, geographical location and various other aspects. Fish farming is basically subject to the same economic laws as traditional farming, capture fisheries deal with certain quite specific and deep-rooted economic problems. The most fundamental of these problems are the inherent dynamic complexities of wild fish stocks and the traditional common property arrangement of most capture fisheries. The capture fisheries of Asia and the Pacific region are dominated by pelagic species - in many subregions it is small pelagics that are the dominant group. The proportion of low-value/trash fish is increasing in many areas, a trend that is meeting the growing demand for these fish for use as feeds for livestock and aquaculture. Freshwater fish dominate aquaculture production. As expected, there is considerable variation in general trends among the five main subregions of Asia and the Pacific region. Production from China, South Asia and Southeast Asia continues to grow. Capture fisheries production from Japan and Democratic People's Republic of Korea have shown a steady decline over many years, partly because of declining stocks, whereas aquaculture production has remained relatively stable. With rising global incomes, demand for fish products will almost certainly continue to grow. A corresponding increase in supply cannot come from capture fisheries, which are already close to the maximum sustainable production. It can only come from fish farming which is not subject to the same binding constraints. An integrated production and market analysis predicts more than doubling of the global fish production during the current century with virtually all the increase coming from fish farming. Global fish farming production is predicted to exceed that of capture fisheries in the 2020s.

Fisheries & Aquaculture Economics Management covers all aspects on the economics of fisheries and aquaculture. It provides the prospect on current state of the global fishery and aquaculture and points out certain key features of these production activities. It discusses key issues such as how to finance and plan new aquaculture business, how to monitor and evaluate economic performance, and how to manage capital, labor, and business risk, the book equips aquaculture professionals, students, researchers, or anyone interested in this subject.

Table of Contents

CHAPTER 1

Improving Food Security in an Eco-Friendly Manner through Integrated Aquaculture

Olawale Ahmed Onada[1], Oluniyi Solomon Ogunola[2]

[1]*University of Bremen, Bremen, Germany*

[2]*University of Ibadan, Ibadan, Nigeria*

ABSTRACT

As the global population increases, demand for food, most especially protein, will increase. The increasing production from agriculture, including forestry and fisheries, has, however, contributed significantly to global food security but most often at the expense of environmental deterioration. Integrated aquaculture (IA) can guarantee and sustain adequate food security in environmentally friendly manner. Integrated fish farming is a diversified and coordinated way of farming with fish as the main target along with other farm produces. The aim of integrated fish farming is to create a mutually beneficial system that will lead to maximization of productivity through optimum resource use. The major features of this system include: multiple production that leads to increased productivity from the combination of different farming system, recycling of waste or by-product in which the waste of one system becomes the input of another system thereby reducing the amount of waste generated and input resources, efficient utilization of farm space for multiple production, provision of steady source of income all year round to the farmer; this comes from various farm products. In any integrated system, the interrelationships are many; crop byproducts are fed to animals, while fish and animal manures are returned to the crops and fish in the ponds. The fish may feed on insects and weed in the rice field planted inside the pond and this in turn can increase the available nutrients to the crop. IA has ample capacity of making more food available thus enhancing food security. Aside from production enhancement, integrated fish farming also gives a platform for managing environmental integrity through waste recycling and utilization. It is important therefore to encourage farmers and other food production sectors to engage in a production system that will ensure food security in an eco-friendly manner.

Keywords: Fish Production, Agriculture-Aquaculture System, Eco-Friendly Aquaculture

Subject Areas: Aquaculture, Fisheries & Fish Science

1. INTRODUCTION

As the global population increases, demand for food, most especially protein, will increase. Production from agriculture, including forestry and fisheries, is crucial for food security in the face of current population increase. In order to secure global food security, it is imperative that more food needs to be produced so as to strike balance between increasing human population and food availability. In this regard, sustenance of increased productivity must emphasize on the development of strategies aimed at improving yield through rational utilization of resources and proper environmental management, such as strategies abound in integrated aquaculture (IA). Integration aquaculture with crop production and animal husbandry is one of the indubitable ways to increase the level of food production and simultaneously maintain equitable use of available land and human resources to meet human food demands in an environmentally sustainable manner. This is because integrated aquaculture productivity has been confirmed to relatively reduce cost of production [1] and thus the sustenance of increased food production must increase farmers' income, nutrition and also entrench the development of strategies aimed at overall welfare [2]. In order to encourage food security of human population, there is the need for farmers to engage in a result oriented farming system that can guarantee and sustain adequate food security in environmentally friendly manner. This review emphasizes the food security improvement contribution of integrated aquaculture in an environmentally friendly manner.

2. RATIONALE OF INTEGRATED AQUACULTURE (IA)

Integrated aquaculture is a method of farming which is generally believed to help the rural poor to utilize resources at their disposal towards maximizing production and making more profit from their investment. IA reduces waste generated from different sub-systems on the farm and also increases water use efficiency through the utilization of waste in one sub-system as resource or input in another sub-system within a system of production, recycling and re-use of water in one sub-system for irrigation in another sub-system. This will, however, reduce the negative impact of farming on the environment due to zero tolerance to waste discharge to the environment while at the same time improve economy of production for profit maximization of the farmers.

3. FORMS AND FEASIBILITY OF INTEGRATED AQUACULTURE

Integration of animal rearing, crop production with fish farming is the most complex of all forms of integrated aquaculture and consists of raising animal and plants on or near fish ponds. Each component of the system receives something and gives something to other components. The feasibility of this system has been earlier carried out in NIFFR (National Institute for Freshwater Fisheries Research) by [3] where he found that integrated combination of fish with poultry and crop is feasible with better performance as compared with single production of the individual systems. The research on the integration of animal and fish by [4] divides the system of production under integrated aquaculture into two:

- Semi-integration (raising livestock on land and transporting the manure to the fish pond, using waste water from pond to irrigate crop outside the perimeter of the pond).
- Complete or Vertical integration (raising livestock directly above the fish pond so that animal manure can drop directly into the fish pond, growing vegetable on the dyke or center of the pond where it can easily utilize the pond water).

Conventionally, IA combines animal husbandry and crop production either singly or in combination with fish production in a strategic way that can ensure mutual interrelationship. IA can come in the form of crop-fish-livestock integration, livestock-fish integration, crop-fish integration and crop-livestock-agro-processing integration.

4. BENEFITS OF INTEGRATED AQUACULTURE

The benefits of integrated Aquaculture system are enormous and encompass those of traditional farming system and modern farming system. The benefits of IA include the following.

4.1. Food Security

IA improves food security due to its various combinations of crop, livestock and fish. More food is thus produced in the single production from the various combinations.

4.2. Self Sufficiency

IA is suitable for low income farmers with remarkably low expenditure pattern and continuous low spending for food and other dietary requirement. The system is all encompassing as the time is well utilized in other farming activities depending on the type of integration involved. In fish cum crop production, crops like the vegetables are harvested continuously even in dry season as water from the pond is used to water the farms adequately.

4.3. Production Indices

Field studies show that variable and fixed inputs are well utilized in integrated aquaculture as compared with single system of production as presented in Table 1.

4.4. Regular Source of Income

IA provides platform for steady source of income all year round; this comes from various farm products. For example, in poultry, vegetable cum fish farming, before the harvesting of the fish, which may take some months, the farmer can sell the eggs and vegetables which will generate money for some time.

Table 1. Production indices.

	Integrated Fish Farming				Single Production System			
Fixed input	Pig	Poultry	Rice	Fish	Pig	Poultry	Rice	Fish
Land	R*	R*	R*	R*	R	R	R	R
Cage	-	R	-	-	-	R	-	-
Poultry house	-	R	-	-	-	R	-	-
Pig sty	R	-	-	-	R	-	-	-
Fish pond	-	-	-	R	-	-	-	R
Variable input								
Fish seed	-	-	-	R	-	-	-	R
Poultry feed	-	R*	-	-	-	R	-	-
Fish feed	-	-	-	R*	-	-	-	R
Pig feed	R*	-	-	-	R	-	-	-
Rice seed	-	-	R	-	-	-	R	-
Rice processing	-	-	R	-	-	-	R	-
Irrigation services	-	-	-	-	-	-	R	-
Organic manure	-	-	R	R	-	-	OP	OP
Inorganic fertilizer	-	-	-	-	-	-	R	OP
Labour	R*	R*	R*	R*	R	R	R	R
Health services	R	R	-	R	R	R	-	R

R* = required in a reduced quantity or cost; R = required; OP = optional.

4.5. Waste Recycling

The waste products of one biological system serve as input for other biological system. For example, in the inte- gration of fish and poultry, the droppings from poultry can be used to fertilize fish pond and also serves as maggot generation medium which serves as input to the fish production system, also, when the fish are harvested, the fish offal can be dried and used for poultry feed. The common feature of most integrated aquaculture systems is that livestock excreta are used as fertilizer for crop production; and livestock waste is also used to enhance the growth of various natural planktons in the ponds which may serve as fish feed.

5. WATER USE EFFICIENCY AND PRODUCTIVITY IN IA

The importance of water in agriculture production cannot be over-emphasized. In production from agriculture, including fish production, water is gotten majorly from precipitation, underground water and reservoirs including rivers and lakes. However, water supply from these sources is not entirely reliable. The growing world population requires increased food production, while available water resources may not be sufficient for agriculture. This alarming situation can only be resolved if water is managed more efficiently, so that yield of agriculture produce per unit of water consumption increases. To increase agriculture per water usage, a system that will ensure recycle and re-use of water will need to be entrenched. Such system is obtainable in IA. IA will assure high water productivity in agricultural production system. Water productivity is defined as the ratio of the net benefits from crop, forestry, fishery, livestock, and mixed agricultural systems to the amount of water required to produce those benefits. In its broadest sense it reflects the objectives of producing more food, income, livelihoods, and ecological benefits at less social and environmental cost per unit of water used, where water use means either water delivered to a use or depleted by a use. Put simply, it means growing more food or gaining more benefits with less water. In fish-crop integration, the waste water from aquaculture can be used to irrigate crop production thereby increasing the use to which the water is employed. In this case, water serves dual purpose of usage in fish production and crop irrigation, and also maintained increase productivity of agriculture produce.

6. ECOLOGICAL SIGNIFICANCE OF INTEGRATED FISH FARMING

Sustainable aquaculture depends upon eco-friendly culture system for its survival. One of the interesting features of IA is that it leads us to view systems under integration as synergistic components. IA are comprised of different ecosystems that can be described, modeled, analyzed and compared; they are guided by principles, and the parts that make them up and the way these parts are related [5].

In any integrated system, the interrelationships are many; crop byproducts are fed to animals, while fish and animal manures are returned to the crops and fish in the ponds. The fish may feed on insects and weeds in the rice field planted around the pond, and when the pond water is used to irrigate the crop, this in turn will increase the available nutrients to the crop. The ecological efficiency of an integrated fish farm is very paramount to the success of the entire farming enterprise. The synergistic and mutually beneficial approach of IA is what makes it different from several other systems of agriculture production.

7. IMPORTANCE OF NUTRIENT CYCLING IN IA

Nutrient recycling in IA provides the opportunity of managing organic waste to an extent that close to zero tolerance is given to waste discharge to the environment; also, the concept of turning waste to resources will bring about increase productivity and provide huge economic benefits. This is important to sustainable aquaculture, and reduces expenses on feed and fertilizer to a large extent. Outside the possibility of sustainable aquaculture production, environmental sustainability is also one of the importances of IA. It is however important to be able to determine and quantify nutrient flow in such a way that will strike balance between demand and supply of bio-resources in and out of a system. For effective cycling of nutrients in IA, a balance of bio-resources must be established within each system, strict ecosystem monitoring and management must be ensured, taking into cognizance the type and level of integration involved.

In time immemorial, poultry droppings have been used as fertilizer in fish pond to encourage the growth of phytoplankton and zooplanktons, and also to supply undigested feed in the dropping for consumption of fish, this poultry dropping also act as organic fertilizer to planted crop when applied directly. Recently, nutrient rich pond water has also been used for irrigation of crops most especially vegetables and rice.

8. ECOLOGICAL SUSTAINABILITY OF INTEGRATED AQUACULTURE

All modern, large scale food systems have discernible environmental and social impacts. Even the sustainability of modern, large scale, organic agriculture has been questioned. The sustainability of ecological parameters in any agro-ecosystem is very important in determining the success of the farming venture. The Broundtland Commission [6] defined sustainable development as the ability to meet the needs of the present without compromising the ability of the future generation to meet their own needs. Some forms of integrated aquaculture development have caused severe social and environmental impacts [7]. The degraded state of most aquatic ecosystems combined with public concerns about adding "new" sources of aquatic pollution to the already overburdened ecosystems will require aquaculture to develop new ecosystem approach and sustainable operating procedures [8]. Poorly managed integrated systems usually have high nutrient loading leading to deleterious effect of cyanobacterial bloom [9], reverse is the case if the system is well managed. In a well-managed system, there is relatively low to zero output of waste. However, ecologically sustainable aquaculture is the development of aquatic farming systems that preserve and enhance the forms and functions of the natural and social environments in which they are situated. It involves realistic ecological approaches that will develop aquaculture production techniques for various integration systems, by using local resources, effective recycling of wastes and materials that can degrade natural ecosystems and proper planning for job creation and marketing strategies that will be widely accepted as meeting both economic and ecological considerations [10].

9. COMPARISON OF PRODUCTIVITY ON INTEGRATED AQUACULTURE AND CONVENTIONAL SYSTEM

The productivity of integrated aquaculture as compared with conventional system of production of agricultural produce has been compared by [11] as presented in Table 2. The advantage of integrated aquaculture is revealed in the research which showed increased productivity.

10. CONSTRAINT OF INTEGRATED FISH FARMING

As it has been observed through several field practices, poly-culture of Tilapia and catfish integrated in a rice paddy area is inefficient in productivity, this is because tilapia has the capacity to destroy rice bed by digging the rice bed to prepare nest for reproduction, thereby reducing the space available for rice production. Synergistic characteristics of IA needs to be properly understood before combination, this is because through waste recycling, the waste from one system that should serve as input in the other can end up becoming a threat, for example, poultry dropping that should fertilize a pond can become a threat to the fish if over-fertilized. Link in different sub-systems needs to be adequately quantified so that resources can be allocated in the right quantity and proportion.

Table 2. Summary of comparative performance of fish, rice and poultry production of integrated system and conventional system in terms of growth, yield and egg production.

	Fish			Rice			Pig	
	CS	IS		CS	IS		CS	IS
IFW (g)	42	10	RS/60 m^2	1200	1000	IW	15.26	14.1
Av feed/fish	639	221	yield (kg)/60 m^2	12.67	23	FW	41.24	41.8
AWG	395	179	yield (ton/hec)	2.08	3.8	WG	26.0	27.7
FCR	1.7	1.25					4.2	3.6

IFW = Initial fish weight; RS = Rice stand; IW = Initial weight; AWG = Average weight gain; FCR = Feed conversion ratio; CS = Conventional system; IS = Integrated system; FW = Final weight; WG = Weight gain.

11. CONCLUSION AND RECOMMENDATION

IA has ample capacity of making more food available thus enhancing food security. Aside from production enhancement, integrated fish farming also gives a platform for managing environmental integrity through waste recycling and utilization. It is important therefore to encourage farmers and other food production sectors to engage in a production system that will ensure food security in an eco-friendly manner.

REFERENCES

1. Ugwumba, C.O.A. and Orji, E.C. (2006) Traditional Farming System and Its Effect on Farm Cash Income in Old Njikoka Local Government Area of Anambra State. Int. J. Agric. Res. Politics, 1, 21-24
2. Okoye, F.C. (2004) Integrated Rice-Fish Farming in Nigeria: Its Feasibility and Economic Viability. 18th Annual Conference of the Fisheries Society of Nigeria (FISON), Owerri, 8-12 December 2003, 168-172. http://aquati ccommons.org/872/1/FISON2003_168%2D172.pdf
3. Otubusin, S.O. (1986) Modern Aquaculture Practices for Increased Fish Production in Nigeria. 3rd Annual Conference of the Fisheries Society of Nigeria (FISON), Maiduguri, 22-25 February 1983, 89-104
4. Barash, H., Planik, I. and Moar, R. (1982) Integration of Duck and Fish Farming: Experimental Results. Agriculture, 27, 129-140
5. Altieri, M.A. (1987) Agro Ecology the Scientific Basis of Alternative Agriculture. Intermediate Publications, London, 45-48
6. WCED (World Commission on Environmental and Development) (1987) Environment Our Common Future. Oxford University Press, Oxford, 1-4
7. Black, K.D. (2000) Environmental Impacts of Aquaculture. CRC Press, Boca Raton, 36.

8. Akpan, E.R. and Okafor, N. (1997) On Organic Fertilization and Plankton Development in Two Experimental Fresh Water Ponds of Nigeria. J. Aqua. Trop, 12, 147-154
9. Pearl, H.W. and Tucker, O.S. (1995) Ecology of Blue-Green Algae in Aquaculture Ponds. Journal of the World Aquaculture Society, 26, 109-131. http://dx.doi.org/10.1111/j.1749-7345.1995.tb00235.x
10. Costa Pierce, B.A. (2002) Farming Research and Extension Methods for the Development of Sustainable Aquaculture Ecosystems. In: Costa Pierce, B.A., Ed., Ecological Aquaculture, CRC Press, Boca Raton, 103-123. http://dx.doi.org/10.1002/9780470995051.ch5
11. Onada (2014) A Research Dissertation for the Fulfillment of Master Degree in the Department of Aquaculture and Fisheries Management. University of Ibadan, Ibadan

CHAPTER 2

A Concept of Bayesian Regulation in Fisheries Management

Noél Michael André Holmgren[1*], Niclas Norrström[1], Robert Aps[2], Sakari Kuikka[3]

[1]*Systems Biology Research Centre, School of Bioscience, University of Sko¨vde, Sko¨vde, Sweden,*

[2]*University of Tartu, Estonian Marine Institute, Tallinn, Estonia,*

[3]*Fisheries and Environmental Management Group, Department of Environmental Sciences, University of Helsinki, Helsinki, Finland*

ABSTRACT

Stochastic variability of biological processes and uncertainty of stock properties compel fisheries managers to look for tools to improve control over the stock. Inspired by animals exploiting hidden prey, we have taken a biomimetic approach combining catch and effort in a concept of Bayesian regulation (BR). The BR provides a real-time Bayesian stock estimate, and can operate without separate stock assessment. We compared the performance of BR with catch-only regulation (CR), alternatively operating with *N*-target (the stock size giving maximum sustainable yield, MSY) and *F*-target (the fishing mortality giving MSY) on a stock model of Baltic Sea herring. *N*-targeted BR gave 3% higher yields than *F*-targeted BR and CR, and 7% higher yields than *N*-targeted CR. The BRs reduced coefficient of variance (CV) in fishing mortality compared to CR by 99.6% (from 25.2 to 0.1) when operated with *F*-target, and by about 80% (from 158.4 to 68.4/70.1 depending on how the prior is set) in stock size when operated with *N*-target. Even though *F*-targeted fishery reduced CV in pre-harvest stock size by 19–22%, it increased the dominant period length of population fluctuations from 20 to 60–80 years. In contrast, *N*-targeted BR made the periodic variation more similar to white noise. We discuss the conditions when BRs can be suitable tools to achieve sustainable yields while minimizing undesirable fluctuations in stock size or fishing effort.

INTRODUCTION

Fisheries managers are challenged with two widely permeated properties of their study system, uncertainty [1]–[3] and variability [4]–[6]. These can confound each other, e.g. imprecise spawning stock size estimates can generate apparent variability in the stock – recruitment relationship [7]. Reduced inter-annual variability in effort has socio-economic benefits with yields matching the capacity of the processing industry [8], a more stable job market, and fewer years with over-dimensioned fleets [9]. Often there is a trade-off between maximizing yield and stabilizing yield and fishing effort that makes management objectives ambiguous [10]–[12]. There are also concerns that fishing increases the temporal variability of harvested stocks, [4], [13]–[15].

The problem with temporal variability in fisheries has led to an increased interest in the performance of alternative harvest control rules [16]–[18]. Stephenson et al. [19] argue that the total allowable catch (TAC) could be used to prevent unsustainable use, but is insufficient to control spatial and temporal variability, and cannot be used to achieve socioeconomic objectives. In a reflection over proposed alternative harvest controls, May et al. [20] conclude that further mathematical refinement is probably not as important as developing “robustly self-correcting strategies that can operate with only fuzzy knowledge about stock levels and recruitment curves”. If our belief of the stock size is a probability function (in contrast to a point estimate), Bayes’ theorem postulates that harvesting information can be used to calculate a conditional probability function [21]. Here we propose Bayesian regulation (BR) that uses catch- and effort data from the ongoing fisheries to make real-time estimates of stock size and fishing mortality. The years can be linked by using the posterior distribution as a prior in the sequential year, and hence exploitation is combined with stock size assessment. Real-time assessment of BR would be advantageous to management routines relying on forecasts of stock abundances. Given that most commercial fisheries with CR-based control of biomass and fishing mortality (e.g. TAC) routinely monitor effort and catch, this additional information is surprisingly poorly utilized during the CR fishing season. We suggest a methodology where this information can be used for updating population size estimates to facilitate in-season management decisions.

The BR is derived from Bayesian foraging theory in behavioral ecology, which describes a giving-up rule for patch foraging animals [22]–[24]. The rule is a relaxation of the full information assumption of the marginal value theorem [25], with search time and number of prey caught as information variables. It applies to patches in which the prey are hidden, such as a woodpecker feeding on pupae under bark [26], [27]. Given some prior information available to the forager, e.g. experience from foraging in patches of its territory, a Bayesian posterior distribution can represent the forager's continuously changing belief about the prey density in the current patch [28]. The animal maximizing its intake rate would thus leave the patch for a new one when the anticipated intake rate in the current patch drops below the one expected from patches on average in the territory. The expected number of remaining prey in a patch can be expressed with a fairly simple equation derived by Iwasa et al. [29], but the decision to leave should include a discounting of the value of further bits of information [30].

For fisheries applications, we have modified Iwasa's equation for Bayesian prey density estimation for recurrent exploitation of one population. Both are cases of Bayesian update of mean and variance of a population size estimate. In the original scenario, the exploiter is informed by the mean and variance of several exploited patches within its foraging area. In the fishery scenario, the exploiter (manager) is informed by previous exploitations and surveys in the same area, followed by quantitative assessment and stochastic forecast simulations. We compare CR with BR, and show how they perform in relation to management objectives and targets, by simulating fishing on a model of the main basin Baltic Sea herring (Appendix S2) [31]. We present levels and temporal variability in yield, fishing mortality, stock abundance, spawning stock biomass (SSB), and finally how frequently it surpasses the MSY related reference point B_{trigger} [32].

METHODS

We look at three levels of decision-making in fisheries management as they are described in the common fisheries policy of the Council of the European Union [33]: (i) The *management objective*, which is the ultimate goal of fisheries management. The objective can be simple or more complex, for example weighing incompatible goals [34], [35]. We have chosen the maximum sustainable yield (MSY) because it is the current objective of fisheries management in the EU [36]. (ii) The *target* of exploitation, which should be given in a quantitative unit that relates to the management objective. We compare the efficiency of two targets: fishing mortality (F_{MSY}) and post-harvest stock size ($\acute{N}_{\text{MSY}}$). (iii) The *harvest rules,* which "[lay] down the manner in which annual catch and/or fishing effort limits are to be calculated and provide for other specific management measures, taking account also of the effect on other species" [33]. On this level we choose to use other terms. We use the term *regulation type* to denote the use of catch and/or effort regulation, whereas the term *harvest control rule* (HCR) is used for the combined objective, target and regulation type.

The harvest control rules under development by the International Council for the Exploration of the Sea (ICES) involve a fishing mortality target (F^*) and the spawning stock biomass reference point $B_{trigger}$ [32], [37]. If the SSB goes below $B_{trigger}$ the fishing mortality is reduced below target, a situation that should warrant a reinvestigation of the stock condition and the harvest control [32]. Based on a stock assessment and a short-term forecast, the fishing mortality target is recalculated as a TAC, which is how the recommendation is presented to the European Commission [38]. ICES harvest control rules use precautionary reference points below which fishing mortality is reduced [32], something we do not use in our simulations. We have focused on the principle differences between CR and BR, e.g. the elaborate data collection and assessment procedure is simplified by using the actual size of the simulated stock with a random error.

HARVEST CONTROL

Altogether we compared eight different harvest controls achieved by the combinations of two exploitation targets, two regulation types and two stock assessment modes (Table 1). The Bayesian regulation $BR(\gamma, f|\mu, \nu)$ is a function of catch (γ) and effort (f) given the prior information of the estimated stock size before harvest, μ, and the uncertainty of that estimate given as variance, ν. The Bayesian estimate of the number of remaining individuals (r) in a harvested stock is:

$$r = \frac{\mu - \gamma\left(1 - \frac{\nu}{\mu}\right)}{(e^{qf} - 1)\frac{\nu}{\mu} + 1} \tag{1}$$

where q, the catchability, is defined as the fraction of the population captured by one unit of fishing effort and is the scalar between catch per unit effort indices and average population abundance, hence $q = \frac{\gamma}{fN}$. When fishing is closed by fulfillment of the target condition, $f = f^*$, then the fishing mortality $F = q\,f^*$. This is a generalization of Iwasa's [29] Bayesian estimates of remaining prey population, see Appendix S1 for details on how Equation (1) is derived from some specific probability distributions. With a known prior probability distribution and a given catch and effort, a posterior distribution can be calculated with *r* being the mean. It is very unlikely that the prior mean is equal to the actual population size. The posterior mean will approach the actual value with accumulating catch and effort, but there will always be a bias towards the prior [39]. For large deviations of the prior mean in relation to the real population size, it may take more than one year to track the population size more closely. Hence, there can be temporal correlations in estimate biases. This depends on the random perturbations displacing the population size from the prior mean and the harvesting information making the posterior mean approach the real value.

Table 1. Characteristics and settings for the evaluation of eight different harvest controls: CR being a simple catch-only regulation; BR is the proposed regulation combining catch and effort.

Harvest control	Target	Supervision	Prior μ	Prior v
CR_F^s	F_{MSY}	Supervised	$N+\varepsilon$	0
CR_F^u	F_{MSY}	Unsupervised	N_{MSY}	0
CR_N^s	N'_{MSY}	Supervised	$N+\varepsilon$	0
CR_N^u	N'_{MSY}	Unsupervised	N_{MSY}	0
BR_F^s	F_{MSY}	Supervised	$N+\varepsilon$	$var(N_{MSY})$
BR_F^u	F_{MSY}	Unsupervised	N_{MSY}	$var(N_{MSY})$
BR_N^s	N'_{MSY}	Supervised	$N+\varepsilon$	$var(N_{MSY})$
BR_N^u	N'_{MSY}	Unsupervised	N_{MSY}	$var(N_{MSY})$

The subscript denotes type of target and the superscript the existence of supervision. The target column denotes the targets used: F_{MSY} is the fishing mortality giving highest yield in the MSY-analysis of the SOM, and N'_{MSY} is the corresponding post-harvest stock size. The management is either supervised with a separate assessment or unsupervised (see methods). For unsupervised management we use the mean, μ, and the variance, v, of the pre-harvest stock size from MSY-analyses: N_{MSY}. In supervised management μ is instead the actual pre-harvest stock size, N, with an added randomly generated error term ε drawn from the normal distribution, (mean = 0, var = $var(N_{MSY})$ where $var(N_{MSY})$ is taken from an MSY-analysis).

When Equation (1) is solved for catch as a function of *f*, and *r* is given a target value, it can be visualized as harvest control curves (Fig. 1A). These curves can be calculated prior to the fishing season and define the combination of total catch and total effort at which the Bayesian information indicates that the fishery has reached its target. The real-time accumulation of catch and effort during the fishing season is responsive to over- and under-estimation of stock size and will develop different trajectories. The intersection of the real-time accumulation of catch and effort with the BR-curve gives the total catches when an initial over- or under-estimation is compensated for (Fig. 1B). Note that if we have no uncertainty in our prior estimate, i.e. $v=0$, Equation (1) simplifies to $r=\mu-\gamma$, in other words the remaining individuals in the population is our prior estimate minus the catch size. This means that catch-only regulation is a special case of BR (Fig. 1A). To be used in fisheries, Equation (1) needs to account for the simultaneous removal by predators (M), which is done by introducing two scaling parameters α and β:

$$r = \frac{\mu - \beta\gamma\left(1 - \frac{v}{\mu}\right)}{\left(e^{\alpha\hat{q}f} - 1\right)\frac{v}{\mu} + 1} \qquad \textbf{(2)}$$

where parameter $\alpha = 1 + M/\hat{q}$ simply scales up the effort with the total mortality in proportion to . $\hat{q}$ The notation $\hat{q}$ denotes the assumed value of the catchability. As default $\hat{q}$ is constant and unbiased, but we make separate runs to explore the impact of biases and stochastic errors in $\hat{q}$. Catchability is difficult to determine accurately and can be affected by aggregation behavior of fish and technical enhancement of fishing gear [40]. Parameter β scales the catch γ to the total number of casualties from fishing and natural mortality:

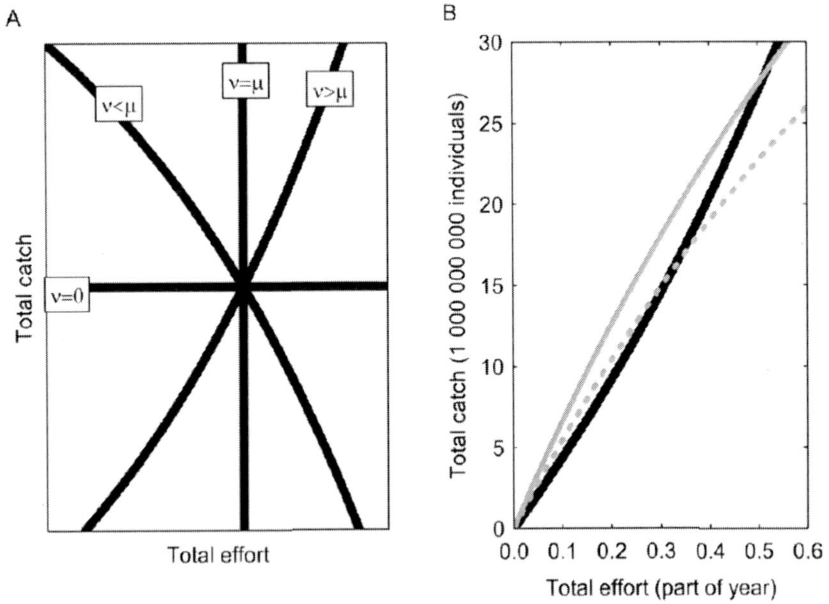

Figure 1. (A) Bayesian regulation curves indicating the combined total catch (y-axis) and total effort (x-axis) when fishing should be terminated for the season. The curves are given by equation 1 when r is set to the target number of fish, and solved for total catch. Hence, the curves show when the fishery target is achieved in catch-effort space, given by the ratio of variance (v) to mean stock size (μ). There are four qualitative cases of the v-μ ratio: when the ratio is zero, the fishery is regulated by catch only because there is no uncertainty around the mean. When the ratio equals 1, regulation is by

effort only. Apart from these special cases, the fishery should be regulated by both catch and effort. If the variance is lower than the mean, the posterior mean will decrease with accumulating catches and effort. If the variance is higher than the mean, the posterior mean will increase with increasing catches but decrease with effort. These effects can be deduced from analyzing equation 1. (B) Here the solid curve denotes the Bayesian regulation when $v>\mu$, which is determined prior to the fishing season (see Methods). The grey curves are idealized trajectories of how cumulative catch and effort develops from the origin during the fishing season. When the trajectories cross the regulation curve, the Bayesian posterior is on target, and the fishery should be closed for the season. The hatched grey line indicates when the initial stock size is at the prior mean (μ), the solid grey line when the stock size is at the mean plus one standard deviation (see F_{MSY} -analysis). The BR thus allows larger catches when the stock size has been underestimated, and vice versa.

$$\beta = \frac{1 - e^{-(M+\hat{q})f}}{1 - e^{-\hat{q}f}} \tag{3}$$

Note that β depends on effort such that Equation (2) can work as a real-time estimator of the population size. The Bayesian estimator is used with the different targets (Table 1). When the target is formulated as a post-harvest population size (N'_{MSY}), harvesting should stop when the Bayesian estimate of remaining stock size (r) equals target N'_{MSY}:

$$re^{-(1-f)M} = N'_{\mathrm{MSY}} \tag{4}$$

The target N'_{MSY} denotes the stock size at the end of the year, we therefore need to take into account the removal of individuals due to natural mortality for the remaining part of the year after fishing has ended, $e^{-(1-f)M}$. When our target is defined as fishing mortality, F_{MSY}, harvesting is aborted when.

$$\frac{\gamma}{\beta\gamma + r} = 1 - e^{-F_{\mathrm{MSY}}} \tag{5}$$

Note that the prior μ cannot be used in the denominator, because the estimated size of the pre-harvest population changes as we receive information from catch and effort. When the conditions in Equation (4), or (5) are fulfilled, we extract the fishing effort (f^*) from Equation (2). Equations (2), (4) and (5) together define the control rules both for CR and BR.

Assessment and priors

Estimates of the pre-harvest stock size, the prior μ, are produced in two ways: by supervised assessment and unsupervised assessment (Table 1). Supervised assessment represents the current annually revised assessment practiced by ICES and other stock assessors. In this case, μ is calculated by adding an assessment error to the actual pre-harvest stock size, N. The error is a random value drawn from a normal distribution with mean=0, and variance$_v$which is equal to the variance of the pre-harvest stock size in the F_{MSY}-analysis (see below). The coefficient of variance in the F_{MSY}-analysis is 21%. This represents an ideal assessment where the error solely stems from the yearly variation of the population when fished at F_{MSY}. The unsupervised assessment uses μ equal to the average, and v the variance of the pre-harvest stock size from the F_{MSY} -analysis (i.e. when fishing at F_{MSY}; see below). The rationale of using a constant μ is that the Bayesian information of stock size when a fishery has been closed in the foregoing year indicates N'_{MSY} at the end of the year when using N-target (Equation 4). Similarly for an F-targeted fishery, fishing is closed every year when at estimated F_{MSY}, which is associated with the mean $N'=N'_{MSY}$. The variance is used for the BR, whereas v=0 for the CR.

The stochastic operating model

We have used a stock model of the herring population in the main basin of the Baltic Sea (ICES catch area subdivisions 25–27, 28.2, 29 and 32) for harvest control evaluation. The model is a stochastic operating model (SOM) parameterized from statistical analyses of ICES catch data and outputs from ICES XSA runs (Appendix S2).

MSY analysis

We performed an MSY-analysis on the SOM by stepping the fishing mortality in steps of 0.01, and for each F-value simulating 40,000 years and rejecting the first 500 to minimize the effects of initiation values. In contrast to simulated management, this algorithm executes perfectly-controlled constant fishing mortality. The relationship between yield and fishing mortality was used to identify the F_{MSY} and its associated average post-harvest stock size. These were used as targets for F-targeted and N-targeted management, respectively. The average pre-harvest stock size, given fishing mortality F_{MSY}, was used as the prior (μ) in the unsupervised simulations, and the variance in the pre-harvest stock size was used as a measure of the uncertainty (v) of μ in the BR. The lower 2.5% percentile of the associated SSB was used as the $B_{trigger}$. Targets, priors and assessment modes were used in defining the eight different HCRs, all aiming at achieving MSY (Table 1).

Evaluation of harvest control rules

The eight HCRs were applied to Monte Carlo simulations of the SOM. A simulation started with a number of years of controlled fishing with F_{MSY} to get away from the initial population size and age-structure. After the initial period, data was collected for a number of years of applied HCR. We ran shorter time series with an initiation period of 1,000 years and 100 years of applied HCR. The time period of 100 years reflects a reasonably long time-horizon for management. We also ran longer series with an initiation period of 100 years and 20,000 years of applied HCR, in order to establish more accurate estimates of mean yields. We collected annual data on yield, fishing mortality, pre-harvest and post-harvest population size, and SSB. Cod SSB and year-specific growth were kept at a mean level (C_y=100,000 tons, k_y=0) during the simulations with the addition of random noise, the size of which was extracted from historical data after removing long-term changes [31]. Technically, fishing on the SOM is performed by applying the effort, f^*, from the solution of the quitting rules in Equation (4) and (5). f^* is calculated by halved-distance iterations until the estimate (the left side of Equation 4 and 5) differed from the target with less than 0.1‰. The population is harvested by stepping the SOM one year with the catch being determined by Baranov's catch equation:

$$\gamma = \frac{qf^*}{M+qf^*}\mu\left(1-e^{-(M+qf^*)}\right) \quad (6)$$

in which we use the actual q of the SOM.

Sensitivity to error in catchability

We keep the estimated catchability $\hat{q}$ in Equation (2) and (3) constant and unbiased in our base runs, but fishery dependent catchability may in reality be biased and vary over time. Such uncorrelated errors and biases in catchability will contribute to error or bias in the assessment. Our base model already has assessment error (CV=21%, as described in *Assessment and priors*) for supervised HCRs, but unsupervised BRs do not (Table 1). Unsupervised BRs do not use separate assessments, but on the other hand they will be affected by error or bias in $\hat{q}$. We chose to compare supervised F-targeted CR and unsupervised F-targeted BR because the former is only affected by the error in the assessed stock size, μ, whereas the latter is affected only by the error in $\hat{q}$. We also explored the sensitivity to correlated biases in $\hat{q}$ and μ by changing their values ±20%. Simulations ran for 41,000 generations and data from the last 40,000 were used for the analyses.

Statistical analyses

Unsupervised CR expectedly led to population crashes within a few decades or less. We therefore excluded them in the statistical analyses, which were performed using STATISTICA software (Statsoft, Tulsa, Oklahoma). General linear models (GLM) were used for three-way ANOVAs, testing the effects of supervision mode, regulation type and target type on yield, fishing mortality, spawning stock biomass, and post-harvest population size. Fourier analyses were performed on 1,000 year data series, tapered by 15% and padded to the length power of 2. The period length in years with the highest spectral density was identified after applying Hamming weights with a data window of size 7. Test of significant deviation from white noise was performed using Kolmogorov-Smirnov deviation (*d*) statistics, Table Y, ρ=0.5, n>100 in Rohlf & Sokal [41].

RESULTS

Yield

Mean yield in 100-year simulations is significantly higher with BR than with CR although the differences are small (Fig. 2, Table 2, 3). With an increasing length of the time series the confidence intervals narrow (Fig. 3). Analysis of 20,000 years reveals more clearly that BRs operated with *N*-target give the highest average yields, even though the largest differences in means are less than 10% (Table 2, 4). Supervised and unsupervised BRs give higher yields (262 thousand tons) than the reference MSY from the F_{MSY}-analysis (254 thousand tons; Table 2, 3, 4, 5). Contrary to the small differences in mean yields, the CV in yields is clearly affected by the choice of HCR. In general, using *F*-targets gave less temporal variation than using *N*-targets (Fig. 2, Table 2, 3, 4, 5). There is also an effect of regulation type with BR, giving less variation than CR (Fig. 2, Table 3, 4). However, supervised BR and unsupervised BR have the same CV regardless of target (Table 5).

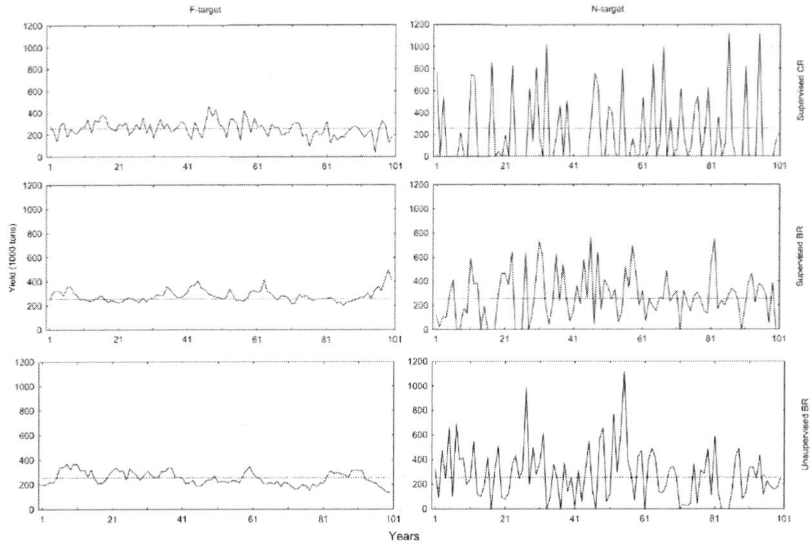

Figure 2. Yield over 100 years for different HCRs. The dashed reference line is the MSY=254.4 thousand tons obtained from the F_{MSY}-analysis during controlled *F*.

Table 2. Mean, standard deviation and CV (%) of annual yield.

	100 yrs			20,000 yrs		
	mean	**SD**	**CV**	**mean**	**SD**	**CV**
CR_F^S	256.6	72.0	28.1	254.3	83.9	33.0
CR_N^S	220.3	322.4	146.3	244.5	343.0	140.3
BR_F^S	282.8	50.6	17.9	258.6	58.9	22.8
BR_N^S	272.5	198.8	73.0	262.3	195.2	74.4
BR_F^U	251.8	55.0	21.9	253.5	57.6	22.7
BR_N^U	281.9	218.9	77.7	262.1	196.5	75.0

Two time series, one of 100 years and one of 20,000 years are presented. Mean values are given in thousands of tons. MSY from the F_{MSY}-analysis is 254.4 thousand tons.

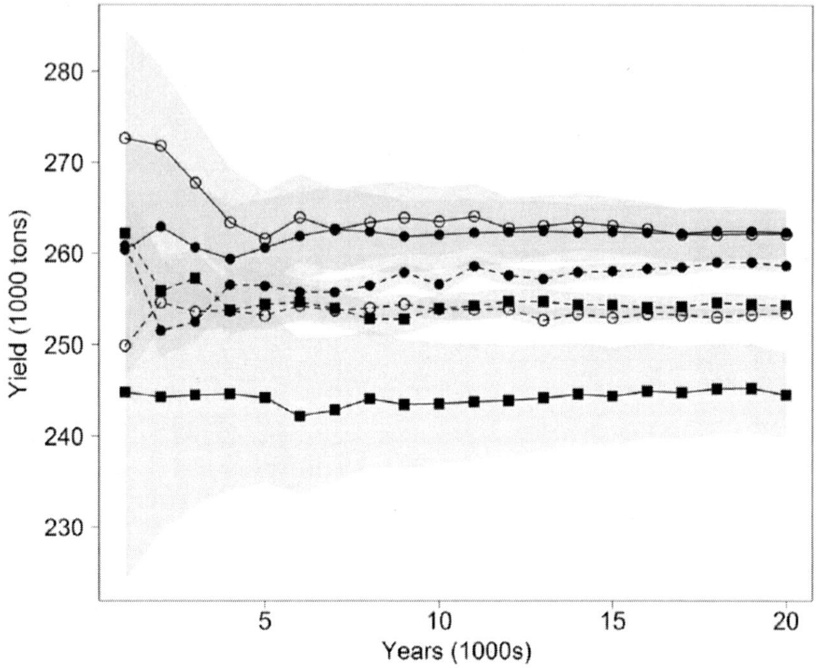

Figure 3. Average yields and confidence intervals as functions of increasing length of simulation in years. Filled symbols denote supervised and open symbols unsupervised harvest control. Squares denote CR and circles BR. Solid lines denote *N*-target management whereas hatched lines denote *F*-targeted management.

Table 3. ANOVA-table showing effects of supervision, regulation type, target type, and target type interacting with supervision/regulation type on annual yield.

100 yrs	SS	DF	MS	F	p
Supervision	11,691	1	11,691	0.35	0.556
Regulation type	153,773	1	153,773	4.56	0.033
Target type	2,831	1	2,831	0.08	0.772
Superv.×Target	40,995	1	40,995	1.22	0.271
Regul.×Target	16,783	1	16,783	0.50	0.481
Error	20,018,836	594	33,702		

Data is from model simulations over 100 yrs.

Table 4. ANOVA-table showing effects of supervision, regulation type, target type, and target type interacting with supervision/regulation type on annual yield.

	SS	DF	MS	F	p
Supervision	141,961	1	141,961	4.16	0.041
Regulation type	1,582,157	1	1,582,157	46.40	<0.001
Target type	58,951	1	58,951	1.73	0.189
Superv × Target	120,612	1	120,612	3.54	0.060
Regul × Target	1,582,157	1	1,582,157	46.40	<0.001
Error	4,091,815,000	119,994	34,100		

Data is from model simulations over 20,000 yrs.

Table 5. Pair-wise test of CV in yield for the 100 year series, applying the F-statistics of testing differences in the variance of annual yield.

	CR_F^s	CR_N^s	BR_F^s	BR_N^s	BR_F^u	BR_N^u
CR_F^s		<0.001	<0.001	<0.001	0.007	<0.001
CR_N^s	27.205		<0.001	<0.001	<0.001	<0.001
BR_F^s	2.455	66.785		<0.001	0.024	<0.001
BR_N^s	6.764	4.022	16.605		<0.001	0.268
BR_F^u	1.647	44.809	1.490	11.141		<0.001
BR_N^u	7.661	3.551	18.807	1.133	12.619	

F-values are presented in the lower left triangle and p-values in the upper right.

Fishing mortality

There is no significant difference in mean fishing mortality (F) due to any effect over 100 years (Table 6, Table 7). All HCRs reach the F-target of 0.17 as their mean, except supervised CR with N-target, which has a mean F of 0.20. Although the mean Fs are very similar across HCRs, the differences in CV are more pronounced. Supervised CR with N-target exhibits the highest CV of 158% (Table 6). F-targeted supervised CR has a CV of 25% (Table 6, Fig. 4). Since catchability is constant here, F is proportional to fishing effort. It is therefore not surprising that F-targeted HCRs result in less variation than N-targeted HCRs. F-targeted BRs, supervised and unsupervised, show very high precision in reaching the target (Fig. 4), the CV being as low as 0.1% (Table 6). As expected, N-targeted BRs exhibit higher CVs (68% unsupervised and 70% supervised; Table 6). In addition, it is only the CV between supervised and unsupervised BRs that does not differ significantly from each other (Table 8).

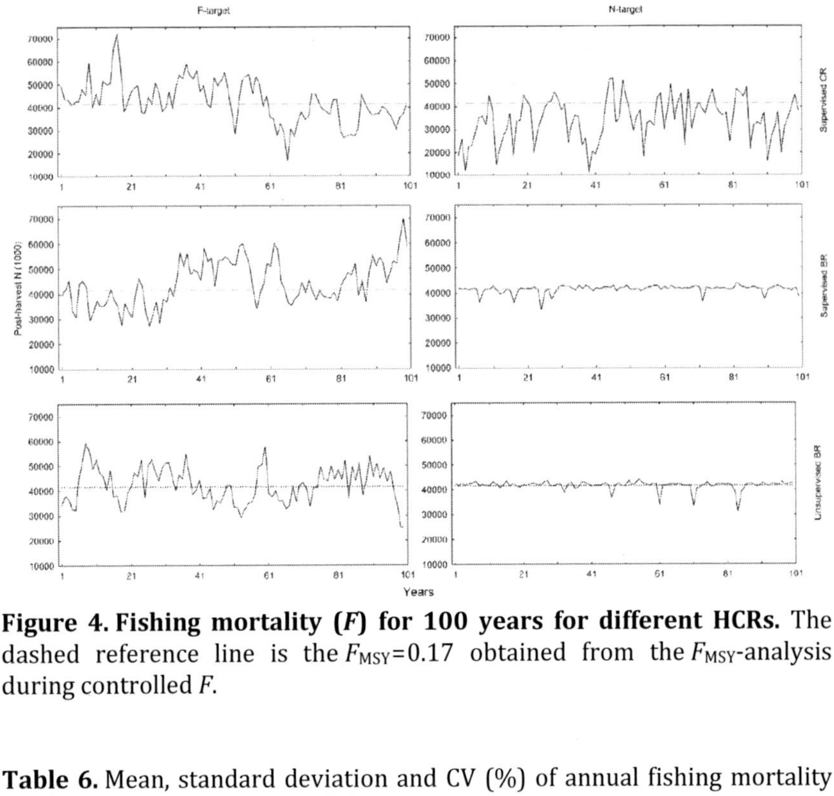

Figure 4. Fishing mortality (F) for 100 years for different HCRs. The dashed reference line is the F_{MSY}=0.17 obtained from the F_{MSY}-analysis during controlled F.

Table 6. Mean, standard deviation and CV (%) of annual fishing mortality (F) of 100 years of simulated fishing.

	mean	SD	CV
CR_F^s	0.17	0.043	25.2
CR_N^s	0.20	0.325	158.4
BR_F^s	0.17	0.000	0.1
BR_N^s	0.17	0.118	70.1
BR_F^u	0.17	0.000	0.1
BR_N^u	0.17	0.119	68.4

F_{MSY} from the F_{MSY} -analysis is 0.17.

Table 7. ANOVA-table showing effects of supervision, regulation type, target type, and target type interacting with supervision/regulation type on annual fishing mortality.

	SS	DF	MS	F	p
Supervision	0.00047	1	0.00047	0.02	0.886
Regulation type	0.03166	1	0.03166	1.40	0.236
Target type	0.02945	1	0.02945	1.31	0.254
Superv × Target	0.00046	1	0.00046	0.02	0.886
Regul × Target	0.03265	1	0.03265	1.45	0.229
Error	13.39279	594	0.02255		

Data is from model simulations over 100 yrs.

Table 8. Pair-wise test of CV-values applying the F-statistics of testing differences in the variance of annual fishing mortality.

	CR_F^S	CR_N^S	BR_F^S	BR_N^S	BR_F^U	BR_N^U
CR_F^S		<0.001	<0.001	<0.001	<0.001	<0.001
CR_N^S	39.520		<0.001	<0.001	<0.001	<0.001
BR_F^S	67,742	2,677,131		<0.001	0.066	<0.001
BR_N^S	7.731	5.112	523,678		<0.001	0.404
BR_F^U	49,987	1,975,470	1.355	386,425		<0.001
BR_N^U	7.361	5.369	498,613	1.050	367,930	

F-values are presented in the lower left triangle and p-values in the upper right

Post-harvest population size and SSB

Looking at post-harvest population size (*N'*), the harvest controls operating on *N*-targets are aiming for 41.5 billion individuals at the end of the year. The two *N*-targeted BRs reach this target with little variation between years (4% CV; Fig. 5, Table 9, 10, 11), whereas *N*-targeted CR leads to over-fishing in the sense that *N'* is on average 17% below target. In contrast to the *N*-targeted regulation types, there is a small difference in *N'* between *F*-targeted BR and CR. This explains the significant interaction between regulation type and target type in the ANOVA (Table 10). Supervision has no significant effect on the mean *N'* (Table 10), nor the CV (Table 11).

Table 9. Mean (Million), standard deviation and CV (%) of annual post-harvest population size, *N'*, of 100 years of simulated fishing.

	mean	SD	CV
CR_F^S	42,620	9,201	21.6
CR_N^S	34,282	9,470	27.6
BR_F^S	44,391	8,798	19.8
BR_N^S	41,508	1,616	3.9
BR_F^U	42,270	7,248	17.1
BR_N^U	41,569	1,840	4.4

N-target from the F_{MSY}-analysis is 41,491 million.

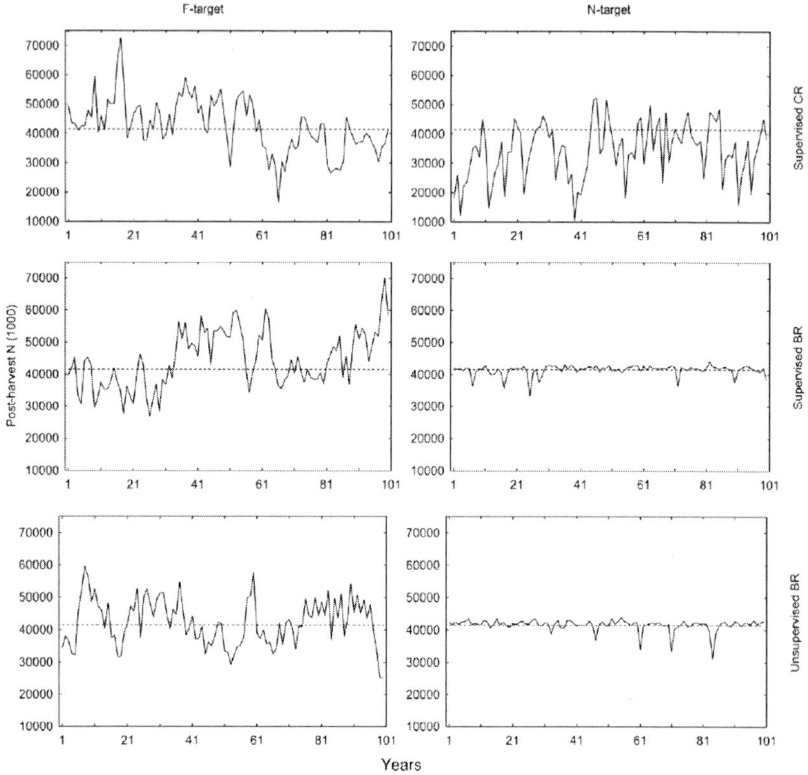

Figure 5. Post-harvest *N*-values for different HCRs in simulations over 100 years. The dashed reference line is the *N*-target=41,491 million obtained from the F_{MSY}analysis during controlled *F*.

Table 10. ANOVA-table showing effects of supervision, regulation type, target type, and target type interacting with supervision/regulation type on annual post-harvest population size.

	SS	DF	MS	F	p
Supervision	1.06E+08	1	1.06E+08	2.1	0.153
Regulation type	2.02E+09	1	2.02E+09	39.1	<0.001
Target type	2.44E+09	1	2.44E+09	47.2	<0.001
Superv ×Target	1.19E+08	1	1.19E+08	2.3	0.130
Regul ×Target	7.44E+08	1	7.44E+08	14.4	<0.001
Error	3.07E+10	594	5.17E+07		

Data is from model simulations over 100 yrs.

Table 11. Pair-wise test of CV-values applying the F-statistics of testing differences in the variance of annual post-harvest population size.

	CR_F^s	CR_N^s	BR_F^s	BR_N^s	BR_F^u	BR_N^u
CR_F^s		0.007	0.198	<0.001	0.011	<0.001
CR_N^s	1.637		0.001	<0.001	<0.001	<0.001
BR_F^s	1.187	1.943		<0.001	0.076	<0.001
BR_N^s	30.736	50.318	25.903		<0.001	0.103
BR_F^u	1.585	2.595	1.336	19.387		<0.001
BR_N^u	23.800	38.964	20.058	1.291	15.012	

F-values are presented in the lower left triangle and p-values in the upper right.

The mean SSBs differ significantly by regulation type, harvest type and supervision (Table 12). BR SSBs are on average larger than CR SSBs (Fig. 6, Table 13, 13). Supervised CR with *N*-target has the lowest mean and is below $B_{trigger}$ in 50% of the years (Table 13, Fig. 6). Supervised BRs has on average larger SSBs than unsupervised, and the BRs operating with *N*-targets surpass the $B_{trigger}$ less frequently than when operating with *F*-targets (Table 13, Fig. 6). The lower CV for BRs operating with *N*-target (Table 13, 14) seems to be the main reason for this result.

Table 12. ANOVA-table showing effects of supervision, regulation type, target type, and target type interacting with supervision/regulation type on annual SSB.

	SS	DF	MS	F	p
Supervision	1 054 971	1	1 054 971	19.8	<0.001
Regulation type	6 757 878	1	6 757 878	126.9	<0.001
Target type	2 008 542	1	2 008 542	37.7	<0.001
Superv × Target	164 153	1	164 153	3.1	0.080
Regul × Target	2 271 205	1	2 271 205	42.6	<0.001
Error	31 642 602	594	53 270		

Data is from model simulations over 100 yrs.

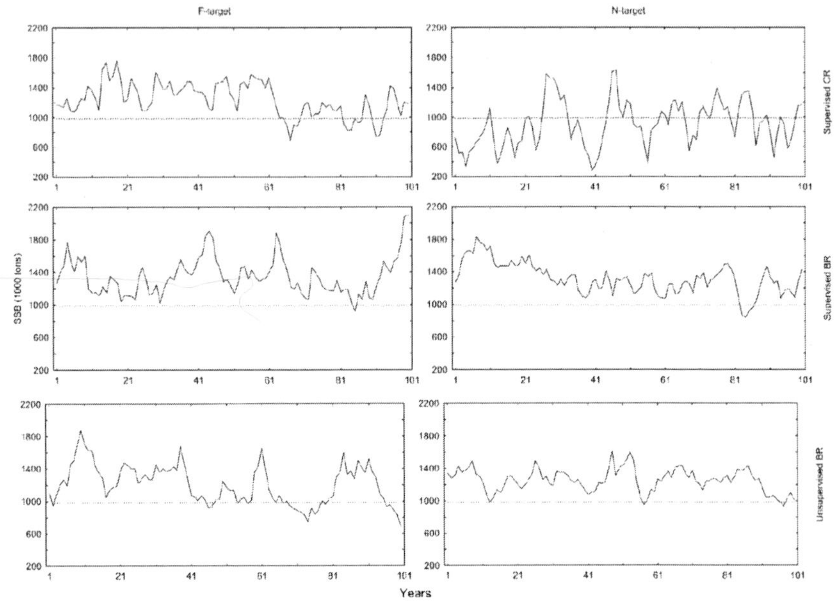

Figure 6. Spawning stock biomass (SSB) for different HCRs in simulations over 100 years. The dashed reference line is the $B_{trigger}$=988 thousand tons. It is the 2.5% lower percentile of SSB obtained from the F_{MSY}-analysis during controlled F.

Table 13. Mean in thousand tons, standard deviation and CV (%) of annual SSB of 100 years of simulated fishing.

	mean	SD	CV	$<B_{trigger}$ (%)
CR_F^S	1 240	230.0	18.6	26.1
CR_N^S	906	308.6	34.1	49.9
BR_F^S	1 349	232.8	17.3	18.3
BR_N^S	1 317	192.4	14.6	5.6
BR_F^U	1 206	245.1	20.3	20.6
BR_N^U	1 255	142.3	11.3	5.9

The percentage of years the SSB surpassed the $B_{trigger}$ ($B_{trigger}$=998 thousand tons), is given in the rightmost column.

Table 14. Pair-wise test of CV in yield for the 100 year series, applying the F-statistics of testing differences in the variance of annual SSB.

	CR_F^S	CR_N^S	BR_F^S	BR_N^S	BR_F^U	BR_N^U
CR_F^S		<0.001	0.236	0.009	0.183	<0.001
CR_N^S	3.368		<0.001	<0.001	<0.001	<0.001
BR_F^S	1.156	3.894		0.049	0.053	<0.001
BR_N^S	1.614	5.434	1.396		0.001	0.006
BR_F^U	1.200	2.807	1.387	1.936		<0.001
BR_N^U	2.678	9.018	2.316	1.660	3.213	

F-values are presented in the lower left triangle and p-values in the upper right.

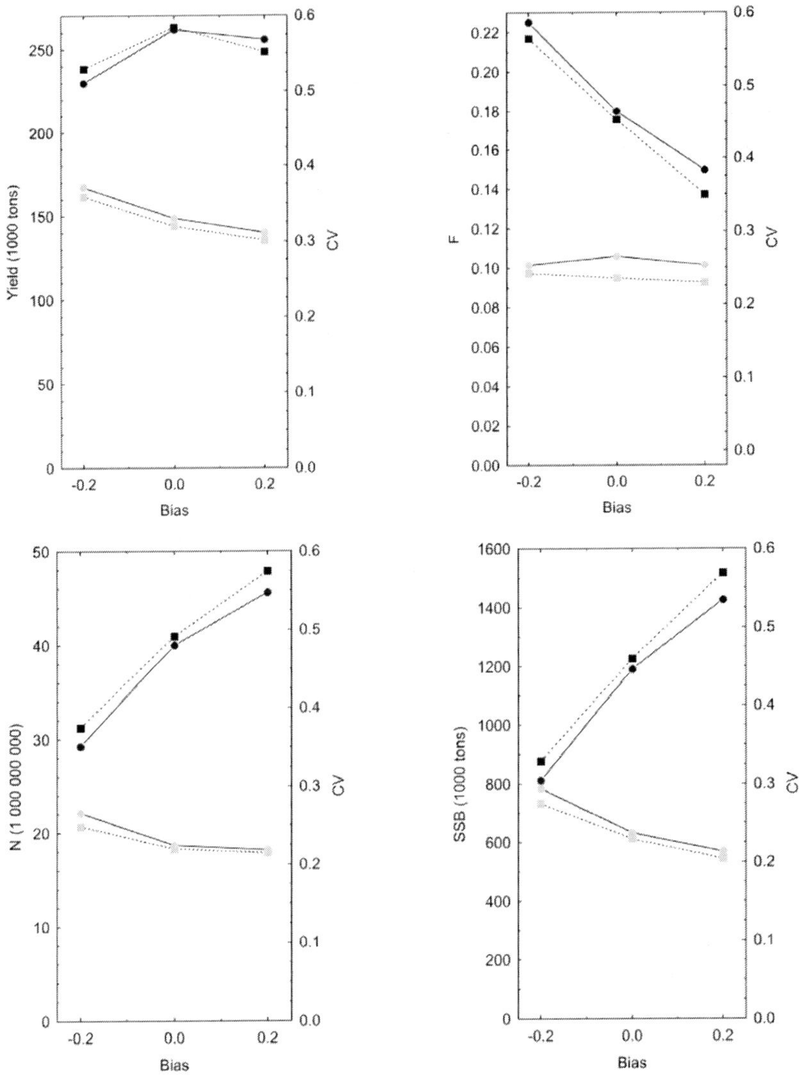

Figure 7. A comparison of yield, fishing mortality (*F*), post-harvest stock size (*N*′) and spawning stock biomass (SSB) when random errors of the same CV are applied to both the stock size assessment (μ) and the estimated catchability parameter ($\hat{q}$). Relative positive and negative biases applied to $\hat{q}$ and μ are denoted on the x-axis. Circles denote results from *F*-targeted unsupervised BR and squares *F*-targeted supervised CR. Black symbols denote the value (left axis) and grey symbols the CV of the effect-variable (right axis).

Bias and error in harvest rates

So far we can summarize the difference between CR and BR as being BR's ability to reduce the variance in the target variable (F or N'; Tables 3, 4). Yield is partially dependent on F, and SSB on N', and thus the variance of these variables is also reduced, although to a lesser degree (Tables 2, 5). These results are based on the assumption of a constant and un-biased estimate of catchability ($\hat{q}$). When random error is added to the catchability (q) of the same magnitude as the error in μ, the CV in γ and F for the unsupervised F-targeted BR become similar to supervised F-targeted CR (Fig. 7). The two F-targeted harvest rules also perform very similar in response to the off-set bias in μ and $\hat{q}$. Hence, supervised CR and unsupervised BR behave similarly when error and bias in μ and q are of the same relative magnitude

Temporal variation in population size

We found that all HCRs decrease the amplitude of temporal variation in pre- and post-harvest population numbers, except for post-harvest numbers under supervised N-targeted CR, at which the variation is indifferent from the unexploited population (Table 15). The BR is effective in reaching the N-target, and hence reduces the CV markedly in the post-harvest population size (Fig. 8, Table 15). The effect is still evident in the pre-harvest population size, in which the inclusion of recruits from the foregoing year increases the variation (Table 15).

Table 15. Coefficient of variation (CV) of pre- and post-harvest population numbers compared with the unexploited population.

	Pre-harvest N			Post-harvest N		
	CV	F-value	p	CV	F-value	p
CR_F^S	21.6	1.52	<0.001	22.0	1.42	<0.001
CR_N^S	24.5	1.18	<0.001	26.1	1.00	n.s.
BR_F^S	21.2	1.57	<0.001	21.2	1.53	<0.001
BR_N^S	14.8	3.25	<0.001	4.6	33.01	<0.001
BR_F^U	20.8	1.64	<0.001	20.7	1.60	<0.001
BR_N^U	15.1	3.12	<0.001	4.7	30.58	<0.001
Unexpl. Pop.	26.6			26.1		

CV is from simulations of 10,000 years. F-value is the ratio of squared CVs (i.e. comparison of standardized variances).

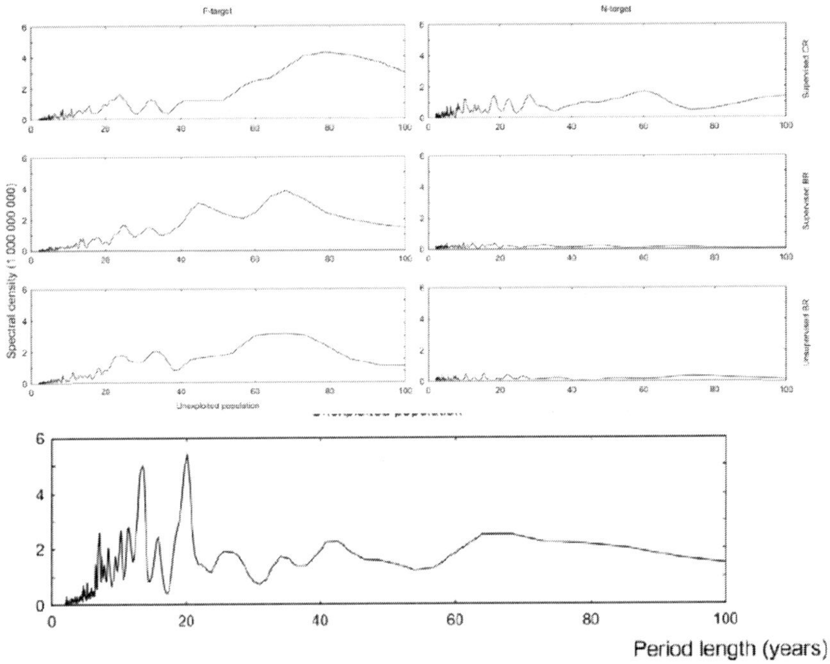

Figure 8. Spectral densities of Fourier analysis of 1,000 years of pre-harvest population size under various HCRs. Period lengths from 0 to 100 years are shown (depicted on the abscissa).

The unexploited population exhibited a very pronounced periodicity of about 20 years (Fig. 8, Table 16). There is only uncorrelated variation in our operating model, which means that any periodicity is due to the demographic parameters and structure of the population. The BRs with *N*-target remove much of this periodicity and puts it closer to white noise (Lower K-S d-values, Table 16). HCRs with *F*-target increase the dominating period length to about 70–80 years (Table 16). Supervised CR with *F*-target, a simplified version of the HCR currently used, reduces the amplitude of variation in both pre- and post-harvested population size (Table 15). At the same time, it increases the dominating period length from 20 years in the unexploited population to about 80 years in the exploited (pre-harvest population size, Table 16).

Table 16. Fourier analysis of pre-harvest population numbers in series of 1,000 years of fishing with alternative HCRs.

HCR	Period max	Density	K-S d	p
CR_F^S	78.769	4.29E+09	0.5013	<0.01
CR_N^S	60.235	1.71E+09	0.3152	<0.01
BR_F^S	68.267	3.84E+09	0.5370	<0.01
BR_N^S	12.488	4.62E+08	0.0929	<0.01
BR_F^U	68.267	3.13E+09	0.5368	<0.01
BR_N^U	15.515	5.10E+08	0.0716	<0.01
Unexpl. Pop.	20.078	5.42E+09	0.4995	<0.01

Period length with maximum spectral density is given in years along with density value. Kolmogorov-Smimov d-value for deviation from white noise is given with its p-value.

DISCUSSION

We have suggested a new approach to fisheries regulation, which we name *Bayesian regulation* (BR), and made a basic evaluation of its properties in relation to catch-only regulation. The idea of BR is adopted from Bayesian foraging theory, in which patch-foragers are assumed to rely on a giving-up rule based on prior information and sampling information [29], [30], [42]. The Bayesian foraging theory has been studied extensively, including the derivation of the probability function used and that cumulative catch and effort are sufficient information variables [29]. The BR uses the uncertainty of the stock size estimate as a parameter, which is rarely the case in current management [43]–[45], and which has been requested for some time [46]. The general applicability of the BR is emphasized by the fact that catch-only regulation, (CR), is a special case of BR — i.e. when there is no uncertainty in the prior estimate of the stock size from an assessment or survey. We have analyzed three aspects of harvest controls: (i) the relevance of supervision (with or without stock size assessment), (ii) operational target type (constant fishing mortality, *F*, or constant stock size, *N*), (iii) type of regulation (catch only=CR, or Bayesian evaluation of combined catch and effort=BR), and their effect on mean and variance in yield and stock size.

The effect of fishing on stock dynamics

There is an ongoing discussion about whether fishing leads to increased temporal variability in stock size compared to unexploited populations [4], [14]. Observations from California, USA, corroborate the view that fishing magnifies fluctuations in fish abundance, driven by increased intrinsic growth rates as a response to higher mortality [4]. Models without age-structure show an increased variability in population abundance as a consequence of exploitation [14]. In a structured model, fishing can either increase or decrease the stock size variability [15]. We find that the CV of population size has been reduced in all investigated HCRs, and especially with *N*-targeted BR. The different effects of fishing between our and the previous studies may depend on our implementation of errors in assessment, whereas earlier studies implemented perfect fishing mortality.

The comparison of Bayesian- and catch-regulation

We expected *N*-targeted BR to give the highest mean yields, because of its ability to maintain the population in the most productive state. It turned out to return the highest mean yield, but the differences between HCRs are small. The CV of fishing mortality and the population size is, however, much more affected using BR compared to CR. The advantage of *F*-targeted BR is seen in a reduced variation in fishing mortality and yield. Similarly, the *N*-targeted BR is associated with less variation in post-harvest population size and SSB. Under some conditions CR can be a blunt harvest control instrument; populations inevitably go extinct if not supervised, and it exhibits the largest CV in yield, *F* and SSB of all harvest controls when operated with *N*-target. For the better, *F*-target supervised CR is a commonly applied harvest control, e.g. within the European Union, and exhibits a smaller CV than *N*-targeted CR in all fisheries and stock variables investigated. Choosing BR, a manager has to consider the trade-off between the pros and cons of an *F*- versus an *N*-targeted fishery, since targeting one variable pushes the variability to the other [47]. From a fisheries perspective, *F*-targeted BR can appear as the most preferable HCR: low variability in effort (costs) and yield (income) enables a more stable economy, and an even process load is more manageable for the food industry [47]. At the same time, single-stock *F*-targeted BR can potentially yield a conflict of interest between fishers, and may be less desirable when taking into account the ecological interaction between different species of fish. *N*-targeted BR, on the other hand, gives low variation in population size, which means lower risks of extinction and smaller indirect effects of variability on prey and predator species, which may also be subject to commercial fishing. The stable post-harvest population size of *N*-targeted BR, in combination with the inherent variation in the productivity of the population causes a highly variable yield.

Our results show that *F*-targeted BRs come close to resolving the trade-off between a high long-term yield and the stable fishing effort and yield acknowledged by some authors [20], [35]. Although giving a slightly lower average yield than BRs with *N*-target, the yield is higher than with CR and the fluctuations in the yield are reduced considerably.

The ability of the BR to control the target variable depends on the reliability of the estimated fishery-dependent catchability. We have shown, as a rule of thumb, that if the CV of the error in catchability is as large as the error CV in the stock size assessment, the unsupervised BR performs similar to the supervised CR. Lack of knowledge makes it difficult to say in general which of the errors is the largest, in fishery-dependent catchability or in assessments. In the case of the Newfoundland cod, CPUE increased when stock abundance declined due to hyper-aggregation of the fish [40]. Such deceitful behavior of CPUE warrants close attention, and neither BRs nor CRs are immune to it. It is possible to estimate catchability accurately when we know the fishing gear, the fishers' behavior, and the fish dispersal pattern. One could say that a transfer from CR to BR would require a shift of focus by managers from surveys and assessments to gear efficiency, behavior of fish and fishing fleets.

HCRs' effects on period lengths of stock size variation

Not only the amplitude of population variation, but also the frequency of the variation has consequences for the population [48]. The induced periodic population variability in predator-prey interactions is well known in ecology [49], but the result of the dynamic interaction between the fish stock and the harvest control is much less studied [15]. It has been hypothesized that populations truncated by fishing have a shorter inherent periodicity, and track environmental fluctuations more closely. Given the assumption that unharvested populations fluctuate at longer frequencies than the environment, this could be the reason for increased amplitude of population variation [4]. Since populations have inherent dynamics controlled by age-specific demographic parameters, one can conclude that populations truncated by harvest should exhibit fluctuations of shorter period length than unexploited ones. In our study, this is true for the N-targeted BRs, but for the other harvest controls the period length of the fluctuations increase. For F-targeted fisheries, stock size deviations from equilibrium delays return times because of a positive feed-back: enlarged populations face reduced mortality and reduced populations are subject to increased mortality. If stocks exhibit long-wave fluctuations with period lengths of 60–80 years due to fishing as indicated with F-targeted CR fisheries, the stress it imposes on ecosystems warrants more attention in future studies and consideration in the choice of method for harvest control.

Potential application of Bayesian regulation

The Bayesian regulation of fisheries modeled here is highly idealized. Real, well-informed management processes are far more complex, and may have sources of error not accounted for. We want to emphasize the potential of BR and stimulate ideas in how it might be implemented. The concept of BR can be developed further, for example to include a Bayesian estimate of the catchability parameter in the probability function (Equation 1). BR has the advantage that, within a year, it can give much higher precision in reaching the management target compared to CR. For instance, stock control with quantitative assessment and forecast can utilize a Bayesian HCR for the harvest year. New assessment methods that calculate a confidence interval for stock size estimate, such as SAM (Nielsen and Berg unpublished), is ideal as they provide priors for a Bayesian regulation.

Another useful property of BR is the ability to work without other assessments, since it is an assessment in itself. It can be an option for the more than 99% of fished species which lack assessments because of the costs and other requirements for collecting the necessary data [50]. An application of BR would still require daily reporting of catches, effort and gear used. But with such a system in place, Bayesian estimates of stock size and fishing mortality could be provided daily, with fishers and managers more up-to-date with current stock status.

ACKNOWLEDGMENTS

Dr. Michael Errigo and an anonymous reviewer gave valuable comments on the manuscript. Oskar MacGregor provided improvements on the language.

REFERENCES

1. Virtala M, Kuikka S, Arjas E (1998) Stochastic virtual population analysis. ICES J Mar Sci 55: 892–904.
2. Mäntyniemi S, Kuikka S, Rahikainen M, Kell LT, Kaitala V (2009) The value of information in fisheries management: North Sea herring as an example. ICES J Mar Sci 66: 2278–2283.
3. Polasky S, Carpenter SR, Folke C, Keeler B (2011) Decision-making under great uncertainty: environmental management in an era of global change. Trends Ecol Evol 26: 398–404.
4. Anderson CNK, Hsieh C, Sandin SA, Hewitt R, Hollowed A, et al. (2008) Why fishing magnifies fluctuations in fish abundance. Nature 452: 835–839.
5. Jonzén N, Ripa J, Lundberg P (2002) A Theory of Stochastic Harvesting in Stochastic Environments. Am Nat 159: 427–437.

6. Getz WM (1985) Optimal and Feedback Strategies for Managing Multicohort Populations. J Optimiz Theory App 46: 505–514.
7. Walters CJ, Ludwig D (1981) Effects of Measurement Errors on the Assessment of Stock-Recruitment Relationships. Can J Fish Aquat Sci 38: 704–710.
8. Hjerne O, Hansson S (2001) Constant catch or constant harvest rate? Baltic Sea cod (*Gadus morhua* L.) fishery as a modelling example. Fish Res 53: 57–70.
9. Aps R, Fetissov M, Holmgren N, Norrström N, Kuikka S (2011) Central Baltic Sea herring: effect of environmental trends and fishery management. In: Villacampa Y, Brebbia CA, editors. Ecosystems and Sustainable Development VIII. Boston: WIT Press Southampton. pp. 69–80.
10. Hightower JE, Grossman GD (1985) Comparison of constant effort harvest policies for fish stocks with variable recruitment. Can J Fish Aquat Sci 42: 982–988.
11. Koonce JF, Shuter BJ (1987) Influence of Various Sources of Error and Community Interactions on Quota Management of Fish Stocks. Can J Fish Aquat Sci 44: s61–s67.
12. Murawski SA, Idoine JS (1989) Yield Sustainability under Constant-Catch Policy and Stochastic Recruitment. T Am Fish Soc 118: 349–367.
13. Hsieh C, Reiss CS, Hunter JR, Beddington JR, May RM, et al. (2006) Fishing elevates variability in the abundance of exploited species. Nature 443: 859–862.
14. Shelton AO, Mangel M (2011) Fluctuations of fish populations and the magnifying effects of fishing. PNAS 108: 7075–7080.
15. Wikström A, Ripa J, Jonzén N (2012) The role of harvesting in age-structured populations: disentangling dynamic and age-truncation effects. Theor Popul Biol 82: 348–354.
16. Kell LT, Pilling GM, Kirkwood GP, Pastoors M, Mesnil B, et al. (2005) An evaluation of the implicit management procedure used for some ICES roundfish stocks. ICES J Mar Sci 62: 750–759.
17. Kraak SBM, Kelly CJ, Codling EA, Rogan E (2010) On scientists' discomfort in fisheries advisory science: the example of simulation-based fisheries management-strategy evaluations. Fish and Fisheries 11: 119–132.
18. Rochet M-J, Rice JC (2009) Simulation-based management strategy evaluation: ignorance disguised as mathematics? ICES J Mar Sci 66: 754–762.
19. Stephenson R, Peltonen H, Kuikka S, Pönni J, Rahikainen M, et al.. (2001) Linking Biological and Industrial Aspects of the Finnish Commercial Herring Fishery in the Northern Baltic Sea. In: Funk F, Blackburn J, Hay D, Paul AJ, Stephenson R et al.., editors. Herring: Expectations for a New Millennium. Fairbanks: University of Alaska Sea Grant. pp.741–760.
20. May RM, Beddington JR, Horwood JW, Shepherd JG (1978) Exploiting natural populations in an uncertain world. Mathematical Biosciences 42: 219–252.
21. Hilborn R, Mangel M (1997) The ecological detective. Princeton: Princeton University Press.
22. Dall SRX, Giraldeau L-A, Olsson O, McNamara JM, Stephens DW (2005) Information and its use by animals in evolutionary ecology. Trends Ecol Evol 20: 187–193.
23. McNamara JM, Green RF, Olsson O (2006) Bayes' theorem and its applications in animal behaviour. Oikos 112: 243–251.
24. Valone TJ (1991) Bayesian and prescient assessment: foraging with pre-harvest information. Anim Behav 41: 569–577.
25. Charnov EL (1976) Optimal foraging, the marginal value theorem. Theor Popul Biol 9: 129–136.
26. Olsson O, Holmgren NMA (1999) Gaining ecological information about Bayesian foragers through their behaviour. I. Models with predictions. Oikos 87: 251–263.
27. Olsson O, Wiktander U, Holmgren NMA, Nilsson SG (1999) Gaining ecological information about Bayesian foragers through their behaviour. II. A field test with woodpeckers. Oikos 87: 264–276.
28. Green RF (1980) Bayesian birds: a simple example of Oaten's stochastic model of optimal foraging. Theor Popul Biol 18: 244–256.

29. Iwasa Y, Higashi M, Yamamura N (1981) Prey distribution as a factor determining the choice of optimal foraging strategy. Am Nat 117: 710–723.
30. Olsson O, Holmgren NMA (1998) The survival-rate-maximizing policy of Bayesian foragers: wait for good news! Behav Ecol. 9: 345–353.
31. Holmgren NMA, Norrström N, Aps R, Kuikka S (2012) MSY-orientated management of Baltic Sea herring (Clupea harengus) during different ecosystem regimes. ICES J Mar Sci 69: 257–266.
32. ICES (2011) Report of the Workshop on Implementing the ICES Fmsy Framework (WKFRAME-2), 10–14 February 2011, ICES, Denmark. ICES CM 2011/ACOM 33: 110 pp.
33. European Commission (2002) On the conservation and sustainable exploitation of fisheries resources under the Common Fisheries Policy. Council regulation (EC) No 2371/2002.
34. Marchal P, Horwood J (1998) Increasing fisheries management options with a flexible cost function. ICES J Mar Sci 55: 213–227.
35. Pelletier D, Laurec A (1992) Management under uncertainty: defining strategies for reducing overexploitation. ICES J Mar Sci 49: 389–401.
36. ICES (2009) Report of the ICES Advisory Committee, 2009. ICES Advice, 2009. Book 8, 132 pp.
37. ICES (2012) Report of the Workshop 3 on Implementing the ICES Fmsy Framework, 9–13 January 2012, ICES, Headquarters. ICES CM 2012/ACOM 39: 1–33.
38. ICES (2012) Report of the ICES Advisory Committee 2012. ICES Advice, 2012. Book 8, 158 pp.
39. Valone TJ, Brown JS (1989) Measuring Patch Assessment Abilities of Desert Granivores. Ecology 70: 1800–1810.
40. Rose GA, Kulka DW (1999) Hyperaggregation of fish and fisheries: how catch-per-unit-effort increased as the northern cod (Gadus morhua) declined. Can J Fish Aquat Sci 56: 118–127.
41. Rohlf FJ, Sokal RR (1995) Statistical tables. New York: W.H. Freeman & Co.
42. Green RF (1984) Stopping rules for optimal foragers. Am Nat 123: 30–43.
43. Patterson KR (1999) Evaluating uncertainty in harvest control law catches using Bayesian Markov chain Monte Carlo virtual population analysis with adaptive rejection sampling and including structural uncertainty. Can J Fish Aquat Sci 56: 208–221.
44. Prager MH, Porch CE, Shertzer KW, Caddy JF (2003) Targets and limits for management of fisheries: a simple probability-based approach. N Am J Fish Manage 23: 349–361.
45. Shertzer KW, Prager MH, Williams EH (2008) A probability-based approach to setting annual catch levels. Fish Bull 106: 225–232.
46. Patterson K, Cook R, Darby C, Gavaris S, Kell L, et al. (2001) Estimating uncertainty in fish stock assessment and forecasting. Fish and Fisheries 2: 125–157.
47. Getz WM, Francis RC, Swartzman GL (1987) On Managing Variable Marine Fisheries. Can J Fish Aquat Sci 44: 1370–1375.
48. Ripa J, Lundberg P (1996) Noise colour and the risk of population extinctions. Proc R Soc London B 263: 1751–1753.
49. Yodzis P (1991) Introduction to theoretical ecology. New York: Harper & Row.
50. Costello C, Ovando D, Hilborn R, Gaines SD, Deschenes O, et al. (2012) Status and Solutions for the World's Unassessed Fisheries. Science 338: 517–520.

CHAPTER 3

Water Budgets for Freshwater Aquaculture Ponds with Reference to Effluent Volume

K. K. Sharma*, B. C. Mohapatra, P. C. Das, Bikash Sarkar, S. Chand

Central Institute of Freshwater Aquaculture, Bhubaneswar, India;

ABSTRACT

This paper presents the consumptive water use for freshwater pond aquaculture for semi-intensive carps farming practices. The consumptive use of water includes evaporation loss, seepage loss and water exchanges requirements. The water requirement has been estimated to be 10.3 m^3/Kg of fish production under present study for semi-intensive culture and with supplemental feeding. Out of which 7.6 m^3/Kg of fish production is system associated requirement. On an average the evaporation loss from the pond is 1498.3 mm/year and seepage loss per year is about 1182.60 mm/year. Seepage and water exchange losses recharge the ground water aquifers and if they are treated and recycled, the water use in aquaculture can be reduced significantly. A further reduction in fresh water use in pond aquaculture is possible through development of intensive and superintensive culture systems and aqua feeds.

Keywords: Water Budgets; Freshwater Aquaculture; Ponds; Carps; Seepage; Pond Evaporation

1. INTRODUCTION

Indian aquaculture has demonstrated a six and half fold growth over the last two decades, with freshwater aquaculture contributing over 95 percent of the total aquaculture production. The production of carp in freshwater and shrimps in brackish water form the major activity of the areas. The three Indian major carps, namely catla (*Catla catla*), rohu (*Labeo rohita*) and mrigal (*Cirrhinus mrigala*) contribute the bulk of production with over 1.8 million tonnes [1]; followed by silver carp, grass carp and common carp forming a second important group. Average national production from still water ponds has increased from 0.6 tonnes/ha/year in 1974 to 2.2 tonnes/ha/year by 2001-2002 [2], with several farmers even demonstrating production levels as high as 8 - 12 tonnes/ha/year. The technologies of induced carp breeding and polyculture in static ponds and tanks virtually revolutionized the freshwater aquaculture sector and turned the sector into a fast growing industry.

Fish culture is a water-intensive endeavor and requires much more water than conventional agriculture [3]. The current expansion of freshwater aquaculture in the India may require large quantity of freshwater either groundwater or surface water. There already exists concern of adequate water supplies for traditional agriculture. Further, uncertainty in monsoon rain, scare and limited availability of freshwater resource have forced in rethinking wise-use of freshwater in aquaculture sector to increase water productivity. Nowaday's water is increasingly becoming less available and costly to procure. World in general and India in particular, the freshwater supply and reserve are now under threat due to increased population following by increased demand of water in agriculture, aquaculture, industry and domestic sectors. Unplanned wasteful use of water in aquaculture is limiting further development of this sector. Therefore, more research is badly needed on water requirements in pond aquaculture and how to minimize the water use. Till date in India, very little work has been carried out in this regard to quantify optimum water requirement for hatchery, nursery, rearing and grow-out culture of carps. A few researchers outside [3-6] have worked out water budgets based on pond measurements for different type of systems/ponds and also in different climatic conditions. Reference [7] developed water budget model as a general methodology that can be adopted to predict water requirements for new locations and [8] has developed water budgets for portable carp hatchery. Because of increasing demand to produce more per unit volume of water resource, the quantification of water requirement for fish culture assumes great importance in view of proper planning [9] for judicious use of freshwater resource of the country. Reference [10] compared the water use in aquaculture with water used for terrestrial animal production. Water budgeting is very important for estimating the total water requirements of ponds, flow through facilities, hatcheries etc. and also to estimate and predict the fish culture potential for different regions.

This paper presents the consumptive water requirements for freshwater pond aquaculture at study site. The hydrology of pond is depending on local conditions, soil type, construction methods, seepage, evaporation, rainfall and other criteria. However the knowledge or protocols about estimating water use for pond aquaculture developed in this study could be applied in the same agroclimatic region of the country.

2. MATERIALS AND METHODS

2.1. Experimental Site

The experimental site is the farm of Central Institute of Freshwater Aquaculture, Bhubaneswar (CIFA) (Latitude: 20°11'06"N; Longitude: 85°50'52"E and msl: 33 m), Orissa, India, consists of around 350 ponds ranging from 0.01 ha to 4 ha in area (total about 50 ha cultured water body) with depth running from 1 m to 4.5 m. Ponds have been constructed by making earthen embankments around the storage area. These ponds are either rectangular or square in shape. The ponds are mainly divided into two categories, one is where there is ground water contribution during few months and other where these are built in swampy area and retain water throughout the year. The water in ponds is generally maintained by supply of the water either from nearby canal or reservoirs. Such ponds are having outflow pipes to drain off excess water/rainfall in a regulated manner. These ponds are also receiving ground water contribution during some of the month of a year, when surrounding water table is high because of flow in nearby canal. The hydrology of these ponds mainly depends on soil type, water table depth, flow in canals, methods of construction, seepage, evaporation and rainfall etc. The soil composition at Study site-I was sand 88%, silt 8% and clay 4%. Similarly the soil composition at Study site-II was sand 82%, silt 10% and clay 8%.

The farm is located between Latitude: 20°11'06"N; Longitude: 85°50' 52"E and msl: 33 m. The study area is characterized as warm and humid climate, where humidity is always very high and in summer it reaches up to 99%. The summer temperature rises to 44°C and winter lowest temperature reaches 4°C. The averages annual rainfall is about 1120 mm of the region. The rainy season generally occurs from May to September and then there is a post monsoon rainfall also.

2.2. Experimental Designs and Hydrological Measurements

The water budgets study was conducted in three numbers of rearing fishponds initially for one-year period (2007-2008). These ponds were stocked with rohu, catla and mrigal fingerlings of size ranging 4 mm to 6 mm at stocking density of 7000 per hectare. Ponds were fed with floating pelleted feed twice a day at the rate of 2% of body weight of fish. In order to study the effect of special variability and confirm the results obtained, next year same study was conducted in three different ponds at CIFA farm. The ponds are rectangular in shape of size 50 m × 20 m of each and the depths of the pond were about 1.5 m. The water level in ponds was maintained ranging 1.0 m to 1.2 m by providing out let pipes of 15 cm diameter to each pond. These ponds were also provided with inlet pipe of 15 cm diameter from the supply channel for introducing water in the ponds from reservoirs or nearby irrigation canal.

To measure water level changes in ponds, a water level recorder was fabricated which consist of a standpipe and glass cylinder with scale for measurement. The water level is measured inside the measuring cylinder to eliminate the effect of wind wave on stage. These instruments were installed inside the each pond and data on stage change of pond water were taken on daily basis between 7.30 to 8.00 hr. In the next year, these pond water level stage recorders were installed in other nine numbers of ponds for conformity of the results obtained.

A class A Evaporation pan was installed on dyke of pond for measurement of evaporation losses. A hook facilitated accurate estimates of water level changes in the pan. A standard rain gauge was provided in the farm area to measure the rainfall during the study period. The catch of the rain gauge also allowed corrections of water level changes in the evaporation pan so that evaporation could be measured accurately by subtracting rain gauge catch water.

Pond evaporation was estimated by using pan coefficient. The pan-pond coefficient was determined by measuring the stages in the Reinforced Cement Concrete (RCC) tank of 10 m × 5 m of size and with 1.2 m depths with effective water depth at around 1.0 m available in the farm. A staff gauge was fixed inside the tank and the stage at the top of the gauge was fixed. The data on water stages were recorded on daily basis by maintaining the water level in the tank by adding the evaporation losses.

Regulated inflow and out flow from the ponds were measured by providing the inlet and outlet pipes of 15cm diameter of PVC materials. Flow meters were installed in these pipes for measuring the regulated flows.

Runoff was estimated by a curve number method [11]. The soil was classified to a particular hydrologic group based on texture; hydrologic soil group and vegetative cover were then used to obtain curve number. The area from estimating the runoff contributions to each pond was taken as 50% of the pond dyke area because these ponds are levee ponds.

The data as discussed above were collected regularly for one-year period starting from April to March. The pond-pan coefficient, pond evaporation and pond seepage were calculated as follows:

Pond-pan coefficient = Class A pan evaporation/Evaporation from RRC tank

$$\text{Pond-pan coefficient} = \frac{\text{Class A pan evaporation}}{\text{Evaporation from RRC tank}} \quad (1)$$

$$\text{Pond Evaporation} = \left(\text{Pond-pan coefficient} \times \text{Class A pan evaporation}\right) \quad (2)$$

Pond Seepage:

Case 1: When there is no rainfall

$$\text{Pond seepage} = \left\{\left(\text{Stage at time 1} - \text{Stage at time 2}\right) - \text{Pond evaporation}\right\} \quad (3)$$

Case 2: When there is rainfall

$$\text{Pond seepage} = \left[\left\{\left(\text{Stage at time 1} + \text{Rainfall} + \text{Runoff}\right) - \text{Stage at time 2}\right\} - \text{Pond evaporation}\right\}\right] \quad (4)$$

Case 3: When there is heavy rain and overflow

$$\text{Pond seepage} = \left[\left\{\left(\text{Stage at time 1} + \text{Rainfall} + \text{Runoff} - \text{Overflow}\right) - \text{Stage at time 2}\right\} - \text{Pond evaporation}\right\}\right]. \quad (5)$$

3. RESULTS AND DISCUSSION

3.1. Determination of Pond-Pan Coefficient

The pond-pan coefficient was determined for calculation of pond evaporation from the pond water stage data. The daily evaporation data from class A pan evaporimeter and RCC tank of 10 m × 5 m size were scrutinized to get the weekly data. These weekly data were plotted and fitted with straight line (Figure 1). The slope of the line is taken as the pond-pan evaporation coefficient. The pond-pan coefficient for the present case has been estimated as 0.68. This coefficient was later used to separate the pond evaporation and pond seepage from the daily stage data.

3.2. Pond Evaporation Losses

Significantly amount of water can be lost from the pond via evaporation, whether it may be either small or larger ponds. Evaporative water loss is primarily a function of ambient air temperature, relative humidity, solar radiation and wind velocity. The daily data of pan evaporation were scrutinized to get the weekly pan evaporation. These weekly pan evaporation data were multiplied by pond-pan coefficient to get the pond evaporation. The daily pond evaporation on an average varied from 0.20 ± 0.05 mm to 5.96 ± 0.75 mm during different months of the year. Mean weekly variation of pond evaporation is presented in Figure 2. Weekly pond evaporation averaged 35.65 ± 6.78 mm. On an average the evaporation loss from the pond is 1498.3 ± 8.61 mm/year. High evaporation rates during the dry months were influenced by lower humidity, strong weed speed and higher solar radiation. But, in case of rainy months, the pond evaporation was lower due to presence of higher humidity.

3.3. Pond Seepage Losses

High seepage rate is not only the results in water losses but also affects the pond fertility, pond dynamics and environment. Pond water loss or gain by seepage depends on primarily on the soil porosity, methods of pond construction, and structural changes over the period of time and pond management. The daily data of pond seepage were separated from daily pond water stage changes data recorded by installing the pond water level recorder in each pond by deducting the pond evaporation data calculated above using pond-pan coefficient. The daily seepage rate on an average varied from 1.1 ± 0.20 mm/day to 14.26 ± 1.50 mm/day during different months of the year. Mean weekly seepage for the present study is shown in Figure 3. The mean weekly seepage rates ranged from 8.28 ± 2.15 to 84.41 ± 3.60 mm. On average the seepage loss is about 1182.60 ± 10.16 mm/year. The rates of seepage in rainy months were almost 20% - 25% less than the dry months. Seepage losses are generally lower during rainy months, which may be due to rainwater infiltrated through pond dykes and that decreased the net seepage losses.

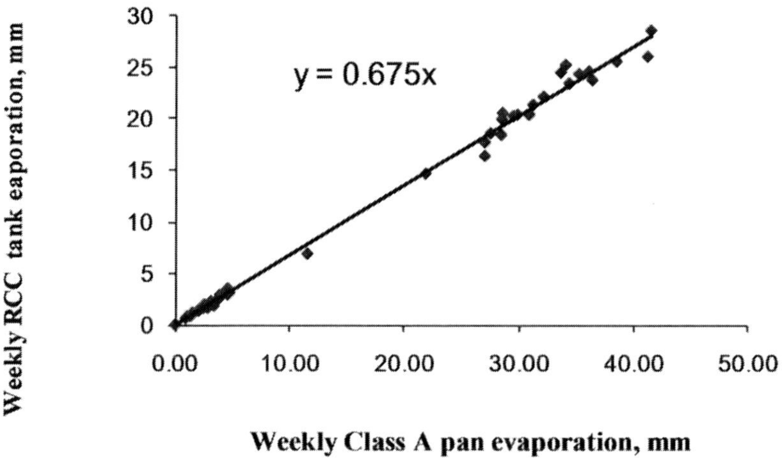

Figure 1. Relationaship between weekly open PanReinforced Concrete Cement tank evaporation.

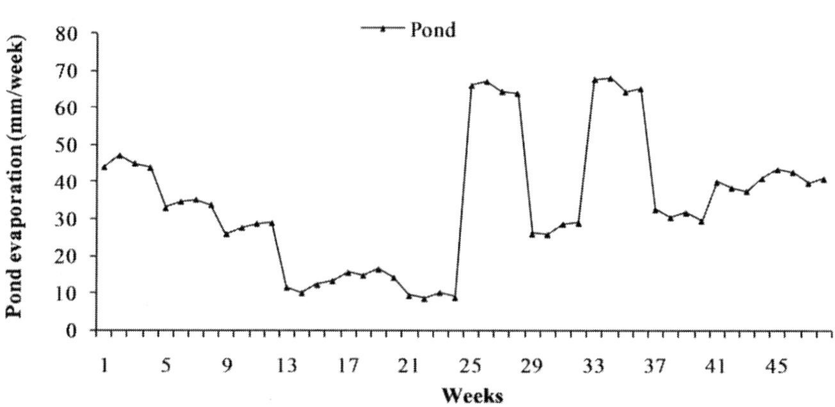

Figure 2. Mean weekly variation of pond evaporation at Study site.

3.4. Rainfall

Measured mean monthly rainfalls during 2006-2009 were shown in Figure 4. In the warm and humid climate rainfall started from early June and continue up to October. During rainy months, the pond water compensated by rainfall. While remaining months, regulated inflow of water was required for each month to replace water losses to pond evaporation during the study period.

3.5. Water Budgets for Pond Aquaculture

The average water budgets have been prepared and summarized in Table 1. From table it is evident that in almost all the months during the year, water had to be added to meet the requirement of water losses from fishponds except during rainy season where in water losses were compensated by gain in ponds either directly from rain or ground water contribution. The monthly average gain was highest in the month of June and lowest in April. Whereas inflow was quite higher in the month of April and lower in July. Total water gain, pond evaporation, seepage and water addition for a year is 1761.8 ± 5.84 1498.3 ± 8.61, 1182.60 ± 10.16 and 2431.2 ± 15.47 mm for the present study.

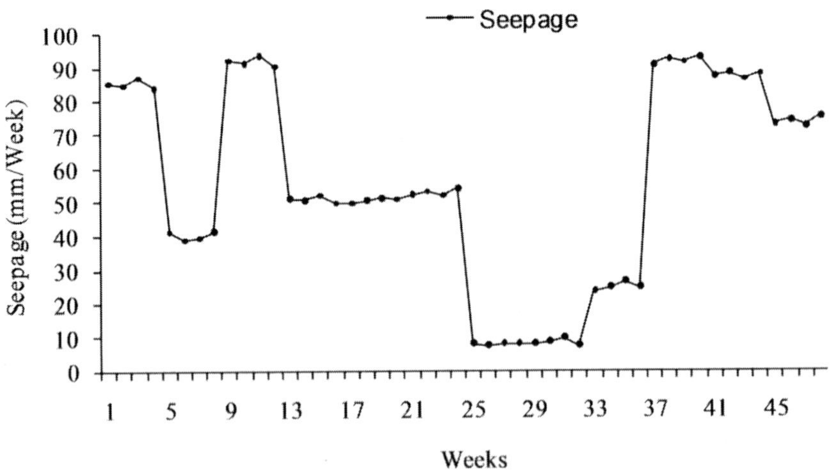

Figure 3. Mean weekly variation of seepage at study site.

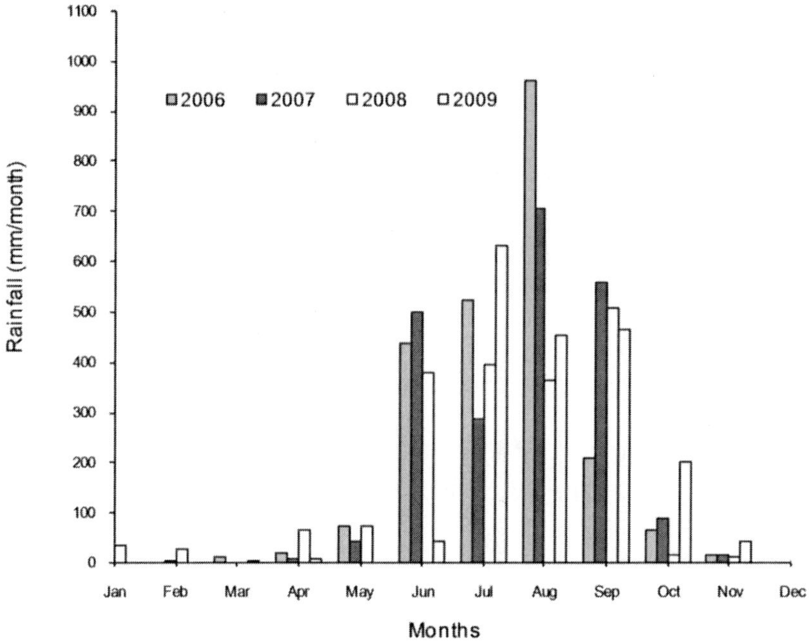

Figure 4. Monthly rainfall on ponds in different years.

Table 1. Mean monthly water budgets of fishponds at CIFA farm.

Months	Water gain (Rain + Run-off + ground water), mm	Pond Evaporation (mm)	Pond seepage, (mm)	Water addition (mm)
April	9.0 ± 1.2	(−)181.3 ± 12.6	(−)237.6 ± 21.4	230.0 ± 18.6
May	0.80 ± 0.2	(−)139.3 ± 6.5	(−)160.9 ± 15.6	300.1 ± 20.5
June	35.5 ± 2.3	(−)110.7 ± 8.1	(−)169.2 ± 18.5	444.5 ± 32.5
July	546.3 ± 19.8	(−)50.7 ± 7.3	(−)42.0 ± 5.8	20.40 ± 2.0
August	434.0 ± 15.2	(−)63.5 ± 6.2	(−)38.5 ± 4.5	207.5 ± 16.8
September	453.0 ± 14.5	(−)36.6 ± 4.5	(−)35.8 ± 4.2	60.0 ± 7.4
October	201.0 ± 10.2	(−)161.8 ± 9.1	(−)53.5 ± 5.8	15.20 ± 2.4
November	44.0 ± 3.5	(−)111.1 ± 7.8	(−)38.8 ± 7.6	106.0 ± 12.6
December	34.0 ± 2.2	(−)132.2 ± 8.4	(−)85.3 ± 8.9	237.2 ± 17.9
January	0.60 ± 0.1	(−)161.0 ± 10.2	(−)93.2 ± 9.7	257.5 ± 15.5
February	0.45 ± 0.1	(−)170.0 ± 10.5	(−)102.0 ± 9.8	243.2 ± 18.3
March	5.0 ± 0.8	(−)179.9 ± 12.3	(−)125.5 ± 10.2	345.2 ± 21.2

The detail of contribution of different components in term of percentage is given below:

Water gain: 35.55% (Rainfall + Run off and Ground inflow)

(Rainfall: 75%; Run off 2% - 3%; Ground inflow 22% - 23%)

Regulated inflow: 64.45% Evaporation: 51.25% and Seepage loss: 48.75%.

3.6. System Associated Water Use in Freshwater Aquaculture

The system associated water use in freshwater aquaculture includes evaporation loss, seepage loss and water exchange for water quality management. In the present study it was assumed that there is no need for water exchange, as the water quality of fishpond did not deteriorate because of extensive aquaculture practices. The on an average the evaporation loss from the pond is 1498.3 mm/year and seepage loss per year is about 1182.60 mm/ year. Consequently, the water requirements for the production of Indian Major carps are presented in Table 2. The water requirement for a production target of 3 - 4 ton/ha/year comes to 7650 litre or 7.6 m^3 of water per Kg of fish production and water requirement decreases with the intensification of aquaculture or increase of production (Table 2). If seepage losses are considered to be green water [12] then total water consumption is about 4.1 m^3 per Kg of fish production. About 850,000 ha pond area is under carp cultivation in India [13] and hence water requirement is very high for freshwater aquaculture sector.

Table 2. Water requirements of Indian Major Carps for production of one Kilogram fish under different production technologies.

Production (Ton/ha/year)	Water requirement at CIFA Study Site in m^3/kg
3 - 4	7.65
6 - 8	3.82
10 - 12	2.43

3.7. Feed Associated Water Use in Aquaculture

The water use associated with the production of bone or blood meal, fishmeal and fish oil can be considered negligible. Animals fed formulated diets indirectly consume large quantities of water. Globally, about 1.2 m^3 of water is needed to produce 1 kg of grain used in animal feeds. Fish or crustaceans require less than 2 kg of grain concentrate for each kg produced, making them the most efficiently producing animals in terms of feed-associated water use. This means 2.4 m^3 of water is also being consumed to produce 1 Kg of fish if the feed is being used above that of 4.1 m^3system associated water requirement. Hence, total water used to produce 1 Kg fish is 6.5 m^3 if the seepage losses are considered as green water. Otherwise the total water used would be about 10.3 m^3 per Kg of fish produced in the present study.

3.8. Freshwater Resources Availability and Utilization in India

The average annual precipitation of the country is about 400 Mham. However the utilizable surface water resources and ground water resources are about 70 Mham and 35 Mham respectively. Additionally 2.36 Mha area of ponds and tanks are also available to be used for various purposes such as domestic use, irrigation, aquaculture etc. The future requirements of water resources have been predicted to meet the growing demand by different sector [14].

About 70% of total water requirement would be going for agriculture through irrigation in 2025. The total utilizable resources are about 105 Mmam and 2.36 Mha area of ponds and tanks after full development of ground water as well as surface water resources for utilization but about 77 Mham would be supplied to irrigate agriculture only to feed the growing population. In conesquence, the future growth of agriculture including freshwater aquaculture would be constrained by the freshwater availability. For development of freshwater aquaculture other water resources like Canal, reservoirs and Oxbow lakes & derelict waters should be fully exploited.

3.9. Comparison of Water Requirement of Freshwater Aquaculture with Crops

The water productivity for freshwater aquaculture, which has been estimated in the present study, was compared with the other crops water productivity Table 3 [15]. It is clear from the table that the freshwater aquaculture needs more water to produce same amount of fish than the grains. Among the crops compared rice is the least water productive with water productivity 3.7 Kg/ha mm and Freshwater aquaculture is even lesser water rice also having 1.50 Kg/ha/mm water productivity.

3.10. Options to Reduce the Use of Water in Freshwater Aquaculture

The water use (7.6 m^3) per Kg of fish production was very high in case of semi-intensive pond culture where production of fish was 3 - 4 ton/ha/year. If the intensive pond culture with supplementary feeding and aeration, the production of 10 - 15 tonnes/ha/yr could be achieved, the water productivity would be almost doubled. The focus should be on using intensive and super intensive culture practices for aquaculture production. The main water consumption in pond aquaculture is in the form of evaporation and seepage losses. Hence our efforts should be such that these losses are minimized during the process. The seepage loss may be controlled by adopting different control measures in pond aquaculture [16]. In intensive or super intensive culture systems, focus should be on assuring the treatment of effluent water and reuse so that these should no longer be considered as a loss. RAS have got very good potential for increasing water productivity very high; they should be tried wherever possible.

On farm water management should also be practiced to reduce the water wastage and make available more water for aquaculture. Water storage capacity of ponds with extensive areas of shallow water can be improved if they are drained and deepened. The shallowest area in a pond should be at least three feet deep. This depth of water will slow growth of aquatic plants, which can take up a large amount of pond volume, effectively reducing the water storage capacity of the pond. To conserve water and reduce energy costs of pumping water in the pond, maintain water levels below the maximum allowed to the standpipe. This practice allows rainfall to be collected in the ponds. How much below maximum level the pond is filled depends upon season, weather patterns, evaporation rates, location, and water holding capacity of the pond, size of the fish in the pond and weight of fish.

Table 3. Water productivity for different crops including freshwater aquaculture.

Crops (New Strains)	Water requirements (Cm)	Yield Kg/ha	Productivity of water Kg/ha/mm
Rice	120	4500	3.7
Sorghum	50	4500	9.0
Pearl millet	50	4000	8.0
Maize	65.5	5000	8.0
Wheat	40.0	5000	12.5
Freshwater aquaculture (Carp culture)	**268**	**4000**	**1.50**

4. CONCLUSION

This study presents the water budgets for freshwater aquaculture ponds for semi-intensive carp culture and also discusses means and ways to reduce the water use in freshwater aquaculture. The total freshwater requirement has been estimated to be 10.3 m^3/Kg production under present study for semi-intensive culture and with supplemental feeding, out of which 7.6 m^3/Kg is system associated requirement. The system-associated requirements are mainly evaporation and seepage losses from the pond during fish culture period. On an average the evaporation loss from the pond is 1498.3 mm/year and seepage loss per year is about 1182.60 mm/year. Water conservation and reuse have become a major issue in aquaculture in recent years. Concern over increased demand on ground water, cost of operating wells, environmental impact of aquaculture effluent and the desire to increase production efficiency continues to drive advances in technology and management practices. Seepage and water exchange losses recharge the ground water aquifers and if they are treated and recycled, the water use in aquaculture can be reduced significantly. A further reduction in freshwater use in pond aquaculture is possible through development of intensive and super-intensive culture systems and aqua feeds.

5. ACKNOWLEDGEMENTS

The authors are thankful to Dr. A. E. Eknath, Director, Central Insti- ute of Freshwater Aquaculture for providing facilities to carry out this work and also for his constant guidance and encouragement.

REFERENCES

1. FAO (2005) Aquaculture production, 2003. Yearbook of Fishery Statistics, Vol.96/2, Food and Agriculture Or-ganization of the United Nations, Rome
2. Tripathi, S.D. (2003) Inland Fisheries in India. Fish for All National Launch, 18-19 December.
3. Boyd, C.E. (1982) Hydrology of small experimental fishponds at auburn, Alabama. *Transactions of the American Fisheries Society*, **111**, 638-44. doi:10.1577/1548-8659(1982)111<638:HOSEFP>2.0.CO;2
4. Boyd, C.E. (1986) Influence of evaporation excess on water requirements for fish farming. *Proceeding of the Conference on Climatic and Water Management*, American Meteorological Society, Boston, 62-64
5. Teichert-Coddington, D.R., Stone, N. and Phelps, R.P., (1988) Hydrology of fish culture ponds in Gualaca, Pa-nama. *Aquacultural Engineering*, **7**, 309-320. doi:10.1016/0144-8609(88)90013-1

6. Green, B.W. and Boyd, C.E. (1995) Water budgets for fish ponds in the dry tropics. *Aquacultural Engineering*, **14**, 347-356. doi:10.1016/0144-8609(95)00002-V
7. Nath, S.S. and Bolte, J.P. (1998) A water budget model for pond aquaculture. *Aquacultural Engineering*, **18**, 175-188. doi:10.1016/S0144-8609(98)00029-6
8. Mohanty, B.B., Mal, B.C., Sharmaand, K.K. and Moha-patra, B.C. (2009) Water requirements of a portable FRP carp hatchery for Rohu spawning and eggs hatching. *Fishing Chimes*, **29**
9. Kathpalia, G. (1991) Water budgeting and planning. Wa-ter pollution and management, edited by C. K. Varshney, 25-35.
10. Verdegem, M.C.J., Bosma, R.H. and Berreth, J.A.J. (2006) Reducing water use for animal production through aqua-culture. *International Journal of Water Resource Devel-opment*, **22**.
11. United States Soil Conservation Service (1971) Ponds for water supply and recreation. United States Department of Agriculture, Agriculture Handbook 387
12. Verdegem, M.C.J. and Bosma, R.H. (2009) Water with-drawal for brackish and inland aquaculture, and options to produce more fish in ponds with present water use. *Water Policy*, **11**, 52-68
13. Ayyappan, S. (2006) National aquaculture sector over-view: India. National Aquaculture Sector Overview Fact Sheets, FAO, Rome
14. Irrigation Commission, Ministry Of Irrigation and Power, Government of India (1972) Report of the Irrigation Commission
15. Michael, A.M. (1997) Irrigation: Theory and practices. Vikas Publishing House Pvt Ltd, Delhi
16. Jayanthi, M., Rekha, P.N., Muralidhar, M. and Gupta, B.P. (2004) Seepage reduction in brackish water ponds with different materials. *Ecology, Environment and Conservation*, **10**, 257-260

CHAPTER 4

Causalities between Price, Pond Area and Employment in Aquaculture Production

Nik Hashim Nik Mustapha[1], Azlina Abd Aziz[1], Nik Mohd Hazrul Hashim[2]

[1]*Department of Economics, Faculty of Management and Economics, University of Malaysia Terengganu, Kuala Terengganu, Malaysia;*

[2]*Graduate School of Business, National University of Malaysia, Bandar Baru Bangi, Malaysia.*

ABSTRACT

The role of aquaculture industry is becoming more prominent in order to supplement marine capture in meeting the food need for the growing Malaysian population. In an attempt to minimize depletion of marine fisheries, only traditional vessels are allowed to fish along the coastal area while bigger vessels are relegated to deep-sea fishing. During the 9th Malaysian Plan (2006-2010) aquaculture has been recognized as the engine of growth in the national food sector's development strategy. Future fisheries policy is expected to focus more on aquaculture production, marketing and technological improvement as an alternative to marine capture. This paper investigates the causalities between the selected freshwater fish prices, aquaculture area and production. The study aspires to establish whether or not market price is a key contributor to a rise in the aquaculture area and production. Aquaculture firms comprising the individual culturists are generally motivated by the economic potential of the industry which is reflected in excess of price over cost of production. Our hypothesis is that government policy and initiation rather than prices had give rise to greater participation of culturists and hence augmented the level of employment. However, production increase has a negative implication on environment degradation. Thus there is a conflicting view as regards to the employment opportunity generated by aquaculture undertakings and the need for sustainable development arising from this growing industry. Multivariate time series analysis was used in this investigation.

Keywords: Aquaculture; Unit Root; Cointegration; Vector Error Correction Model; Granger's Causality

1. INTRODUCTION

The popularity of aquaculture practice in Malaysia comes with its commercialization during the last few decades. As such aquaculture is dominated by several fishery species associated with the market potential which is primarily reflected in high prices offered to the breeders. These fisheries are in fact experiencing shortage in supply. Species such as giant prawn (Macrobrachium rosenbergii) and tiger prawn (Penaeus monodon) are delicacies of hotels and restaurants due to the high demand while the supply is limited. Another underlying factor for their high demand is that special fishery species like the catfish (Clarias gariepinus) is believed among the majority of its medicinal healing capability. However, the success of culturing these aquatic resources either freshwater or brackish-water fisheries is always constrained by the economic factors such as high capital outlay to start the business and the expertise knowledge required in their breeding and maintenance. Availability of fingerlings, cost of feeding ingredients, suitability of land area in terms of location and lastly problems related to marketing and distributional system could be the salient factors that need to be considered in aquaculture practices.

In macro perspective aquaculture fishery is an essential component of the economy food security and sustainability. In 2009 the aquaculture statistics on area, production and the number of culturist were 4916.2 km^2, 21,051 metric tons and 70361.5 persons respectively. The retail value accounted for about RM476, 549.45 for 2009 [1]. These statistics show continuous growth trends over the years indicating that the aquaculture industry is escalating perhaps due to the fact investors are attracted to the profitability of the prospective species. With an increase in demand for food owing to ever-growing population and heavy reliance of the present fishery policy on marine capture, the crucial function of aquaculture as supplementary to marine fisheries becomes obvious. By the end of the 9th Malaysia Plan 2010 Malaysia targeted to produce some 507,550 metric tons of aquaculture and aquatic products under the National Agricultural Policy [2]. Aquaculture is seen as an alternative that could reduce the pressure of over fishing on marine fisheries resource. The role of aquaculture in food economy is stipulated to become more important in future in view of depleting marine fisheries. In the light of this changing scenario which the government has agreed to adopt this as a new fisheries policy it deems necessary to investigate the extent to which interactions between pricing, area and employment are intertwined in order to provide economic rationale.

The objective of this study is to investigate the interdependencies of economic policy variables namely prices of freshwater fisheries, area invested with aquaculture ponds and the number of culturists. The general hypothesis is that although aquaculture provides the best alternative to marine capture its growth depends on the economic interplays of fish prices and aquaculture area. In big cities land competes for alternative investments pushing land price extremely high. Since the investment in aquaculture activity generally requires high initial capital, technical knowhow on fisheries rearing, difficulty of getting seedlings, high percentage of fish mortality due to river pollution, investing in aquaculture is almost impossible. Culturally majority of Malaysian favor marine fisheries it is therefore, hypothesized that the probability of a significant increase in aquaculture operators and thus employment is minimal. The results of this investigation are useful for decision makers in formulating strategic fisheries policies. If price is highly significant and correlated with employment and area cultured or otherwise, then strategic policy can be under-taken to further support such relationships.

2. FRESHWATER CATCH AND PRODUCTION

Statistics provided by AQUASTAT [3] show that freshwater catch in 1990 was estimated at 12,995 metric tons about 10 percent of the total catch of 3,783,743 metric tons in Asia (excluding Middle-East). Freshwater catch for 2000 figures were 22,636 metric tons and 5,959,055 metric tons for Malaysia and Asia respectively. The catch difference is about 12 percent but there is an improvement of about 2 percent per annum over the period 1990-2000. The total share of aquaculture for Malaysia in terms of Asian aquaculture production in 1997 was relatively small which accounted for only 5 percent.

One of the more recent developments in aquaculture species of market potential is the red hybrid tilapia (Oreochromis sp.), a genetic improvement with attractive red coloration. In 2007 Malaysia produced 32,258 metric tons (80% red tilapia and 20% black tilapia) with the wholesale value of MYR170.6 million [4]. According to the same source fresh water red tilapia production constitutes about 46 percent of total aquaculture production and 12 percent of this total is from brackish water. In value term the production of tilapia is currently leading all other aquaculture species constituting 49.4 percent, followed by catfish 37 percent and carps 10 percent. The major drawbacks in aquaculture, in particular, the red hybrid tilapia production are soaring feeds cost, shortage of trained workers, diseases resistance fries and the supply of fingerlings has to be bought from commercial firms. The Institute of Tropical Aquaculture of the University Malaysia Terengganu (UMT) is doing research to produce diseases resistant seeds of tilapia.

The major drawback during the earlier part of freshwater development in this country originated from the domination of the marine capture over freshwater species and the consumers' preference for the former. Over the last decade marine capture had experienced drastic changes due to depletion of the resource partly caused by the increased demand as population and income grew. On the supply side depletion of the resource is attributable to several factors, namely the continuous application of efficient fishing methods such as trawlers, the encroachment of efficient vessels from outsiders within the Malaysian 200 nautical miles Exclusive Economic Zone (EEZ), ineffectiveness policing of the fishing grounds owing to insufficient enforcement officers to guard the assigned EEZ. Aquaculture investment had been in Malaysia for a long period of time but had been tried on a relatively small scale because of the high risk and large capital needed for the industry. Hence, with existing requirement for additional supply of fish in addition to marine capture this industry started to flourish partly due to the aquaculture potential for sea food industry.

Experience shows that the success of aquaculture investment is likely to benefit the large scale operators more relative to small operators pertaining to the technical knowhow, culturing and breeding practices. In delivering their products contracts with the local hotels are made and for exporting fish products to destinations outside the country approval and procedures need to be adhered. The small operators with limited working capital, technical knowhow and access to the target market would be greatly at the disadvantage. Alternatively small operators could make this aquaculture undertaking a success through fishermen Association or Cooperatives. Otherwise the government policy that attempts to motivate culturists on an individual basis would be constrained by those farming practices, technicalities of making contracts and the delivery system requirements.

3. MATERIALS AND METHODS

The theoretical discussion of the econometric multivariate model is discussed in Enders [5]. For a general presentation we begin with Pesaran and Pesaran [6] augmented vector autoregressive model of order ρ that is AVAR (ρ)

$$y_t = \alpha_0 + \alpha_{1t} + \sum_{i=1}^{\rho} \omega_i y_{t-1} + \tau x_t + \mu_t \tag{1}$$

where y_t is an n × 1 vector of endogenous (jointly determined) variables, a_0 is a drift, t is a linear time trend, x_t is an m × 1 vector of exogenous variables, and μ_t is m × 1 vector of unobserved disturbance terms. The assumptions for the disturbance terms are the mean is zero, $E(\mu_t) = 0$, not serially correlated $E(\mu_t\mu_s) = 0$, the variance matrix is time-invariant that is it does not vary with time. Since not all time series data representing these variables are stationary they have to be tested using Dickey-Fuller (DF) and augmented Dickey-Fuller (ADF) unit root statistics which is necessary in identifying their long-run cointegration [7]. Moreover, equation one contains a constant drift a_0 and the linear time trend which has to be de-trended and to achieve the necessary condition for cointegration the non-stationary variables have to be differenced. The difference multivariate econometric equations exemplifying the general vector error correction model (VECM) [5] is shown in Equation (2) below

$$\Delta y_t = \kappa_0 + \kappa_1 t + \sum_{i=1}^{\rho-1} \psi_i \Delta z_{t-1} + \gamma w_0 + \mu_t \tag{2}$$

Where Δy_t is an n × 1 vector of endogenous (jointly determined) $I(1)$ variables.

$\Delta z_{t-i} = \begin{pmatrix} y_{t-i} \\ x_{t-i} \end{pmatrix}$ are n_y × 1 vector of endogenous $I(1)$ and n_x × 1 exogenous $I(1)$ variables respectively.

w_t is m × 1 vector of exogenous $I(0)$ variables.

κ_0 & κ_1 are intercepts and deterministic linear trends.

μ_t is n × 1 vector of unobserved disturbance terms.

T = 1, 2, ..., T represents time periods.

Analysis of aquaculture development uses time series data 1976-2007 obtained from the Department of Fisheries annual statistics, the Ministry of Agriculture Malaysia. The objective of this investigation is to identify causalities between the freshwater fish prices namely that of giant prawn, catfish, river carps (Cyprinus carpio, Pangasius sp.) and goby (Gobiidae) with the total area (in cubic meter) of ponds devoted to rearing of these fresh0 water fisheries and the total number of culturists involved in the production activities. The multivariate equations for the model for the long run variables are presented in Equation (3).

An increase in the area of pond will attract more production of aquaculture into the market that acts to bring down the price of river carps and vice versa. Changes in the price of catfish will have both direct and indirect effect on the price of river carps depending on whether catfish compete or complement the river carps. While a rise in the number of culturists would increase aquaculture production that will ultimately reduce the price of river carps. Similar rationales can be used to explain the economic interrelationships for the rest of the equations in (3).

A direct impact on number of culturist a proxy for employment is anticipated from an increase in the area of pond. On the other hand, changes in either price of catfish or river carps will have a meaningful impact in terms of policy implication on the employment attraction to the aquaculture industry if the prices were found to be significantly associated with the number of culturists and vice versa. In the final analysis only the price of catfish (LPCF) was retained while the price of river carps (LPRC) was omitted from the endogenous system of equation for the long run cointegration analysis.

4. RESULTS AND DISCUSSIONS

4.1. Stationary Tests

Table 1 shows results of unit root tests for the selection of variables that will be used in the multivariate equation model both with trend and without trend. At form level $I(0)$ the number of culturists, the price of river carps and price of catfish were found to be stationary without trend. However, when the linear time trend is considered, with the exception of river carps price, all investigated variables were stationary at level $I(0)$. The results at first difference revealed that all selected variables fulfilled the requirement for the stationary test since the estimated values were found to be greater than the Dickey-Fuller or the augmented Dickey-Fuller critical values as shown in **Table 1** (see the Equation (3) below).

where, $LARP_t$ is annual area of freshwater pond in cubic meters at time t.

$LPRC_t$ is annual average price of river carps in MYR per kg at time t.

$$\begin{bmatrix} \Delta LARP_t \\ \Delta LPRC_t \\ \Delta LPCF_t \\ \Delta LCUL_t \\ \Delta LPGP_t \\ \Delta LARP_t \end{bmatrix} = \begin{bmatrix} \alpha_1 \\ \alpha_2 \\ \alpha_3 \\ \alpha_4 \\ \alpha_5 \\ \alpha_6 \end{bmatrix} + \begin{bmatrix} \beta_{11}\beta_{12}\beta_{13}\beta_{14}\beta_{15}\beta_{16} \\ \beta_{21}\beta_{22}\beta_{23}\beta_{24}\beta_{25}\beta_{26} \\ \beta_{31}\beta_{32}\beta_{33}\beta_{34}\beta_{35}\beta_{36} \\ \beta_{41}\beta_{42}\beta_{43}\beta_{44}\beta_{45}\beta_{46} \\ \beta_{51}\beta_{52}\beta_{53}\beta_{54}\beta_{55}\beta_{56} \\ \beta_{61}\beta_{62}\beta_{63}\beta_{64}\beta_{65}\beta_{66} \end{bmatrix} \begin{bmatrix} \Delta LARP_{t-1} \\ \Delta LPRC_{t-1} \\ \Delta LPCF_{t-1} \\ \Delta LCUL_{t-1} \\ \Delta LPGP_{t-1} \\ \Delta LARP_{t-1} \end{bmatrix} + \begin{bmatrix} \gamma_{11}\gamma_{12}\gamma_{13}\gamma_{14}\gamma_{15}\gamma_{16} \\ \gamma_{21}\gamma_{22}\gamma_{23}\gamma_{24}\gamma_{25}\gamma_{26} \\ \gamma_{31}\gamma_{32}\gamma_{33}\gamma_{34}\gamma_{35}\gamma_{36} \\ \gamma_{41}\gamma_{42}\gamma_{43}\gamma_{44}\gamma_{45}\gamma_{46} \\ \gamma_{51}\gamma_{52}\gamma_{53}\gamma_{54}\gamma_{55}\gamma_{56} \\ \gamma_{61}\gamma_{62}\gamma_{63}\gamma_{64}\gamma_{65}\gamma_{66} \end{bmatrix} \begin{bmatrix} \Delta LARP_{t-p} \\ \Delta LPRC_{t-p} \\ \Delta LPCF_{t-p} \\ \Delta LCUL_{t-p} \\ \Delta LPGP_{t-p} \\ \Delta LARP_{t-p} \end{bmatrix} + \begin{bmatrix} \pi_1 \\ \pi_2 \\ \pi_3 \\ \pi_4 \\ \pi_5 \\ \pi_6 \end{bmatrix} \left[ect_t \right] + \begin{bmatrix} \mu_1 \\ \mu_2 \\ \mu_3 \\ \mu_4 \\ \mu_5 \\ \mu_6 \end{bmatrix} \quad (3)$$

Table 1. Unit root tests for the related variables using dickey-fuller.

		Without Trend		With Trend	
		DF	ADF (1)	DF	ADF (1)
At level form	*LCUL*	−4.2311	−3.0532	−4.1422	−2.9732
	LPRC	−3.4457	−3.1533	−2.7205	−1.9461
	LPCF	−5.2088	−3.7823	−5.5283	−3.7478
	LPGY	−1.8526	−1.4778	−3.0975	−2.6430
	LPGP	−0.9844	−0.6692	−4.1355	−3.5040
	LARP	−7.0818	−4.2542	−6.9190	−4.2648
At 1st difference	*LCUL*	−7.4110	−4.7610	−7.5820	−4.7988
	LPRC	−5.9698	−3.9135	−6.3416	−4.5241
	LPCF	−9.4382	−7.3577	−9.2954	−7.5165
	LPGY	−6.5456	−4.7248	−6.4142	−4.6090
	LPGP	−6.2512	−6.0686	−6.0686	−6.7573

Note: Dickey-Fuller statistics at 95 percent critical value are −2.9798 $I(0)$ and −3.6027 $I(1)$ with trend and 2.9850 $I(0)$ and 3.0532 $I(1)$ without trend.

$LPCF_t$ is annual average price of catfish in *MYR* per kg at time *t*.

$LCUL_t$ is annual total number of culturists in the industry at time *t*.

$LPGY_t$ is annual average price of goby in *MYR* per kg at time *t*, and $LPGP_t$ is annual average price of giant prawn in *MYR* per kg at time *t*.

All variables are measured in natural logarithmic values, $\rho > 1$ represents lag length greater than one, and Δ is the difference notation. *US$1* = *MYR*3.1.

α_i are intercepts of the multivariate equations, β_{ij} γ_{ij} for i = j = 1, 2, ..., $n\left(\beta_{ij} \neq \beta_{ji}, \gamma_{ij} \neq \gamma_{ji}\right)$ represent the coefficients of endogenous and exogenous variables of the multivariate equations to be estimated in the model. The terms ect_i refer to the error correction terms, whose coefficients measure speeds of adjustment and are derived from the long-run cointegrating relationships, μ_i represents the disturbance terms, and ρ is the lag lengths for the respective multivariable equations. Causalities tests between these endogenous variables in multivariate analyses using vector error correction model requires that each of these endogenous variables will be used as right hand side (RHS) equation as shown in Equation (3).

4.2. Number of Cointegrating Variables and Order of Lags

Tables 2(a) and (b) show the number of cointegrating variables, *r*, for the case of restricted intercepts and no trends in vector autoregressive (*VAR*) model. Both maximal eigenvalue and trace of the stochastic matrix suggest that at least the number of cointegrating variables, r = 2 for the order of lag, *VAR* = 2 at either 95% or 90% significant level respectively. Pesaran &

Pesaran [5] caution that the figures reported by Microfit differ from those obtained by Johansen and Juselius [8] and also by Osterwald-Lenum [9]. However, results from the model selection Akaike Information Criterion (AIC), Schwarz Bayesian Criterion (SBC) and Hannan-Quinn Criterion (HQC) in **Table 2**(c) suggest the cointegrating variables of r = 3, r = 2 and r = 3 respectively. The cointegrating vectors based on the theoretical formulation of the endogenous variables that was originally intended for this investigation and those suggested using econometric criteria will be tested in order to get the most appropriate results. The lag order is limited to two after several trials were carried out for higher degree of lags had proven to yield no significant results. With higher degree of lags such as three and above the loss in the degree of freedom is somewhat serious given the limitation on the number of observations for this study.

5. AQUACULTURE ERROR CORRECTION MODEL

Freshwater species which is widely raised by the majority of culturists is the catfish. Catfish species are relatively easy to breed and they are available in local markets. Furthermore, the species are believed by majority of Malaysians to have the specialty of healing capacity for bleeding. The price of catfish is relatively low compared to other freshwater delicacies like giant prawn. Hence, there is sufficient reason to believe that catfish prices may have a significance influence in attracting culturists as well as the cultivated area for the long run cointegration analysis. These three policy variables are grouped as endogenous variables in the current study while the other prices of river carps, giant prawn and goby are included as the exogenous variables as they are being influenced by unobservable factors outside the currently available data.

The log-linear error correction models (ECM) for the three endogenous variables namely the area per pond in hectares (dLARP), number of culturist (dLCUL) and average price of catfish (dLPCF) are shown in Tables 3 (a)-(c). These regression equations were first estimated using the ordinary least square (OLS) omitting their dynamic characteristics; that is, the differences and lags of the estimated endogenous variables are of order one I(1). The residuals have satisfied the necessary condition for the existence of long-run relationships.

The short-run characteristics of the ECM coefficients need to be substantiated from the difference equations of **Table 3**(a) thru **Table 3**(c). The Granger representation theorem (GRT) requires that the ECM coefficients are negatives and statistically significant.

Table 2. (a) Cointegration with restricted intercepts and no trends in the VAR maximal eigenvalue of the stochastic matrix 1978-2007; (b) Cointegration with restricted intercepts and no trends in the VAR trace of the stochastic matrix, 1978-2007; (c) Cointegration with restricted intercepts and no trends in VAR using model selection criteria 1978 to 2007.

(a)

Null	Alternative	Statistic	95% Critical Value	90%Critical Value
$r = 0$	$r = 1$	21.7735	22.0400	19.8600
$r <= 1$	$r = 2$	18.6515	15.8700	13.8100
$r <= 2$	$r = 3$	6.4055	9.1600	7.5300

(Order of *VAR* = 2).

(b)

Null	Alternative	Statistic	95% Critical Value	90%Critical Value
$r = 0$	$r >= 1$	46.8305	22.0400	31.9300
$r <= 1$	$r >= 2$	25.0570	20.1800	17.8800
$r <= 2$	$r = 3$	6.4055	9.1600	7.5300

(Order of *VAR* = 2).

(c)

Rank	Maximized LL	AIC	SBC	HQC
$r = 0$	4.1437	−13.8563	−26.4671	−17.8906
$r = 1$	15.0305	−8.9695	−25.7839	−14.3486
$r = 2$	24.3562	3.6438	−23.2606	−9.9194
$r = 3$	27.5589	−2.4411	−23.4590	−9.1649

(Order of *VAR* = 2) AIC = Akaike Information Criterion; SBC = Schwarz Bayesian Criterion; HQC = Hannan-Quinn Criterion.

As apparent ecm1(−1) =−0.81661 (**Table 3**(a)), ecm2(−1) = −1.0587 (**Table 3**(b)) and ecm3(−1) = −0.58028 (**Table 3**(c)) demonstrate that they are statistically significant at 99 percent probability level for the ecm1(−1) and ecm2(−1) respectively. The ecm3(−1) is also significant but at a probability level of 95 percent. Thus, the endogenous vectors used in current aquaculture study are cointegrated in the long-run. This is illustrated in Figures 1 thru 3 showing the impact of generalized impulse responses to one standard deviation shock on the cointegrated variables.

Further investigation of interest to the current study that utilizes the information given in **Table 3**(a) through **Table 3**(c) is to derive Granger's causality test. According to Rosilawati Amiruddin et al. [10] the works of Granger [11], Hendry [12], Engle and Granger [6] and Johansen and Juselius [13] on the extension of cointegration technique had opened up new avenues for evidence of causalities in one and bi-directional which were once uncertain due to spurious regression results.

With cointegration between two or more variables and either one or bi-directional causalities could be carried out on appropriate model specification and this is done with the Wald test using Microfit. The section is discussed in next section under Granger's causalities.

6. GRANGER'S CAUSALITY

Causality between variables in time series multivariate analysis was initially investigated by Granger [11,14]. In general it provides richer information than the single or univariate equation model because several endogenous and exogenous variables could be tested simultaneously. These causalities are obviously essential for identifying national strategies such as that on the current policy of promoting aquaculture to supplement the existing depleting marine fisheries. It looks reasonable to augment the present shortage of marine supply with aquaculture undertakings but it would be wise to further investigate and to ask; are the firms responsive to this strategy? They may not be responsive if the non-aquaculture investments outside the industry such as manufacturing, services and agricultural sector itself are more profitable and suitable to the investors' skills and interests.

Table 3. (a) Log-linear ECM for dependent variable dLARP estimated by OLS based on cointegrating *VAR* (2), 1978-2007; (b) Log-linear ECM for dependent variable dLCUL estimated by OLS based on cointegrating *VAR*(2), 1978-2007; (c) Log-linear ECM for dependent variable dLPCF estimated by OLS based on co-integrating *VAR*(2), 1978 to 2007.

(a)

Regressor	Coefficient	Standard Error	T-Ratio (Prob)
dLARP1	0.32130	0.17621	1.8234 (0.083)
dLCUL1	0.06382	0.19527	0.3269 (0.747)
dLPCF1	0.28802	0.16021	1.7978 (0.087)
ecm1(−1)	−0.81661	0.19549	−4.1774 (0.000)
ecm2(−1)	−0.18977	0.28099	−0.6754 (0.507)
ecm3(−1)	−0.62160	0.22416	−2.7731 (0.011)
LPRC	−0.14416	0.08403	−1.7157 (0.101)
LPGP	−1.07080	0.29160	−3.6721 (0.001)
LPGY	0.56681	0.17696	3.2030 (0.004)

R-Squared	0.58779	R-Bar-Squared	0.43076
S.E. of Regression	0.20525	F-stat.F (8.21)	3.74320 (0.007)
Mean of Dependent Variable	−0.06535	S.D. of Dependent Variable	0.27205
Residual Sum of Squares	0.88470	Equation Log-likelihood	10.28740
Akaike Info. Criterion	1.28740	Schwarz Bayesian Criterion	−5.01800
DW-statistic	2.37310	System Log-likelihood	27.55890

(b)

Regressor	Coefficient	Standard Error	T-Ratio (Prob)
dLARP1	−0.10651	0.19277	−0.552530 (0.586)
dLCUL1	0.06630	0.21363	0.310360 (0.759)
dLPCF1	−0.19197	0.17527	−1.095300 (0.286)
ecm1(−1)	0.16165	0.21386	0.755850 (0.458)
ecm2(−1)	−1.05870	0.30741	−3.444000 (0.002)
ecm3(−1)	0.38988	0.24523	1.589900 (0.127)
LPRC	−0.14773	0.09193	−1.607000 (0.123)
LPGP	−0.01180	0.31901	−0.036998 (0.971)
LPGY	0.06295	0.19360	0.325130 (0.748)

R-Squared	0.50601	R-Bar-Squared	0.31782
S.E. of Regression	0.22455	F-stat.F (8.21)	2.6888 (0.033)
Mean of Dependent Variable	0.01944	S.D. of Dependent Variable	0.27187
Residual Sum of Squares	1.05890	Equation Log-likelihood	7.59180
Akaike Info. Criterion	−1.40820	Schwarz Bayesian Criterion	−7.71360
DW-statistic	1.82090	System Log-likelihood	27.55890

(c)

Regressor	Coefficient	Standard Error	T-Ratio (Prob)
dLARP1	−0.01249	0.18242	−0.06848 (0.946)
dLCUL1	−0.11646	0.20215	−0.57613 (0.571)
dLPCF1	−0.28045	0.16585	−1.69090 (0.106)
ecm1(−1)	−0.24628	0.20237	−1.21700 (0.237)
ecm2(−1)	0.35571	0.29089	1.22280 (0.235)
ecm3(−1)	−0.58028	0.23206	−2.50060 (0.021)
LPRC	0.26741	0.08699	3.07410 (0.006)
LPGP	0.44522	0.30187	1.47480 (0.155)
LPGY	−0.29847	0.18320	−1.62920 (0.118)

R-Squared	0.63170	R-Bar-Squared	0.49139
S.E. of Regression	0.21248	F-stat.F (8,21)	4.50230 (0.003)
Mean of Dependent Variable	0.04446	S.D. of Dependent Variable	0.29795
Residual Sum of Squares	0.94814	Equation Log-likelihood	9.24860
Akaike Info. Criterion	0.24860	Schwarz Bayesian Criterion	−5.01800
DW-statistic	2.21240	System Log-likelihood	27.55890

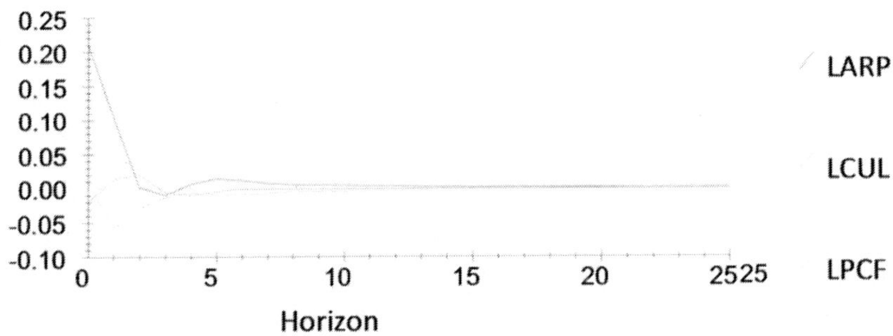

Figure 1. Generalized impulse response(s) to one S.E. shock for LARP.

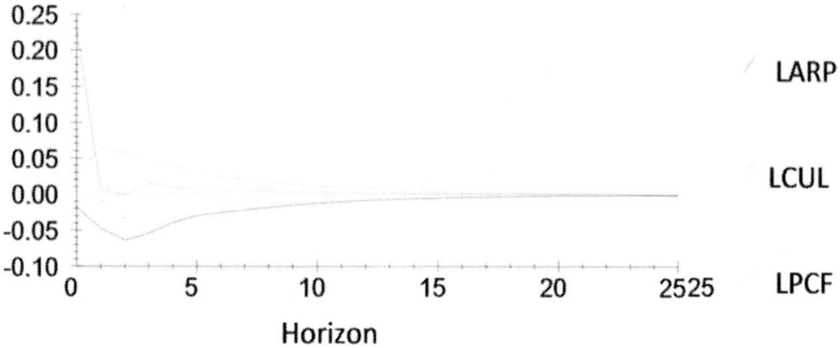

Figure 2. Generalized impulse response(s) to one S.E. shock for LCUL.

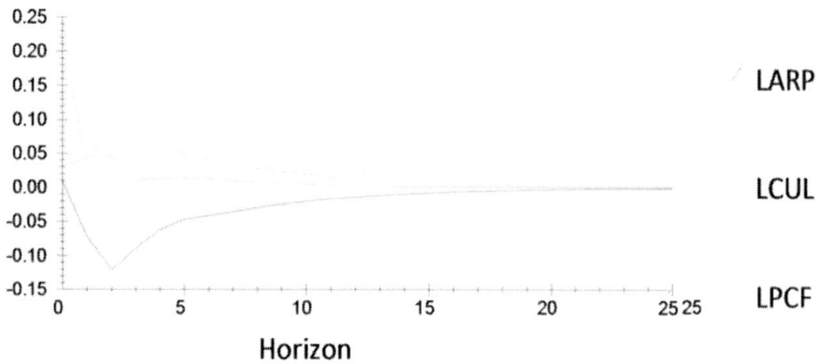

Figure 3. Generalized impulse response(s) to one S.E. shock for LPCF.

Table 4 illustrates the existence of causality among the investigated economic variables in the Malaysian aquaculture industry. There should be either one, bi-directional or nonexistence of causality between these policy variables. The null hypothesis indicated that the size of aquaculture pond area (in ha.) had no causality whatsoever with the number of culturists in the industry and vice-versa. An increase in size of the aquaculture area had not contributed significantly to the course for raising employment. Alternatively, an increase in the number of culturists will not have the potential to increase the size of aquaculture area. However, changes in the one-period lag price of catfish (dLPCF), current prices of river carps (LPRC), giant prawn (LPGP) and current price of goby (LPGY) would draw a direct change in the size of aquaculture pond area. On the other hand, a change in price of catfish will have no effect on inducing one-period lag of size of aquaculture area (dLARP1).

With the exception of direct and current change in price of river carps that have some influence on changing the number of culturists, all other variables in the EVCM model apparently show no significant causality on culturists. Price of river carps may attract those anglers who fish for recreational purpose. Sometimes river carps can fetch a good price for local markets and restaurants and at other times organized competition in sea and river fishing are held. Private individuals also provide special fishing grounds for river carps and selected local freshwater fish species for anglers at reasonable fees.

Table 4. Wald statistical tests for multivariate vector error correction model (VECM), 1978-2007.

	ΔLARP	ΔLCUL	ΔLPCF
ΔLARP1	3.3249	0.3053	0.00469
	(0.068)*	(0.581)NS	(0.945)NS
ΔLCUL1	0.10683	0.09632	0.33192
	(0.744)NS	(0.756)NS	(0.565)NS
ΔLPCF1	3.2322	1.1996	2.8593
	(0.072)*	(0.273)NS	(0.091)*
ecm1(−1)	17.4503	0.5713	1.481
	(0.000)***	(0.450)NS	(0.224)NS
ecm2(−1)	0.45612	11.8611	1.4953
	(0.499)NS	(0.001)***	(0.221)NS
ecm3(−1)	7.6899	2.5277	6.2531
	(0.006)***	(0.112)NS	(0.012)**
LPRC	2.9437	2.5825	9.4502
	(0.086)*	(0.108)*	(0.002)***
LPGP	13.484	0.00137	2.1752
	(0.000)***	(0.970)NS	(0.140)NS
LPGY	10.2592	0.10571	2.6544
	(0.001)***	(0.745)NS	(0.103)*

Notes: Figures in parentheses represent the probability values (p-values) of significant levels. ***denotes significant at 0.01 probability level. **denotes significant at 0.05 probability level. *denotes significant at 0.10 probability level.

The one-period lag catfish price (dLPCF1) has shown some degree of causality with its own current price (dLPCF) at about 90 percent probability level. Similarly, the price of exogenous variable goby (LPGY) has exhibited some degree of causality on the current price of catfish. The degree of causality between the current price of river carps and the difference price of catfish are stipulated to be high at 99 percent probability level. In all of these cases the null hypotheses that support no causality between the prices mentioned above are rejected. It is true that the price of giant prawn has no association with the price of catfish. These species have their own market specialties.

In general, the finding revealed that price of catfish Granger causes the aquaculture area which is only onedirectional since the reverse influence of area on price is not significantly different from zero. Furthermore, the one-period lag aquaculture area Granger causes the current area and the one-period lag price of catfish Granger causes the current price of catfish. This trend is not observable in the case of number of culturists. The exogenous variables of the current prices of giant freshwater prawn, goby and river carps have greatly influenced the increase in size of the aquaculture area. The price of river carps and to a lesser extent the price of goby have a similar influence on the price of catfish. River carps and catfish are substitutes because their prices move in the same direction while goby and catfish are complementary products since their price movements are in the opposite direction.

7. CONCLUSIONS

The multivariate econometric vector error correction model was estimated using Microfit software to investigate the relationship between endogenous and exogenous variables namely the size of pond area, the number of culturists and the prices of the major fresh fish species. The findings revealed that price movements, specifically the price of catfish has caused changes in the size of the aquaculture pond area but the reverse is not true. This indicates that the direction of causality is one way. Furthermore, the potential of the variation in aquaculture area to draw larger investors as indicated by the number of culturists in the industry is most likely not going to happen. The causality between the determining variables is unfortunately insignificant, thus employment wise the promotion for aquaculture projects has not been able to place big hope in terms of employment generation. Possible explanations could be that aquaculture activities require a large amount of initial capital to start, high technical knowhow and it is a risky investment. This is without considering problems relating to marketing of fish harvests, getting supply of fries and the ever-increasing prices of land for fishponds compete with other economically potential investments. In the case of cage culture although it requires high initial outlays competition for water use is less profound; however, water quality needs to be assured from pollution.

The idea of utilizing freshwater and brackish water fisheries to supplement marine catch is nevertheless a worthy strategy that needs to be supported. However, special care for degradation of the environment due to aquaculture development and activities should be taken into consideration by the policy makers. Research and development should be conducted from time to time to find the best way to invest in aquaculture. It would be possible to allow individual households to start a small aquaculture projects to reduce costs and to be self-sufficient in food supply with fishery-based programs in addition to a several large privately owned aquaculture projects with government assistance.

8. ACKNOWLEDGEMENTS

Nik Hashim Nik Mustapha, Azlina Abd Aziz, Nik Mohd Hazrul Hashim (FRGS) vote number 59189 for the duration of April 2009-March 2012. They also wish to thanks the unanimous reviewers of this article for their suggestions and recommendations.

REFERENCES

1. Department of Fisheries Malaysia, "Annual Fisheries Statistics (1987-2007)," 2010. http://www.dof.gov.my/59
2. Government of Malaysia, "Rancangan Malaysia Kesembilan 2006-2010," Unit Perancang Ekonomi Malaysia, 2010.
3. EarthTrends AQUASTAT, "Information System on Water and Agriculture Country Profiles," 2003. http://www.fao.org/waicent/faoinfo/agricult/agl/aglw/aquastat/countries/index.stm
4. P. J. Pradeep, T. C. Srijaya, H. Anuar, S. Faizah and C. Anil, "The Aquatic Chicken Tilapia and its Future Prospects in Malaysia," Institute of Tropical Aquaculture, University Malaysia Terengganu, Terengganu. Prospect, Malaysian Premier Higher Education Magazine Times Guides Sdn Bhd, Pelangi Damansara, 2010, pp. 52-56.
5. W. Enders, "Applied Econometrics Time Series. New Jersey," John Wiley & Sons Inc., Hoboken, 2003.
6. M. H. Pesaran and B. Pesaran, "Working with Microfit 4.0 Interactive Econometric Analysis," Oxford University Press, Oxford, 1997.
7. R. F. Engle and W. J. Granger, "Cointegration and Error Correction: Representation, Estimation and Testing," Econometrica, Vol. 55, No. 2, 1987, pp. 251-276.doi:10.2307/1913236

8. S. Johansen and K. Juselius, "Testing Structural Hypotheses in a Multivariate Cointegration Analysis of the PPP and UIP for UK," Journal of Econometrica, Vol. 53, No. 1-3, 1992, pp. 211-244. doi:10.1016/0304-4076(92)90086-7
9. Osterwald-Lenum, "A Note with Quantiles of the Asymptotic Distribution of the Maximum Likelihood Cointegration Rank Test Statistics," Oxford Bulletin of Economics and Statistics, Vol. 54, No. 3, 1992, pp. 461-472. doi:10.1111/j.1468-0084.1992.tb00013.x
10. Rosilawati, A. H. Shaari, and I. Ismail, "Test on Dynamic Relationship between Financial Development and Growth in Malaysia," Gadjah Mada International Journal of Business, Vol. 9, No. 1, 2007, pp. 1-17.
11. W. Granger, "Developments in the Study of Cointegrated Variables," Oxford Bulletin of Economics and Statistics, Vol. 48, No. 3, 1986, pp. 213-227. doi:10.1111/j.1468-0084.1986.mp48003002.x
12. F. Hendry, "Econometric Modeling with Cointegrated variables: An Overview," Oxford Bulletin of Economics and Statistics, Vol. 48, No. 3, 1986, pp. 201-212.doi:10.1111/j.1468-0084.1986.mp48003001.x
13. S. Johansen and K. Juselius, "Maximum Likelihood Estimation and Inference on Cointegration with Application to the Demand for Money," Oxford Bulletin of Economics and Statistics, Vol. 52, No. 2, 1990, pp. 169-210. doi:10.1111/j.1468-0084.1990.mp52002003.x
14. C. W. Granger, "Investigating Causal Relationships by Econometric Models and Cross-Spectral Methods," Econometrica, Vol. 37, No. 3, 1969, pp. 424-438.doi:10.2307/1912791

CHAPTER 5

A Social Wellbeing in Fisheries Tool (SWIFT) to Help Improve Fisheries Performance

Tracy Van Holt [1,2,*], Wendy Weisman [3], Jeffrey C. Johnson [4], Sofia Käll [1], Jack Whalen [5], Braddock Spear [5] and Pedro Sousa [5]

[1] *Global Economic Dynamics and the Biosphere Program, Royal Swedish Academy of Science, Stockholm SE-104 05, Sweden*

[2] *Stockholm Resilience Center, Stockholm University, Stockholm SE-104-05, Sweden*

[3] *Centre for Social Innovation, New York, NY 10001, USA*

[4] *Department of Anthropology, University of Florida, Gainesville, FL 32611, USA*

[5] *Sustainable Fisheries Partnership, Honolulu, HI 96816, USA*

ABSTRACT

We report on a rapid and practical method to assess social dimensions of performance in small-scale and industrial fisheries globally (Social Wellbeing in Fisheries Tool (SWIFT)). SWIFT incorporates aspects of *security* (fairness and stability of earnings, benefits of employment to local fishing communities, worker protection, and personal safety and health in communities associated with fisheries); *flexibility* (including opportunity for economic advancement); and the fishery's social *viability* (including whether the fishery is recruiting new harvesters and diverse age classes of workers, whether women's participation and leadership in global production networks are on an upward trajectory.). We build on resilience research by conceptualizing wellbeing in terms of *security*, *flexibility*, and *viability*, and assessing wellbeing at individual, community, and system levels. SWIFT makes social performance measures more broadly accessible to global production networks, incorporates an everyday understanding of wellbeing for people involved in the seafood industry, and helps put social sustainability into measurable terms that are relevant for businesses.

Keywords: fisheries improvement; corporate social responsibility; social sustainability; seafood; supply-chain; well-being; global production networks

1. INTRODUCTION

As trends toward fair trade and corporate social responsibility (CSR) continue in many areas of food production, seafood consumers and organizations promoting ecological sustainability have shown growing interest in social wellbeing within global seafood production networks [1,2,3,4]. Interest in wellbeing mirrors a call for ecosystem stewardship that fosters social-ecological sustainability, and identifies strategies to maintain diverse options, enhance social learning, and adapt governance to implement potential solutions [5,6]. In the seafood industry, increasing attention has been paid to human rights frameworks in relation to global production networks in the academic literature [7,8], seafood trade conferences [9,10], and industry-led initiatives [2]. While it is critical to address egregious examples of fisheries worker mistreatment such as those exposed by the media, eradicating those problems is not enough. Seafood industry firms need tools to understand and help improve the everyday wellbeing of people working in their global production networks, and consider their future prospects. We make the case for a rapid tool to help organizations throughout seafood production networks begin the process of assessing social performance.

Social scientists have long studied the social structures, cultures, and dynamics of fishing communities and how they relate to physical environments, but the interest, specifically, to assess wellbeing began to emerge in the late 1990s through policies concerning social impact evaluations of fisheries [11]. Researchers also engaged in social assessment work as a way to understand how and why individual and community wellbeing was declining in historically-strong small-scale fishing regions [12,13], and to meet the needs of a wider movement toward ecosystem-based management approaches that integrate people into the fishery systems concept [14]. Wellbeing has been defined as the degree of individual or community health, happiness, and prosperousness [15], security, health, good social relations, freedoms, and choice [16,17], and in terms of resistance (politics and power), rootedness (sense of place), and resourcefulness (knowledge and learning) [18]. In the resilience literature, it is argued that social-ecological systems that effectively sustain human wellbeing are better prepared to adapt and transform in response to change [19].

1.1. Comparing Social Wellbeing Measures

Our comparison of selected wellbeing measures shows the trade-offs between breadth and depth of information that can be used to assess social performance (Table 1). The need for social data that can only be obtained through detailed field research can drastically inhibit the use of performance measures at global levels by businesses. For example, the Association of Southeast Asian Nations (ASEAN) method has a high standard for evidence for worker fairness using compliance certificates from industry partners, and local research surveys detailing resource conflict. This is feasible for that specific context to the extent that ASEAN's industry-led consortium members are based in a single region and likely have the capacity to track compliance. However, the resource mapping proposed in field sites will be time consuming, expensive, and difficult to obtain globally [20]. NOAA's social scientists generated a way for US decision-makers to consider social impact assessments that incorporate social vulnerabilities and resilience in modeling the impact of fishery policy alternatives [21,22]. The NOAA group's [21] approach has the potential to be accurate in its ability to predict social outcomes related to policy changes at the community level, but it is best suited for the US context because it relies heavily on publically-available data, such as spatially-explicit census data, boating ramp databases, and other data types not applicable to many international fisheries [21,22]. Similarly, the strength of Fair Trade standards is their commitment to ensure updated information, using third party field visits annually to verify components of the production network [23]. However, data from these opportunities will likely not be available to most fisheries in the short term. On the other hand, the Anderson group [24] have created 68 metrics which can be ranked using key informants (rather than on-site field research) and, thus, potentially reduce the burden in uncovering suitable evidence. In this approach rankings are verified through further research if the evaluators find anomalies. Yet, the Anderson group's comprehensive, performance-focused, and carefully-constructed metrics are complex for non-academic audiences, who would likely find it difficult to locate evidence to back up those metrics. Importantly, the Anderson group includes some measures that recognize the basic differences between small-scale and industrial fisheries. This distinction is also explicit in the Food and Agricultural Organization of the United Nations' (FAO) ethical guidelines in small-scale fisheries, which provides a broad overview of human rights themes in the context of fisheries (rather than specific social performance measures) and emphasizes the importance of considering gender, tenure, and power relationships [25]. The rapid method we propose, Social Wellbeing in Fisheries Tool (SWIFT), uses criteria that are appropriate for both industrial and small-scale fisheries. Other key characteristics of SWIFT are that it aims to rely on data that are publically available, or readily-accessible data through desk-based research (reports or online sources), focuses on performance measures that the global production network can likely influence, applies aspects of resilience that likely resonate with both businesses and academics, and uses the fishery as a unit of performance measurement, in keeping with the way fisheries are generally managed.

Table 1. Comparison of six approaches to operationalizing wellbeing for fisheries in terms of the evidence, units of analysis, and context.

Approach	Purpose	Main Focus	Audience	Evidence/Source	Unit of Analysis/Scale
ASEAN [20] *(Industry)*	Measure social performance	Worker fairness and avoiding resource-use conflict	Industry investors, philanthropists	Self-reporting of compliance certificates; mapping community resources on site	Small and large fisheries (Asian Region)
Jepson et al. [21] *and Himes-Cornell* [22] *derived from Pollnac* [13]	Assess social impact	Assess management alternatives' effect on vulnerability, community wellbeing, social disruption	State agencies, fishers	Spatially-explicit census and other datasets	Fishing community (section of a county) (US States)
Fair Trade USA [23]	Set Fair Trade certification standards	Governance structure, empowerment, human rights, wages, working conditions and services	Consumers, fishers, industry	Third-party verifies supply network through fieldwork	Certificate holder/fishery organization (Global/Local data)
Anderson and colleagues [24]	Measure performance of fishery management systems	Ecology, economy, community in harvest and post-harvest sectors	Governments, industry, researchers	Sixty-eight metrics ranked by experts using key informants	Fishery management rather than fish stock (Global)
FAO/United Nations [25]	Provide ethical guidance relevant to small-scale fisheries	Secure tenure, decent work, power relationships, gender equality	Governments, industry, fishers	Document summarizes rights issues in small-scale fisheries)	Small-scale fisheries communities (Local)
Sustainable Fisheries Partnership (SWIFT method proposed here)	Measure social wellbeing performance	Flexibility, security, and viability	Industry, investors, philanthropists	Third parties synthesize public data; performance rankings can be challenged and rebutted with evidence	Small and large fisheries (Global-Fishsource data)

1.2. State of the Art and Contribution to Work in the Area of Wellbeing

SWIFT incorporates resilience-oriented concepts of wellbeing (*security*, *flexibility*, and *viability*) [26,27,28] because these are relevant across disciplinary boundaries to social scientists, ecologists, and those working in the private sector and government. *Security* is the dependability of a continued livelihood. Secure systems are able to maintain themselves and are also well positioned to adapt to change. The profile of a secure fishery would be that workers earn fair wages, are protected, and have a strong, accessible healthcare system. *Flexibility* is freedom of choice, degree of options, and potential mobility within the fishery system. Flexible social systems, often those characterized by diverse economic and livelihood options, are more likely to recover from shocks [28]. The profile of an ideal flexible fishery would be that fishers can sell their catch to whomever they wish, fishery wealth remains local, and young people in the community have a strong educational system that opens up new options. *Viability* refers to whether the fishery and associated global production network is likely to persist from a social perspective, as long as the fishery is well-managed, ecologically. Viability addresses the link between individual wellbeing and the wellbeing of a fishery system as a whole, looking at demographic features and trends in fisheries, rather than their individual participants or communities. A viable fishery demonstrates signs of persistence through demographic patterns that show the new generation and women have attractive opportunities for engagement. Viability helps us grasp the current social and ecological state of the fishery and likelihood the system can persist or transform [29].

2. METHODS

Drawing on the concepts of *flexibility*, *security*, and *viability*, we identified seven social performance dimensions that best met the criteria of being essential, applicable at the fishery level, having publically-available data, and likely to have at least one stakeholder group in the supply network associated with it that has a degree of control to improve performance on the dimension. For each dimension we identified the focus, rationale, operationalization, data sources, and availability. We then ranked two fishery cases to test the method. A fishery was ranked for each of the seven dimensions, often using qualitative data, such as project reports or websites. Each dimension had two or three factors. We used a three-level ordinal scoring system for each dimension: a fishery is ranked 1 if less than half of the factors were present, 2 if at least half of the factors were present, and 3 if all of the factors were present (Table 2). These ordinal variables are not to be totaled together among multiple dimensions to try to obtain a single sum rank. This rule is a characteristic of ordinal variables. The difference between each rank cannot be assessed as they are interval and ratio variables and, therefore, mathematical operations in such cases make little sense. Instead, each dimension should be evaluated by itself to get a sense of where a fishery stands. We did not penalize rankings for missing data, though that would certainly put pressure on those responsible for fisheries performance to seek out information on social wellbeing. To determine ease of finding data, we looked for evidence for the dimensions in five fisheries in as many countries. We categorized the data search difficulty as easy, average, challenging, and very challenging.

Table 2. Social Wellbeing in Fisheries Tool (SWIFT) method for determining social performance in fisheries.

Wellbeing Dimension	Suggested Measures & Rank Ranges		
	For Each Measure, Determine Which Factor(s) Apply and Note the Appropriate Ranking.		
	1	2	3
	Less than half of factors are present	At least half of factors are present	All factors are present
1. Earnings are fair and stable.	1.1 For *wage-based earnings*, harvesting wages are higher than the standard (national minimum wage), OR if *landings-based earnings*, price/weight paid to harvesters is stable or increasing over a five year period of most recently available data. 1.2 Processing sector wages are higher than the standard.		
2. Jobs benefit the communities.	2.1 The majority of the harvesting workforce is comprised of local people. 2.2 The majority of primary processing workforce is comprised of local people.		
3. Workers are protected.	3.1 There is a national level protection of right to strike in place or it is in the nation's constitution. 3.2 There is a national level protection of freedom of collective bargaining, or it is in the nation's constitution. 3.3 Structure is in place to address worker grievances at local level or the global production network uses labor contracts and the terms are transparent/public.		
4. Communities have improving healthcare.	4.1 Made improvement in under-five mortality rate at the relevant fishery level or meets global standard. 4.2 Seafood company has implemented programs to improve healthcare.		
5. Communities have improving education.	5.1 Less than 10% of primary and secondary school age children are out of school. 5.2 Seafood company has implemented programs to improve education.		
6. Fishery will be viable for future generations.	6.1 Harvesters from a range of age classes are represented. 6.2 New fishers are being recruited into the fishery. 6.3 Women are increasingly taking leadership roles in the production network and fishing communities.		
7. Harvesters have economic flexibility.	7.1 Harvesters are free to sell to whomever they wish without retribution. 7.2 Harvesters can access loans from at least two types of lenders at interest rates not exceeding the government rate. 7.3 Formalized training is provided to harvesters concerning how to add value to their landings.		

2.1. The Test Cases

Our two test cases were selected because they represented a small-scale fishery and an industrial fishery that should exhibit wide-ranging rankings (we have anonymized the fisheries for use in this comparison). Industrial work crews are often paid in wages, have large labor unions, and are locked in to one target species, area, fishing gear, and harvest time. In contrast, small-scale fisheries with independent fishers are often paid per catch, local cooperatives and middlemen often provide economic safety nets, and fishers are more readily able to switch gear, species, and harvest time and location. In addition, a global comparison of how trade influences the social-ecological outcomes in small-scale fisheries has shown that strong governance institutions at various levels were a necessary condition for fisheries that ranked high on both social and environmental indicators of success; therefore, a fishery operating in a country with weaker institutions likely would rank lower [30]. Profiles of the two cases are:

- *Small-scale fishery test case:* a fishery with relatively weak fishery management, governance, and labor institutions. Very large number of fishers in the fishery, and the resource is reported to be over-exploited. Small boats operated by individual fishers sell product to local middlemen. The product has a relatively high value per kilo and processed in country for export and sale under several brands. There is a small domestic market for the product. Women are employed in processing.
- *Industrial fishery test case*: a fishery context with relatively strong institutions and legal framework protecting citizens. A large number of fishers are employed in the fishery, with the resource reported to be fairly well managed. Industrial vessels employ workers under a contract. The product has a lower value per kilo and is processed overseas, for sale in a highly-competitive commodity market.

2.2. Social Wellbeing and Fisheries Tool (SWIFT) Dimensions

1) *Earnings are fair and stable.* This focuses on harvesters and processors at the fishery level; these data are of average difficulty to obtain. Wages are a standard economic measure of wellbeing [15]. If either the harvesting (1.1) or processing (1.2) sectors earn less than the standard wage, then social inequality is being perpetuated in the fishery. *Waged-based earning* by country and minimum wages for the fisheries sector are tracked [31]. Local variation in earnings for harvesters or between fisheries in a country is not represented and must be found elsewhere. If both harvesting and processing sectors have average earnings lower than the standard, then the fishery is not providing an acceptable living. The standard is for fishers to obtain higher than the national minimum wage except in countries where wage standards are set by collective bargaining (i.e., Scandinavian countries) or if no minimum wage is available. In these cases, estimates can be made using respected publically-available sources. In *landings-based earnings*, such as small-scale fisheries where fishers work independently or in small groups, landings-based data on stability or increase in price/unit actually paid to fishers for catches over a five-year period is a direct, accessible measure, that can be substituted as a minimum wage (1.1). This information is often collected for fisheries whose products are sold on global markets, and is often collected for fisheries improvement projects (FIPs) seeking to improve marketing opportunities, or implement new policies or practices that may benefit fishers economically [32].

2) *Jobs benefit the communities.* This focuses on harvesters and processors at the fishery level and the data is challenging to find. A reasonable measure for jobs benefiting the community is whether more than half of the harvesting (2.1) and primary processor (2.2) workforce is based in the fishery region (vs. outsourced or migrant labor). An underlying assumption is that a local workforce implies that greater fishery wealth and other benefits remain in a fishing community. Reducing local employment by processing fish in another country may affect local wellbeing indirectly as well as directly, by reducing the tax base and, therefore, reducing political leverage for needed social services. Anderson et al.'s measures of nonresident employment follow a similar rationale [24]. We found no common data source for outsourced or migrant labor for fisheries (2.1 and 2.2). Instead, data on workforce composition is possible to find for fisheries with in-depth reports on trade dynamics. Reporting of non-local fleets is likely more common than reports of local ones, and some fisheries may have no public data.

3) *Workers are protected.* This focuses on institutions at the fisher and national level. These data are average to challenging to obtain. The most direct measures of basic worker protection are the presence of legitimate legal frameworks at the national level that include rights to strike (3.1) and freedom of collective action (3.2) which, for many countries, can be found in online databases, such as wageindicator.org [31]. As Fair Trade USA's first requirement makes clear, if there is no organizational structure for workers, then worker protections are less likely to be a concern [23]. The presence of grievance structures (3.3) can be verified through either active fisher organizations that are common in small-scale fisheries and are often FIP participants or stakeholders (see fishsource.com for specific fisheries) [33]. Another source (3.3) of evidence is a formal structure with employers or local regulations that explain grievance mechanisms. Ideally, we would assess how well fisheries are *actually upholding these rights*, but standardized data would be scarce under the evidence constraints we are working with.

4) *Communities have improving healthcare.* The focus is children at the level closest to the fishing community scale, and the fishing community. These data are easy to obtain. Children are the most vulnerable in a population. Goal #4 of the Millennium Development

Goals of the United Nations calls to reduce the under-five mortality rate by two thirds, between 1990 and 2015 [34]. While few publically-available datasets are global in scope and can be used as a shared standard for fisheries, child mortality under five years old (4.1) is one of these virtually universal measures regularly collected at the national level, and sometimes at regional and finer resolution (preferred). The Demographic and Health Surveys (DHS) Program's Spatial Data Repository funded by USAID provides spatially-explicit health and demographic data [35]. Since the under-five mortality rate may vary across areas of large fisheries that encompass multiple geographic regions, the assessor must take the highest under-five mortality rate of all regions associated with the fishery. We set global standard for under-five mortality as higher than 10, which is in line with many developed countries. Faced with the challenge of this performance measure, global production networks may be encouraged to create new partnerships with public health and other agencies. CSR reports can be used to cite legitimate programs to improve local health that are initiated or supported by the seafood industry (4.2).

5) *Communities have improving education.* The focus is children at the level closest to the fishing community level, and the fishing community. These data are easy to obtain. As with the health dimension, focusing on children and the community (rather than education levels among the workforce) is likely to be more available through desktop research and more indicative of changes. The first component of this builds on the Millennium Development Goal #2 to achieve universal primary education [34], but these alone do not capture enough variation across countries. Instead we focus on whether less than 10% of the children are not enrolled in primary or secondary school (5.1) using UNESCO's global trend on numbers of out-of-school children [36,37]. If data are available at finer resolution they should be used. For 5.2, evidence of legitimate industry-led education-related programs for fishers, as well as others in the community, could be obtained from CSR reports.
6) *Fishery will be viable for future generations.* The focus is the fishery system as a whole, including fishers, women, and newcomers. These data are challenging to very challenging to obtain, as local reports are required. If fishery employment is seen as a viable source of livelihood, women and men value the fisheries. Conversely, fisheries can, and do, extinguish from attrition when the value is seen as low and when resources are overharvested. Women play a critical role in many fisheries and are often the majority of workers in seafood processing [38]. Such work is often poorly paid and low status. Evidence of women in leadership roles (6.1) in workers'

organizations that represent seafood worker interests or seafood companies are shown in FIP, CSR, or other reports. Whether a mix of age groups is represented (6.2) can likely be found in demographic information that are regularly collected at fisheries levels for socio-economic assessments or national statistics of employment by FAO country profiles [39]. Evidence of training of fishers (6.3) can be found in CSR, fisher association, industry association, or government reports.

7) *Harvesters have economic flexibility.* This focuses on harvesters at the fishery level. These data are challenging to very challenging to obtain. Evidence of fishers' freedom to sell to whomever they choose (7.1) is the absence of information indicating that harvesters are locked into one buyer or loan type (which may lead to unfair lending practices and other exploitation by intermediaries leaving them more vulnerable to cope with challenges [40]). Evidence (7.1) can also be the structural dynamics of the fishery, for example, if there are few processors, or fisheries are contracted to sell to a single processor, or if there are very few middlemen in a large fishery system, the harvesters' options and ability to negotiate fair terms are severely limited. Evidence of economic flexibility also includes whether harvesters access multiple loan types at good rates (not exceeding government rate) (7.2), which will need to be obtained through socio-economic reports. We also consider as evidence formalized training such as programs in how to add value to fishers' landings (7.3) that can help harvesters earn more without increasing fishing effort or expenditures. This information can be found in fisher-organization, government, FIP, and CSR reports. We did not include economic fishery analyses, such as cost of entry for fishers, a metric being considered elsewhere [32].

3. RESULTS AND DISCUSSION

Small-scale fishery: The small-scale fishery (with weaker institutional context) ranked lower than the industrial fishery (with stronger institutional context) in most dimensions (Table 3, Figure 1). In the small-scale fishery, the highest rank for any single dimension (3) was for jobs benefitting the communities, because >50% of the fishers and processors were local (community-based). The small-scale fishery ranked medium (2) for earnings and education, and ranked low (1) for worker protection, healthcare, fisheries viability, and economic flexibility. The medium ranking for earnings was assigned because while fishers earned a fair wage in the small-scale example, processors did not. Education ranked medium because less than 10% of primary and secondary school age children were out of school, but there were no company-led education programs. The low (1) rankings for worker protection were assigned because there were essentially no labor protections at the national level, nor any structures in place to address grievances at the fishery level. In the health dimension, the DHS survey data showed >85 deaths per 1000 births, higher than the standard of 10, and there were no records of company-led health initiatives in this area. Regarding fisheries viability, despite having all age-classes of fishers represented in the small-scale example, there was no evidence of either recruitment of new fishers nor of women in leadership positions, resulting in a low ranking. A low ranking for economic flexibility was assigned because loans were obtained mainly from middlemen, and no formal value-added programs were in place.

Large-scale fishery: The large-scale fishery had high rankings (3) in worker protection, healthcare, education, and economic flexibility (Table 3; Figure 1). It ranked medium (2) for jobs benefitting the community and for fisheries viability, and low (1) in earnings. A basic explanation for the high rankings in the industrial fishery case is that the fishery is located in a country with measures in place for human rights and labor protection, and public health and education is strong, overall. Specifically regarding worker protection, strikes were allowed, salaries were collectively bargained, and there was a national fisher's organization where grievances can be addressed. For health, there were <10 deaths per 1000 births and the fishery had documented company-led healthcare

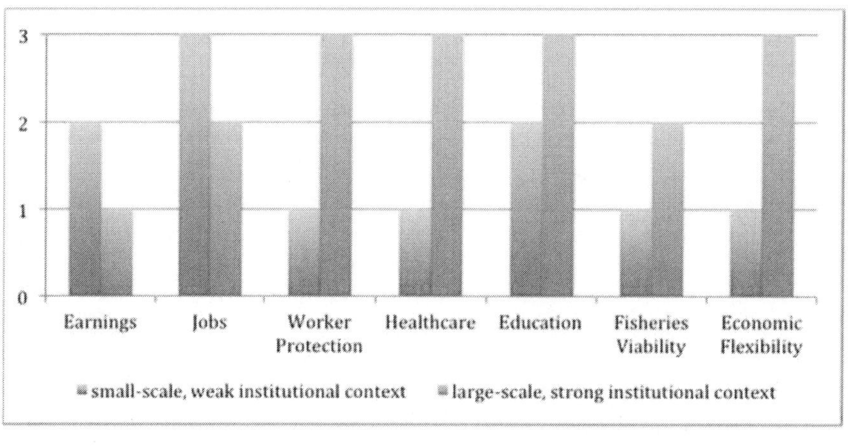

Figure 1. A comparison of SWIFT rankings of a small-scale fishery in a weaker institutional context and an industrial fishery in a stronger institutional context.

Table 3. SWIFT rankings for a small-scale fishery in a weaker institutional context and an industrial fishery in a stronger institutional context, according to criteria in Table 1.

Wellbeing Dimension	Suggested Measures & Rank Ranges	Small-Scale	Rank	Industrial	Rank
1. Earnings are fair and stable.	Harvesting wages are higher than the standard (national minimum wage) OR if no wages price/weight paid to harvesters is stable or increasing over time over a five year period.	Yes >10% national average	2	No Harvesters earn below national average.	1
	Processing sector wages are higher than the standard	No Less than average daily wage in region.		No Outsourced to cheaper labor markets abroad.	
2. Jobs benefit the communities.	Harvesting workforce comprised of >50% local people.	Yes >50% local harvesting crew.	3	Yes National fishing companies own fishing quotas. Strong labor organizations/difficult to employ unregulated workers.	2
	Primary processing workforce is comprised of >50% local people.	Yes >50% local processing workers.		No Processing outsourced to other countries.	
3. Workers are protected.	National level protection: multiple national level protections for right to strike.	No Only provisions in the labor code, not in trade union/collective action.	1	Yes Insufficient data on wage indicator.org but public strikes suggests strikes allowed.	3
	National Level Protection.	No only in the labor code.		Yes Insufficient data on wageindicator.org but salaries are collectively bargained for 80% of workforce.	
	Structure is in place to address worker grievances at local level, or the global production network uses labor contracts and the terms are transparent/public.	No No evidence of fisheries associations to address grievances, & mainly informal contracts.		Yes National fisherman's association addresses grievances, and strong labor protections present.	
4. Communities have improving healthcare.	Made significant improvement in under-five mortality rate at the relevant fishery level or meets global standard (set at 10 deaths per 1000 births).	No DHS Survey data show >85 deaths per 1000 births.	1	Yes <10 deaths per 1000 births or less.	3
	Seafood company has implemented programs to improve healthcare at the relevant fishery level.	No none known to be implemented		Yes CSR program by industry to deliver services such as support for physiotherapist, safe equipment and sponsored sports activities for employees.	

initiatives. For education, less than 10% of primary and secondary school age children were out of school where the fishery is based, and at the harvester level there are documented company-led education initiatives. Regarding economic flexibility, harvesters could obtain government and bank loans at market rates and there were formalized training programs to increase the value of the harvest. For viability, while new fishers were being recruited and all age-classes were represented in the fishery, there were no women leaders in relevant worker or fisher organizations, or board members, resulting in a medium ranking (2).

The industrial fishery ranked higher than the small-scale fishery overall (Table 3), but it ranked somewhat lower than might be expected in some dimensions considering the country's institutional context in which workers have very strong protections in place (e.g., labor contracts, unions, social security, pension, unemployment funds). Specifically, the industrial fishery ranked lower than the small-scale fishery for "earnings are fair and stable" and "jobs benefit communities". In the industrial example, harvesters earned less than the national standard (which was a high standard). Earnings in the small-scale fishery case were higher compared to their national standard. In addition, the business model of the industrial fishery example is to outsource the vast majority of fish processing jobs overseas where labor is cheaper. Thus, despite employing mostly local harvesters, it ranked low (1) because processing jobs did not benefit the fishing communities.

Of the seven variables, the easiest to obtain were those regarding earnings, and community-level education and healthcare. The most challenging were data on viability and flexibility. The local workforce characteristics and worker protection were in the middle. This was not surprising, since concepts such as viability and flexibility were more complex measures of wellbeing and were not yet measured in a standardized way, as opposed to regularly-collected education and health metrics, and economic information on wages.

3.1. Areas for Future Work

Testing SWIFT on two fisheries gave insights on data collection process and availability in order to obtain rankings. The next step is to expand testing to a wider range of fisheries in terms of geography, socio-cultural context, and scale to better understand potential issues of validity. This will provide additional information on how fisheries rank relative to each other and how to improve the methodology. If common challenges are encountered in the data collection or scoring, or if biases are detected across the fisheries tested, adjustments can be made. Another key consideration is how these dimensions resonate with key stakeholders in small-scale and large-scale source fisheries. The results of the tests should be shared with the stakeholders involved in the fisheries that are included. Their understanding of, and response to, SWIFT will inform adjustments to the method.

SWIFT highlights that it is possible for a fishery to meet certain criteria in practice, but to be unable to demonstrate that through publically-available data at the time of assessment. For example, some countries (like Scandinavian countries) may have no legally-set minimum wage but pay rates are agreed through collective bargaining agreements between the industry and the union. In our industrial fishery example 85% of workers in the country were automatically protected through labor unions' agreements, and fishers have their own separate association and contract. It is also possible that evidence used to justify ranks may be outdated by the time it becomes available to the public. Political regimes change, fisheries collapse, but such events will also be public so the public can request that this information be corrected if evidence is supplied, as is the case with other ranks on fishsource.com.

Once additional test cases are analyzed and further improvements are made to SWIFT, opportunities for better understanding the interaction between social and ecological issues in fisheries will be explored. Changes or trends in social performance ranks can be overlaid with ecological ranks to look for insights into causality or indirect effects. For example, if a fishery's rank for dimension 1 (Earnings are fair and stable) drops over the course of a period of a few years and illegal fishing increases over the same timeframe, we can begin to ask questions and design improvement solutions that consider both areas, if a link can be shown.

Ideally, wellbeing should be examined at all levels in the global production network, and across all populations and locations where the product is processed, but for now we focused mainly at the origin of the seafood supply network. A contribution of SWIFT to social performance measures is that it includes, but is not limited to, assessing wellbeing at individual and community levels. Community resilience has been found to be important in disaster preparedness and in social sustainability [41,42]. Wellbeing applies to various levels and sectors in fishery systems (e.g., fishers, seafood processors, women, children, institutions related to fishing, and processing labor, more generally) and they are interrelated. If fishers have a secure livable income, but children in the community do not complete basic education, then fisheries may not be able to capitalize on new opportunities or respond to future change or crises, because the system, as a whole, is not well.

The issue of units of analysis—whether a species, fish stock, fishing community, area defined by a regulatory system, or gear types—is problematic for all wellbeing approaches, including the one proposed. For example, the census-level data that the NOAA group relies on may be much broader than the scale appropriate to a fishery [21]. In some fisheries, information is so scarce that information from a single area and tens of fishers may be applied to a fishery that spans a whole country and many thousands of fishers and processing workers. Personal biases of key informants can also come into play in the Anderson group's approach [24] because it is unclear whether site-, region-, or country-level data has been used to arrive at a rank. Resolving this problem is beyond the scope of this paper, but this still is important to consider.

The difficulty of obtaining data specifically focused on resilience concepts—security and viability demonstrate opportunities to bring these concepts into the social responsibility mainstream, such as those that collect data on education, health, and wages.

While our desktop research approach cannot replace any "bottom-up" participatory method that uses local-level definitions and aspirations of wellbeing, it does help to open up the subject of human wellbeing as an important issue to more seafood industry actors and institutions, which is a much needed step. SWIFT can offer a snapshot of the wellbeing of a fishery, but this does not replace the function of conducting in-depth social research in fisheries, nor the use of comprehensive methods to assess social dimensions of fisheries.

3.2. How Can Global Production Networks Improve Social Performance?

SWIFT may compel companies to look more closely at low-scoring aspects of their source fisheries, or make the effort to seek out legitimate evidence to improve a low ranking. Companies sourcing from high-scoring fisheries may see opportunities for communication and marketing about successful efforts to improve socio-economic aspects of fisheries. However, we recognize that actions of global production network actors are limited in certain ways by their context and the business models underlying them. Likewise, the motivation and potential reward for examining social performance will vary. Types of actions by seafood companies may depend, in part, on whether the production network in question produces a globalized commodity with low profit margins, or a product with higher margins that distinguishes itself by brand/more specialized markets. For example, global production networks (especially those harvesting or processing in countries with a high national wage as in our test case) may be less likely or able to overhaul their wage scales given the low-margin/high volume business model, and might be more likely to invest more heavily in other areas of performance that positively affect worker wellbeing.

The industry can also help foster existing fishing organizations or help develop new ones according to local desires and needs. These entities could co-create appropriate programs, such as more value-added training for fishers, more microfinance options, etc. which also could improve fishery performance ranks. Another example of how seafood companies can use resources and influence to improve social *and* ecological conditions in a fishery might be a landings data collection program that hires and trains interested women and men in the community, including students. Programs of this type that monitor performance can also help recruit individuals into the fishery workforce, creating a virtuous cycle in terms of viability of the fishery system. As this assessment method develops and patterns emerge within certain regions or fishery types, perhaps health and education programs might be designed with the common concerns of fishing communities or regions in mind at larger than local scales, and cooperation or networking among fisheries and across global production networks may be facilitated to address wellbeing issues.

Whatever actions may be available to global production network participants, it is clearly in the seafood industry's interest to be seriously concerned with the relationship between social and economic performance. A goal is for the seafood industry to be able to use some iteration of SWIFT to evaluate fisheries and create a publically-available snapshot of social performance in fisheries. A similar tool implemented by SFP has been successful in engaging industry to make changes to the environmental aspects of global production networks. This assessment methodology may also help seafood companies see the limits and consequences of certain fishery improvement strategies, such as drastic catch reductions and other constraints, and to improve chances of success. The economic hardship fishing communities often suffer when drastically new catch regimes are implemented harms wellbeing, and can doom efforts to failure by undermining fishers' abilities to comply. However, fishery improvement projects using social performance measures at the start have a foundation to design projects that activate the fishery's existing capacity to adapt, and take approaches that may be more step-wise and less risky from a harvester perspective [43], given the uncertain outcomes. We would expect that fishers who are less vulnerable, socially and economically, can better cope with the impacts of such efforts to improve fisheries ecological performance.

4. CONCLUSIONS

The SWIFT method offers a rapid, broad assessment of wellbeing and represents an accessible first step in understanding the situations of people working in global production networks. The proposed methodology provides a snapshot assessment of social wellbeing. The question of why a certain rank for a dimension is high or low will usually require a more detailed explanation or in-depth research. These "why" questions matter greatly for understanding fishery dynamics at the human-ecological system level. In the shorter term, however, SWIFT can help generate questions about causality and relationships between social performance measures for further research. The performance measures are meant to encourage current and future global production network participants, marine resource managers, and socially responsible investors to become more interested in the basic question of how social wellbeing and socio-economic sustainability relate to ecological sustainability.

ACKNOWLEDGMENTS

Funding was provided by the Rockefeller Foundation Ocean and Fisheries Program. The input of Tracy Van Holt was also made possible, in part, by funding from the Erling-Persson Family Foundation through

Global Economic Dynamics and the Biosphere, the Royal Swedish Academy of Sciences.

REFERENCES

1. ANOVA. *Sustainability Report-Fishing and Living: Improving Life in the Fishing Community*; ANOVA Holding: Tampa, FL, USA, 2015.
2. Sustainable Fisheries Partnership. Seafood and Social Development:Engaging the seafood industry in the social development of seafood-producing communities in developing countries. In Proceedings of the Seafood and Social Development Forum, Annapolis, MD, USA, 21–22 September 2015.
3. Brinson, A.; Lee, M.Y.; Rountree, B. Direct marketing strategies: The rise of community supported fishery programs. *Mar. Policy* **2011**, *35*, 542–548.
4. Food & Agriculture Organisation (FAO). *Overcoming Factors of Unsustainability and Overexploitation in Fisheries: Selected Papers on Issues and Approaches*; Food & Agriculture Organisation: Rome, Italy, 2005; Volume 782.
5. Folke, C.; Colding, J.; Berkes, F. Synthesis: Building resilience and adaptive capacity in social-ecological systems. In *Navigating Social-Ecological Systems: Building Resilience for Complexity and Change*; Cambridge University Press: Cambridge, UK, 2003; pp. 352–387.
6. Chapin, F.S.; Carpenter, S.R.; Kofinas, G.P.; Folke, C.; Abel, N.; Clark, W.C.; Olsson, P.; Smith, D.M.S.; Walker, B.; Young, O.R.; et al. Ecosystem stewardship: Sustainability strategies for a rapidly changing planet. *Trends Ecol. Evol.***2010**, *25*, 241–249.
7. Ratner, B.D.; Åsgård, B.; Allison, E.H. Fishing for justice: Human rights, development, and fisheries sector reform. *Glob. Environ. Chang.* **2014**, *27*, 120–130.
8. Allison, E.H.; Ratner, B.D.; Åsgård, B.; Willmann, R.; Pomeroy, R.; Kurien, J. Rights-based fisheries governance: From fishing rights to human rights. *Fish Fish.* **2012**, *13*, 14–29.
9. Pickerell, T.; Newman, M.; MacDonald, S. New tools for seafood business to understand human rights risks and improve social compliance. In Proceedings of the Conference at Seafood Expo North America, Boston, MA, USA, 6–8 March 2016.
10. Greenfield, K. Success, failure, and corporate responsbility. In Proceedings of the Conference at Seafood Expo North America, Boston, MA, USA, 6–8 March 2016.
11. Wingard, J. Community transferable quotas: Internalizing externalities and minimizing social impacts of fisheries management. *Hum. Organ.* **2000**, *59*, 48–57.
12. Miller, M.; Johnson, J. Hard work and competition in the bristol bay salmon fishery. *Hum. Organ.* **1981**, *40*, 131–139.
13. Gaynor, J.L. Flexible fishing: Gender and the new spatial division of labor in eastern indonesia's rural littoral. *Radic. Hist. Rev.* **2010**, *2010*, 74–100.
14. Jentoft, S.; McCay, B.J.; Wilson, D.C. Social theory and fisheries co-management. *Mar. Policy* **1998**, *22*, 423–436.
15. Pollnac, R.B.; Abbott-Jamieson, S.; Smith, C.; Miller, M.L.; Clay, P.M.; Oles, B. Toward a model for fisheries social impact assessment. *Mar. Fish. Rev.* **2006**, *68*, 1–18.
16. Millenium Ecosystem Assessement. *Ecosystems and Human Well-Being: A Framework for Assessment. Millennium Ecosystem Assessment*; Island Press: Washington, DC, USA, 2003.

17. Millenium Ecosystem Assessement. *Ecosystems and Human Well-Being: Synthesis*; Island Press: Washington, DC, USA, 2005.
18. Brown, K. *Resilience, Development and Global Change*; Routledge: London, UK, 2016.
19. Biggs, R.; Schlüter, M.; Schoon, M. (Eds.) *Principles for Building Resilience: Sustaining Ecosystem Services in Social-Ecological Systems*; Cambridge University Press: Cambridge, UK, 2015.
20. Association of Southeast Asian Nations (ASEAN) Steering Committee. *Fishery Improvement Project Protocol for the ASEAN Region Draft for First Public Comment Period*; ASEAN: Jakarta, Indonesia, 2014.
21. Jepson, M.; Colburn, L.L. *Development of Social Indicators of Fishing Community Vulnerability and Resilience in the U.S. Southeast and Northeast Regions*; NOAA Technical Memorandum NMFS-F/SPO-129; U.S. Department of Commerce: Washington, DC, USA, 2013.
22. Himes-Cornell, A.; Kasperski, S. Using Socioeconomic and Fisheries Involvement Indices to Understand Alaska Fishing Community Well-Being. *Coast. Manag.* **2016**, *44*, 36–70.
23. Fair Trade. *Fair Trade USA Capture Fisheries Program*; Fair Trade USA: Oakland, CA, USA, 2014.
24. Anderson, J.L.; Anderson, C.M.; Chu, J.; Meredith, J.; Asche, F.; Sylvia, G.; Smith, M.D.; Anggraeni, D.; Arthur, R.; Guttormsen, A.; et al. The fishery performance indicators: A management tool for triple bottom line outcomes. *PLoS ONE* **2015**, *10*, e0122809.
25. Food & Agriculture Organisation (FAO). *Voluntary Guidelines for Securing Sustainable Small-Scale Fisheries in the Context of Food Security and Poverty Eradication*; Food & Agriculture Organisation: Rome, Italy, 2015.
26. Berkes, F.; Folke, C.; Colding, J. (Eds.) Building resilience for adaptive capacity in social-ecological systems. In *Navigating Social-Ecological Systems: Building Resilience for Complexity and Change*; Cambridge University Press: Cambridge, UK, 2002.
27. Cumming, G.S.; Barnes, G.; Perz, S.; Schmink, M.; Sieving, K.E.; Southworth, J.; Binford, M.; Holt, R.D.; Stickler, C.; Van Holt, T. An exploratory framework for the empirical measurement of resilience. *Ecosystems* **2005**, *8*, 975–987.
28. Gunderson, L. Resilience, flexibility and adaptive management—Antidotes for spurious certitude? *Ecol. Soc.* **1999**, *3*, 7.
29. Adger, W.N. Social and ecological resilience: Are they related? *Prog. Hum. Geogr.* **2000**, *24*, 347–364.
30. Crona, B.I.; Van Holt, T.; Petersson, M.; Daw, T.M.; Buchary, E. Using social-ecological syndromes to understand impacts of international seafood trade on small-scale fisheries. *Glob. Environ. Chang.* **2015**, *35*, 162–175.
31. Wage Indicator. Available online: http//:www.wageindicator.org (accessed on 15 December 2015).
32. Metz, S. *Developing Economic Scores for FishSource*; Sakana Consultants: Brest, France, 2015.
33. Sustainable Fisheries Parnership. FishSource. Available online: http://www.fishsource.com/ (accessed on 27 May 2016).
34. United Nations. *Millenium Development Goals Report 2015*; United Nations: New York, NY, USA, 2015.
35. Demographic and Health Surveys. Spatial Data Repository for the Demographic and Health Surveys (DHS) Program. Available online: http://spatialdata.dhs program.com/home/ (accessed on 27 May 2016).
36. United Nations Educational, Scientific, and Cultural Organization (UNESCO). UNESCO eAtlas of Out-of-School Children. Available online: http://tellmaps.c om/uis/ oosc/ (accessed on 27 May 2016).
37. United Nations International Children's Emergency Fund (UNICEF). *Global Out-of-School Children Initiative Operational Manual Global Initiative on Out-of-School*

Children; United Nations International Children's Emergency Fund: New York, NY, USA, 2015.
38. Griffith, D.; Valdez Pizzini, M. *Fishers at Work, Workers at Sea: A Puerto Rican Journey through Labor and Refuge*; Temple University Press: Philadelphia, PA, USA, 2002.
39. Food and Agriculture Organization of the United Nations (FAO). FAO Country Profiles. Available online: http://www.fao.org/countryprofiles/en/ (accessed on 27 May 2016).
40. Cinner, J.E.; Sutton, S.G.; Bond, T.G. Socioeconomic thresholds that affect use of customary fisheries management tools. *Conserv. Biol.* **2007**, *6*, 1603–1611.
41. Norris, F.H.; Stevens, S.P.; Pfefferbaum, B.; Wyche, K.F.; Pfefferbaum, R.L. Community resilience as a metaphor, theory, set of capacities, and strategy for disaster readiness. *Am. J. Community Psychol.* **2008**, *41*, 127–150.
42. Magis, K. Community Resilience: An indicator of social sustainability. *Soc. Nat. Resour. Int. J.* **2010**, *23*, 401–416.
43. McCay, B.J. Muddling Through the Clam Beds: Co-operative Management of New Jersey's Hard Clam Spawner Sanctuaries. *J. Shellfish Res.* **1988**, *7*, 327–340.

CHAPTER 6

Managing Small-Scale Fisheries in Colombia

Lina M. Saavedra-Díaz[1]*, Robert Pomeroy[2] and Andrew A. Rosenberg[3]

[1]Department of Biology, University of Magdalena, Carrera 32 No. 22 – 08, Santa Marta, Magdalena P O Box 2-1-21630 Colombia

ABSTRACT

The small-scale fishermen of Colombia face a wide range of problems and conflicts. While many problems are shared among individuals from both the Atlantic and the Pacific coasts (bi-coastal), others are unique to a subset of the communities, only occur on one of the coasts (uni-coastal) or in an individual locality. To come to grips with the major problems for these fisheries requires establishing a fisheries management strategy that can prioritize solutions at different levels: national, coastal, and local. This study describes the solutions identified by three sets of stakeholders: fishermen, local leaders and fisheries experts, to improve small-scale fisheries management in Colombia. Some cross-cutting solutions were recommended by all three sets of stakeholders. In other cases, only two of the three stakeholder groups agreed on certain proposals, and some isolated solutions were found in only one type of stakeholder. All three sets of stakeholders recommended that the government put in place fundamental regulatory framework for small-scale fisheries including support for alternative employment opportunities to reduce fishing pressure on the resources. Some but not all groups supported specific measures, such as gear restrictions, closed areas and closed seasons. There was also a clear need to distinguish those truly engaged in fishing as their livelihood from opportunists moving in and out of the sector. Specific recommendations are here presented to reform and restructure governance through co-management, and to develop a consensus among the main government and user stakeholders.

Keywords: Colombia Small scale fisheries Co-management Fishery solutions

INTRODUCTION

Small-scale tropical fisheries studies in the last few decades have strongly benefitted from a multidisciplinary approach that includes ecological, economic, sociological, technological and governance information (Christy 1997; Preikshot and Pauly 1998). Many fisheries frameworks and governance models around the world are based on a combination of government and community efforts. These models could be used as points of departure for discussions with important stakeholders in Colombian small-scale fisheries.

A growing global literature describes lessons learned from implementing small-scale fisheries management (Berkes 2004, 2010; Brown and Pomeroy 1999; Chuenpagdaee and Jentoft 2007; Cinner et al. 2012; FAO 2006; Francis et al. 2007; Garcia and Cochrane 2005; Lam and Pauly 2010; Lebel et al. 2006; Olsson et al. 2004, 2006; Pomeroy et al. 2004; Salas et al. 2011; Tyler 2006). Understanding social interactions within the context of fishing, and the dynamic among fishermen within and among fishing communities (St. Martin et al. 2007) are essential in order to develop effective management strategies.

Moreover, the theoretical context for co-management of small-scale fisheries is an ongoing debate among academics and fisheries management practioners, especially in the context of social-ecological systems (Berkes et al. 2001; Ostrom 1990, 2009; Pomeroy et al. 2004, 2011). Weak governance is one of the main causes of the present poor condition of fisheries and is characterized by (but are not limited to) corruption, lack of stakeholder participation, political will and capacity, weak institutional capacity and capabilities, poor enforcement, and inadequate information (CRC et al. 2006). Small-scale fisheries are extremely diverse, complex and dynamic (Berkes et al. 2001). They operate at many different scales and with many different stakeholders. These characteristics make the challenges and concerns that confront small-scale fisheries governance more difficult. Given the range of actors in small-scale fisheries, with their varying perceptions, agendas and power, there is a need to develop a governance system that can include multiple stakeholders (Bavinck et al. 2005). How best to tap their various interests, agendas and capacities remain a challenge.

In the case of Colombia, small-scale fishermen have not been fully recognized as key stakeholders in the fisheries management process

(Cuello and Duarte 2010; García 2010; Saavedra-Díaz et al. 2015a). It is critically important to involve communities in co-management practices in order for management to be fully effective and to reduce the propensity for overexploitation. In the Caribbean and Latin America, examples of co-management exist in Mexico, Belize, Brazil, Peru and Chile. Due to the high diversity of the fisheries in these areas, as well as poverty, violence, forced displacement, among other situations, many co-management practices need to be implemented in concert in order to be effective. Also, decision makers need to recognize the importance of supporting management decisions with local knowledge (Begossi and Brown 2003). Recently, fisheries co-management was successfully implemented in Chile by dividing responsibilities for the Management Area System between government agencies and small-scale fishers' organizations (Marin and Berkes 2010). This success raised consciousness among fishermen. It avoided the tragedy of the commons by building rapport between the state and fishermen, raising ecological and management awareness, and promoting teamwork between fishermen and scientists (Schumman 2007). The research presented here explores possible cross-cutting solutions of this kind for the particular realities of the under-researched small-scale fisheries of Colombia.

Government institutions and agencies in Colombia have attempted to organize the fisheries sector and articulate policy through the publication of different sets of national documents. However, a disarticulated institutional framework and a thicket of contradictory policies highlight the fragmented nature of Colombian fisheries management. Unfortunately, national planning documents such as the "Colombian Vision 2019" have set unrealistic goals for increases by 30 % in fisheries captures for 2019 (ECOVERSA 2007). Development goals such as this misunderstand the current status of both fishery resources and the management system and have not directly involved local communities in the development and implementation of policy to achieve sustainable fisheries (Hart 2003; McCay and Jentoft 1996; Schumman 2007). The Colombian legal framework that regulates fishery resources at the national level is the Code of Renewable Natural Resources and Protected Environment, created by the Decree/law 2811 of 1974 and Law 13 that created the General Statute for Fisheries in 1990. The Rural Development and Agriculture Ministry-through the Fishery and Aquaculture National Authority - is in charge at the national level of the Marine Small-Scale Fishery sector since 2011. However, over the last decade five different institutions have been in charge of fisheries. Such institutional instability at the national level has being one of the major challenges for successful management of the small-scale fishery sector.

Another challenge to overcome is the difference of social perceptions among stakeholders –i.e., fishermen, local leaders, and fishing experts and their existing relationships within marine ecoregions. According to Saavedra-Díaz et al. (2015b), fishermen perceive destructive fishing methods as the most significant problem with the highest percentage response among all respondents (70 % in the Caribbean coast and 79 % in the Pacific), followed by the lack of regulation enforcement (60 % in Caribbean and 47 % in Pacific) and pollution and industrial contamination. Local leaders identified the low quality of life of fishermen including the lack of public services, lack of potable water and lack of education focused on environmental awareness. They also noted a weakness of fisher's organizations and leadership. In contrast, the majority of fisheries experts believed that the small-scale fishery sector has been largely ignored by governmental actions. The fisheries experts identified the instability of regulatory authorities and lack of control measures, oversight and monitoring by the agencies in charge of artisanal fishery activity due to lack of personnel, and the lack of infrastructure or equipment to facilitate this work. While many problems occur on both coasts, other problems are coastally unique, and further, some affect only a small portion of fishing communities. In consequence, there is a need to establish a fisheries management strategy that can prioritize solutions at different levels: national, coastal, and local.

Using results from the same survey discussed in Saavedra-Díaz et al. (2015b), the objective of this paper is to discuss the solutions identified by fishermen, local leaders and fisheries experts for improving small-scale fisheries management in Colombia. In addition, we present specific recommendations on a framework for small-scale fisheries governance in Colombia.

METHODS

Study area

The Colombian coastline is divided between the Caribbean and the Pacific. The Caribbean coastline is 1642 km in length and the Pacific coastline is 2188 km in length. The Caribbean coast has a population of 2,919,348, while the Pacific coast has a population of 543,594 (Posada and Rozo 2005). Colombia has 12 political and administrative coastal states, eight on the Caribbean coast and four on the Pacific coast. Different from the states, coastal and marine environmental divisions separate the Colombian coast into six Coastal and Marine Ecoregions (CME) on the Caribbean and four CMEs on the Pacific coast (Fig. 1), plus four Insular Ecoregions on both oceans. CMEs are distinguished by different environmental characteristics such as geo-morphology, hydrography, sedimentology, and coastal and marine ecosystems (Posada and Rozo 2005). Nevertheless, state and CME boundaries are relatively similar, in some cases nearly overlapping. Since the present study focuses on environmental conditions, CMEs provide spatial orientation. Although the Caribbean and Pacific Insular Eco-regions (San Andrés and Providence Archipelago, San Bernardo, Malpelo and Gorgona Islands) are not included this study they should be included in future efforts to improve Colombian fisheries management overall.

Figure. 1Map of the selected fishing community (*white dots*) in coastal and marine ecoregions (CME) on the Colombian Caribbean coast and Pacific coast (Taken and modified from Posada and Rozo 2005)

The following parameters were employed in identifying one "typical" community in each CME: (1) it was not located in a marine protected area or on an island; (2) it was historically recognized as a fishing community; (3) it relied on fishing as a primary economic or subsistence activity; (4) there existed some level of fisher's organization (formal); (5) the community was involved previously in research projects; (6) there was a low incidence of violence or drug trafficking (for safety); and (7) it had been considered by Fishery experts.

Caribbean coast fishing communities selected in each coastal and marine ecoregion (CME) for this study (Fig. 1) include *Ahuyama* in the Guajira CME, *Taganga* in the Sierra Nevada de Santa Marta CME, *Las Flores* in the Magdalena CME, *San Antero* in the Morrosquillo and Sinú CME and *El Roto* in the Darién CME. Pacific coast fishing communities selected in each CME are *Bahía Solano* in the Alto Chocó CME, *Pizarro* in the Baudó CME, *Juanchaco* in the Málaga-Buenaventura CME, and *Tumaco* in the Llanura Aluvial del Sur CME.

Data sources

The fishermen, local leaders and fisheries experts were interviewed concerning their solutions to the identified problems that are affecting the small-scale fisheries communities and the marine fisheries resources they depend upon for food and livelihood. In addition, fishermen at community meetings were asked questions about their interest in, readiness for and how to move toward fisheries co-management. Fishermen are defined here as coastal marine small-scale fishermen or fisherwomen, including a great variety of racial and ethnic groups and are directly involved in fishing activities. Local leaders include presidents of local fishing associations or persons recognized as influential members of the fishing community including traditional authorities who fishers look to for leadership and who bring a broader perspective than just of a single fishermen. Finally, fisheries experts are scientists and technicians from Colombian fisheries institutions and administration with expertise and experience working with fishing communities, fishermen and marine resources. Semi-structured interviews of 309 people were conducted. Table 1 shows the number of interviews undertaken for the Caribbean and Pacific coasts per stakeholder group per community.

Table 1 Number of semi-structured interviews and number of fishermen performed in each stakeholder group on each community and eco-region on each coastal region

	Caribbean					Pacific				
Methods	*Ahuyama*	*Taganga*	*Las Flores*	*San Antero*	*El Roto*	*Bahía Solano*	*Pizarro*	*Juanchaco*	*Tumaco*	Total number of participants by method
Fishermen interviews	18	23	31	36	17	14	15	18	23	195
Leaders' interviews	2	3	4	6	1	4	2	3	2	27
Fishery experts' interviews	5	8	5	6	7	2	2	18	4	57[a]
Co-management hearings	17	10 and 18	20	27 and 28	20	35	13	13 and 17	17	235

[a]Thirty interviews performed in 2007 (15 from the Caribbean and 15 from the Pacific) in the feasibility study for the present research have been included in this study, for a total of 87 fishing experts interviews in total

Fieldwork performed in the fishing communities took place from August 2008 to August 2009 and lasted approximately one month within each community. To collect the fishermen and local leaders' perceptions about fisheries management solutions, a representative sample of leaders and fishermen in each small- scale fishing community were chosen (Table 1). The great variety of the fishing communities imposed important methodological limitations. For instance, fishing communities vary in population from villages such as *El Roto* with 50 Fishermen, to municipalities such as *Tumaco* with 4000 fishermen. In high population fishing communities such as *San Antero* on the Caribbean and *Tumaco* on the Pacific community leaders identified neighborhoods populated mainly by fishermen, for whom fishing activity and proximity to landing places determine where they live. Some municipal fishing communities also cover a much larger area than others (*Tumaco* extends over 167 counties). Therefore, the number of *Tumaco* fishermen involved in the study was low compared to the total number of fishermen in the community, and most came from the municipal center so that peripheral areas were under-represented.

Small communities with fishing populations of around 200 were simpler to sample. The lead author lived approximately three weeks to one month in each community and developed a greater understanding of the situation those fishermen faced. Recognizing these limitations is important in understanding the results of this research.

Community leaders were interviewed first upon arrival at each community. Local leaders identified the principal fishing gears used locally and explained the general economic, social, and cultural conditions in the community. The interview contained 15 standard questions focused on general subjects pertaining to the local artisanal fishing community. Questions were divided into five main subjects: (1) the role of local leaders in the community; (2) overview of the condition of marine artisanal fisheries; (3) fisheries problems; (4) proposed solutions; and (5) fisheries management.

Given the time and resources available, at least two fishermen from each community were interviewed for every fishing method (i.e., gillnets, mainline, longline, beach seine net with bag, harpoon, surrounding net, bottom trawl, among others). While representativeness of such a survey is always challenging, every effort was made to capture a wide range of perspectives. The interview consisted of 89 questions focused on seven main subjects: (1) personal information; (2) demographics and quality of life (family, education, living conditions, among others); (3) current fishing activity (i.e., time spent fishing, reasons for fishing, technology employed, fishing locations, marketing, among others); (4) long term changes in fishing activity and fisheries problems; (5) proposed solutions; (6) fisheries management; and (7) information regarding the community (i.e., relationships with community members, vulnerability to violence or drug traffic, happiness being a fisherman, among others).

In addition to interviews, we also utilized data from public meetings in each community related to the strategy of co- management. The meetings were held where fishermen brainstormed about top-down and bottom- up fisheries management strategies. Exchanging information and opinions, and discussing the co- management process took place in two main steps:

- Fishermen learned about fisheries management strategies, in particular, about co-management. The basic concepts and features of the major types of top-down and bottom-up fisheries management were explained. Examples of traditional management were analyzed and then contrasted with co-management to show the benefits and drawbacks of this process. This basic information allowed them to understand how fishery management has been applied in other countries, what alternatives exist, the benefits communities might gain from co-management, and the importance of working with other fisheries actors (stakeholder groups) in this process.
- An open discussion focused on the following questions:

What opinions do the fishermen have about management in general and co-management in particular?

What weakness and strengths within the artisanal fisheries community might affect the success of co- management?

What first steps could the community take to start the co-management process?

On the Caribbean coast, 141 community members participating in the interviews, while 81 on the Pacific coast were participated (Table 1). In addition, 140 community members participated in the hearings on the Caribbean coast and 95 on the Pacific coast. Even though it was planned to have one meeting per community, in some communities it was necessary to hold two hearings (*Taganga* and *Juanchaco* communities) because of low fishermen participation.

Data analysis

Qualitative information from semi-structured interviews and fieldwork activities was coded into quantitative data then organized and content analysis performed using computer-assisted qualitative data analysis software (CAQDAS), N-Vivo/8 software (García-Horta and Guerra-Romos 2009; Ozkan 2004; Saldaña 2009; Thayer et al. 2007; Yin 2003). Based on Miles and Huberman (1994) and Yin (2003), coding was done by the meaning of phrases, following the elemental method and incorporated in N-vivo. Then, structural coding and the elemental method were used to pre-code questions by creating main categories of common subjects that allow different opinions to be combined in a single category. Sixteen main categories were created corresponding to the common subjects under which the codes are aggregated (Miles and Huberman 1994). Chi-Square tests were used to distinguish between the perceptions of fisheries solutions and the three groups of fisheries stakeholders. Differences in the perceptions of the main categories of fisheries solutions among stakeholders were analyzed using a Kruskal-Wallis test. In addition, Mann-Whitney tests were performed in order to measure differences in perceptions of the same main categories between the Pacific and Caribbean regions. Finally, to explore the relationships between stakeholders' perceptions of fisheries solutions and the marine eco-regions, a redundancy analysis (RDA) was performed. Descriptive statistics (percentages) were used to analyze the results of each stakeholder group and among all stakeholders.

RESULTS

Given the range of actors in Colombian small-scale fisheries, with their varying perceptions, agendas and power, it can be expected that there will be differences in the proposed solutions to the challenges and problems facing the fisheries. Fishermen are focused on the fishing activity, leaders on governing, and experts on technical aspects of management. The theoretical debate over fisheries governance is how to develop a consensus among stakeholders for a governance system that can allow for including multiple stakeholders.

The results are grouped into cross-cutting solutions (identified by the fishermen, local leaders and experts at the same time), inter-group solutions (identified by only two of the three stakeholders), and isolated solutions (identified by only one stakeholder). In addition to the interviews, the results from public meetings in each community related to the strategy of co-management are presented.

Cross-cutting solutions

Fishermen from both coasts identified 133 solutions, of which 25 were represented by more than 3 % of respondents. Leaders identified 121 solutions, of which 38 were represented by more than 3 % of respondents. Experts identified 397 solutions, of which 135 were represented by more than 3 % of respondents. The responses from the three groups generated 22 cross-cutting solutions in eight categories. The Chi Square Test was used to establish the significance of these cross-cutting solutions to the three participant groups (Table 2).

Table 2: The perception of importance of cross-cutting solutions ranked by the fishermen, local leader and fisheries experts groups and analyzed using the Chi square test

Categories	Cross-cutting solutions	Code	Fisheries experts (%)	Fishermen (%)	Local leaders (%)	X^2	*p*-value
Aquaculture	Explore aquaculture as an option	Aqua	22.50	13.33	37.04	10.66	0.05
Fishermen and communities	Changes in Community attitude	Comm	2.50	5.64	7.40	1.56	0.46
	Changes in	Fish	41.25	14.35	3.70	30.2	<0.00

	Fishermen attitude					8	1
Fishing equipment	Open waters fishing depth sea fish – *pesca de altura*	OpenW	23.75	13.84	14.81	4.09	0.13
Fishing methods	Use friendly gears and recommendati ons	GearsOk	7.50	9.23	14.81	1.28	0.53
Goverment-Administrati on	Control over	Control	7.50	4.61	3.70	1.09	0.58
	Find or generate other jobs	Jobs	13.75	14.87	3.70	2.52	0.28
	Government support to real Fishermen	RealFish er	1.25	8.20	3.70	5.16	0.08
	Loan program with clear rules	Loan	11.25	1.02	3.70	15.5 4	<0.00 1
Industrial fishing activity	Restrictions on Industrial fishing	IndusRe st	12.50	7.69	7.40	1.69	0.42
Organizatio n of Fishermen	Strengthen Fishermen associations	F Asso	25.00	6.66	3.70	20.7 7	<0.00 1
	Strength Leaders	Leaders	5.00	0.51	3.70	6.31	0.04
Regulations	Create regulations	Regulate	61.25	50.76	62.96	3.34	0.19
	Create Protected Marine Areas	PMA	10.00	0.51	3.70	15.9 6	<0.00 1
	Fishing Zoning plan	Zoning	10.00	10.25	18.51	1.75	0.42
	Regulate gears	Gears	11.25	39.48	14.81	24.8 0	<0.00 1
	Increase mesh size	Mesh	7.50	6.66	3.70	0.47	0.79
	Prohibit gill nets	Not-Gill	7.50	19.48	3.70	9.36	0.01
	Change unfriendly	C-	11.25	40.00	3.70	31.9	<0.00

	methods for environmental friendly	Unfriend				9	1
	Establish minimum size per spp.	Min-size	10.00	2.05	7.40	8.62	0.01
	Veda-Time closed season	Veda	23.75	17.43	37.03	6.07	0.05

Results show that 11 cross-cutting solutions had a *p-value* less than or equal to 0.05, and thus were ranked at different levels of importance to all three groups, while 10 cross-cutting solutions had similar importance. Support for proposed solutions in the category of Government-Administration were not significantly different among the groups with the exception of the need for a loan program for fishermen based on clear rules, which experts felt would be particularly valuable.

All groups perceived with a similar sense of urgency that the government must regulate the fishery sector, that other jobs must be generated to replace fishing and decrease pressure on fragile resources and ecosystems, and that the administration needs to target support for "real fishermen" rather than opportunists (participants used the term "real fishermen" to distinguish between legitimate fishermen like themselves and pretenders who show up when aid is being handed out).

In the category of Regulations, only fishermen placed high importance on the need to regulate the use of gear, to prohibit the use of gillnets, and to exchange damaging gear for methods that are environmentally friendly. Experts placed great importance on creation of marine protected areas and establishing minimum catch size. For local leaders, the creation of close seasons (*veda*) was particularly important. There was significant agreement in the need to create regulations for fishing. In fact, this solution had the highest representation among the three groups. The category, Organization of Fishermen, was assigned a different level of importance by each group, with only experts emphasizing the need to strengthen fishermen associations.

Non-significant differences in importance were found in solutions in the following categories: (1) Changes in community attitude (Fishermen and Communities); (2) The use of vessels that encourage open waters and deep sea fishing (Fishing Equipment); (3) The need to use environmentally friendly gear (Fishing Methods); (4) Restricting industrial fishing (Industrial Fishing Activity); and (5) The need to establish fisheries zoning in each community, and to increase gill net mesh size (Regulations).

To relate the ranked importance of the perceived solutions with different stakeholders and regions, a redundancy analysis (RDA) was carried out (Fig. 2). In Fig. 2, the first two factors of the RDA explained 63 % of the total variance. The x-axis (F1) represents 44 % of the variation of perceived solutions between fishing communities on the Pacific (positive scores) versus fishing communities on the Caribbean (negative scores). Meanwhile, the y-axis (F2) is showing a gradient of perceived solutions represented 20 % of the variation between the perceived solutions by fishers (negative scores) versus experts (positive scores). Local leaders have low representation, being located almost in the middle of the other two stakeholders.

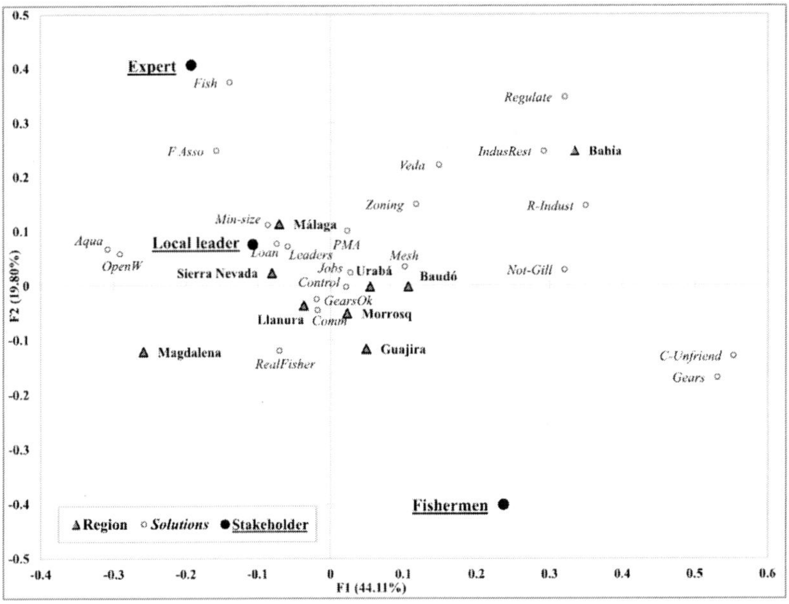

Figure. 2 Redundancy analysis of primary solutions (Cross-cutting) obtained through semi-structured interviews related to ecoregions and stakeholders (explaining variables). The x-axis (F1) represents 44 % and the y-axis (F2) 20 % of the variation

While solutions proposed by Caribbean communities differed from solutions proposed by communities on the Pacific, differences are not as great as was the case with the problems (refer to Saavedra-Díaz et al. 2015b). The eco-regions of *Alto-Chocó* (*Bahía Solano* fishing community) and *Baudó* (*Pizarro*) are representative of the fishery situation on the

Pacific, whereas the eco-region of *Magdalena* (*Las Flores*) is representative of the fishery situation on the Caribbean coast. Conversely, the Pacific eco-regions of *Málaga-Buenaventura* (*Juanchaco*) and *Llanura Aluvial del Sur* (*Tumaco*) correspond more closely to the Caribbean situation in terms of perceived solutions. Interestingly the eco-region of *Sierra Nevada de Santa Marta* (*Taganga*) was closer in viewpoint to Pacific communities. On the Pacific coast, the *Bahía Solano* community shared opinions with other communities, but their strong tradition of group deliberation and awareness of the fisheries situation resulted in a clarity of thought that produced more solutions and a wider spectrum of options.

Pacific fishing communities are more aware of the need for regulations (restricting industrial fishing, prohibiting gill nets, establishing closed fishing seasons, fisheries zoning, minimum mesh sizes net, ecologically friendly fishing methods, among others), while Caribbean communities proposed fewer solutions over a narrower range. They focused primarily on the need for oversight and control of fishing activity, change in community attitude, identifying "real fishermen", as well as for strong fishermen's associations, change in fishermen's attitude, exploration of aquaculture and access to open water vessels.

The y-axis (F2) shows how the opinions about solutions of fishers were far different from the opinions of experts. This result supports the impression that the opinions of fishers are not represented by experts. In fact, need for changes in fishermen's attitude (Fish), strengthen fishermen associations (F Asso), and the creation of marine protected areas (MPA) are perceived as important solutions mostly by experts. In contrast, fishers identified solutions that were not considered important by the other two stakeholder groups, such as identifying "real fishermen" (RealFisher), and access to loans (Loan).

All three stakeholder groups identified regulation as the most important solution (fishermen 51 %, experts 61 %, and leaders 63 %). All groups agreed on five additional solutions, but differed significantly on the degree of importance. They are presented in order from the most to the least different. The change from unfriendly (environmentally damaging) to friendly (less damaging) fishing methods ($X^2 = 31.99$; p- value <0.001) was most important to fishermen (40 %). Changes in fishermen's attitude ($X^2 = 30.28$; p- value <0.001) was most important to experts (41 %). Gear regulation ($X^2 = 24.80$; p- value <0.001) was also most important to fishermen (39 %). The fourth and fifth solutions that differ in importance, the need to strengthen fishermen associations ($X^2 = 20.77$; p- value <0.001), and to create marine protected areas ($X^2 = 15.96$; p- value <0.001), were most important to the group of experts.

Inter-group and isolated solutions

There are 37 inter-group solutions which were proposed by the stakeholder groups. These solutions are separated into nine categories (coastal uses and infrastructure, fishermen and communities, fishing resources, government-administration, institutions, marine ecosystems threatened, marketing, organization-fishermen association, regulations). The top solutions in each category are presented in Table 3.

Table 3 Main inter-group solutions shared by two stakeholder groups

Main intergroup solutions	% both coasts		
	Fishermen	Local leaders	Fisheries experts
Coastal Uses and Infrastructure			
Tourism		25.90	2.40
Fishermen and Communities			
Invest in Fishermen Education-read and write		11.11	16.67
Assume responsibility as Fishermen and improve attitude	14.36		7.14
Fishing Resources			
Find new target spp.	0.51		10.71
Goverment-Administration			
Recognition of the importance of Fishermen role and the job-chain involved		18.52	13.10
Empower community	7.20		17.86
Institutions			
Academia important actor		3.70	13.10
Establish Real Fishery institution	3.60		56.95
Marine Ecosystems Threatened			
Recover mangroves and breeding zones		7.40	3.60
Marketing			
Have or improve their own selling and		14.81	5.95

marketing process			
Improve product manipulation and quality		7.41	5.95
Organization - Fishery Association			
Promote all Fishermen to be associated and the importance of it		3.70	8.33
Support Fishermen in organizing their F.A.		3.70	8.33
Regulations			
National Fishery policy-regulations		22.22	16.67
Work and control fish buyers or merchant		7.41	4.76

In addition to these inter-group solutions, there were 37 isolated solutions separated in 9 categories (aquaculture, coastal activities and infrastructure, fishermen and communities, government- administration, institutions, national situation, marketing, organization-fishermen association, regulations). The largest number of inter-group solutions were proposed by experts. These included improved fishery information, more research, co-management, and learning about successful management from other countries. Two of these 37 solutions contain sub-solutions. Establishing a real fishery institution was described in 10 sub-solutions, and creating a national fishery policy was identified in 22 sub-solutions.

In order to understand differences at the category level, solutions in the first (cross-cutting solutions), second (inter-group) and third (isolated) orders were combined for analysis. The Kruskal–Wallis test (Kruskal and Wallis 1952) was used here to evaluate differences among categories with respect to the three interviewed groups (Table 4). No significant differences were found in the categories of Fishing Equipment, Fishing Methods, Industrial Fishing Activity, and Regulations; all stakeholders viewed these categories with the same level of importance. Leaders and experts contributed to categories such as Marketing and Fishermen Associations. The former gave great feedback about solutions related to the category of Coastal Uses and Infrastructures. The latter brought substantial investigative experience to the national fishery situation through their knowledge in the categories of Institutions and Government Administration Fishermen were not highly represented in any category compared with the other two stakeholders.

Table 4 Kruskal – Wallis test by the categories of solutions through the three stakeholders groups

Categories	Fisheries experts	Fishermen	Local leaders	X^2	*p*-valor
Acuaqulture	0.225	0.133	0.370		
	(0.420)	(0.341)	(0.492)	10.621	0.005
Coastal Uses and Infrastructure	0.100	0.056	0.370		
	(0.302)	(0.231)	(0.492)	26.86	<0.001
Fishermen and Communities	0.438	0.190	0.444		
	(0.499)	(0.393)	(0.438)	21.35	<0.001
Fishing Equipment	0.263	0.159	0.185		
	(0.443)	(0.367)	(0.396)	3.960	0.138
Fishing Methods	0.075	0.092	0.000		
	(0.265)	(0.290)	(0.000)	2.783	0.24
Fishing Resources	0.125	0.000	0.074		
	(0.333)	(0.000)	(0.267)	24.066	<0.001
Goverment-Administration	0.713	0.333	0.593		
	(0.455)	(0.473)	(0.501)	34.948	<0.001
Industrial Fishing Activity	0.125	0.077	0.000		
	(0.333)	(0.267)	(0.000)	4.388	0.111
Institutions	0.625	0.077	0.037		
	(0.487)	0.267	0.192	105.148	<0.001
Threatened Marine Ecosystems	0.063	0.036	0.148		
	(0.244)	(0.187)	(0.362)	6.132	0.047
Marketing	0.238	0.036	0.148		
	(0.428)	(0.187)	(0.362)	26.471	<0.001
National Situation	0.038	0.000	0.000		
	(0.191)	(0.000)	(0.000)	8.381	0.015
Organization of	0.300	0.077	0.333		

Fishermen	(0.461)	(0.267)	(0.480)	27.77	<0.001
Regulations	0.613	0.508	0.630		
	(0.490)	(0.501)	(0.492)	3.333	0.189
Small Scale Fishing	0.050	0.005	0.074		
	(0.219)	(0.072)	(0.267)	8.409	0.015

Mean relative value (and standard deviation between parenthesis) for each category and stakeholder is shown

The Mann-Whitney test was used to evaluate differences among the categories with respect to the Caribbean and Pacific coastal regions (Table 5). These results show that category responses on one coast are not significantly different from the other, as was shown in the Redundancy Analysis (RDA).

Table 5 Mann-Whitney test by categories of solutions, comparing Colombia's Caribbean and Pacific coasts

Mann-whitney test				
Categories	Caribbean	Pacific	*U*	*p*-valor
Acuaqulture	0.214	0.122	11743.50	0.043
	(0.411)	(0.328)		
Coastal Uses and Infrastructure	0.112	0.070	11212.0	0.222
	(0.317)	(0.256)		
Fishermen and Communities	0.262	0.304	10297.5	0.427
	(0.441)	(0.462)		
Fishing Equipment	0.166	0.226	10104.0	0.194
	(0.373)	(0.420)		
Fishing Methods	0.086	0.070	10924.5	0.619
	(0.280)	(0.256)		
Fishing Resources	0.032	0.052	10536.5	0.387
	(0.177)	(0.223)		
Goverment-Administration	0.444	0.478	10382.5	0.561

	(0.498)	(0.502)		
Industrial Fishing Activity	0.043	0.148	9623.0	0.001
	(0.203)	(0.356)		
Institutions	0.193	0.261	10017.5	0.164
	(0.395)	(0.441)		
Threatened Marine Ecosystems	0.064	0.035	11068.5	0.270
	(0.246)	(0.184)		
Marketing	0.059	0.165	9608.5	0.003
	(0.236)	(0.373)		
National Situation	0.011	0.009	10774.0	0.868
	(0.103)	(0.093)		
Organization of Fishermen	0.144	0.183	10342.5	0.379
	(0.352)	(0.388)		
Regulations	0.503	0.617	9519.0	0.052
	(0.501)	(0.488)		
Small Scale Fishing	0.037	0.000	11155.0	0.036
	(0.190)	(0.000)		

Mean relative value (and standard deviation between parenthesis) for each category and stakeholder is shown

However, exceptions are seen in categories such as Aquaculture, with higher representation on the Caribbean coast, and Industrial Fishing and Marketing, with greater feedback on the Pacific. This supports the conclusion that solutions on both coasts are basically the same, but that differences in expectations must be part of any management framework.

Fisheries management

The three stakeholder groups – fishermen, local leaders and experts – were asked a number of questions concerning the concept of fisheries management and co-management (bottom-up management).

Fishermen

Only one percent of the fishermen responded that they knew what fisheries management is. These fishermen used a variety of words or phrases to explain the concept, such as organization and order, closed fishing zones, temporary closures, fishermen's association, marketing, control, changing fishing gears, processing fish products, aquaculture, and managing fishing equipment. There is a lack of understanding among fishermen of what fisheries management is.

When asked if their community was ready for bottom-up management, 37 % answered positively and negatively at the same time (42 % Caribbean vs. 28 % Pacific), only 31 % emphatically answered positively (31 % Caribbean vs. 30 % Pacific), 9 % answered negatively (8 % Caribbean vs. 11 % Pacific), and 23 % did not know or were unsure. Fishermen's answers revealed that they believe fisheries management is necessary because otherwise fishing will get worse (8 %), they need order and rules (7 %), and the community should try and see if it works (4 %). Fishermen reported that some communities have implemented rules by themselves. For example, *Pizarro* established a rule to use gillnets with mesh size larger than 3 in.. *El Roto* established a minimum mesh size of 3 ½ inches for gill nets. The fishermen reported that while most efforts at implementing their own rules have been unsuccessful, they are optimistic that it is possible for fishing communities to come together to establish rules. They felt that it would be possible for the community to implement fisheries management under certain conditions, such as: (1) If all fishermen agree to it; (2) If management is controlled by both community and government; (3) If training in fishery management comes first; and (4) If a subsidy mitigates a fishing restriction. However, fishermen believed that establishing bottom-up management is difficult because: (1) Fishermen's attitude prevents it; (2) It is difficult to get fishers to agree; (3) Some attempts at management had failed in the past; (4) Fishermen cannot stop fishing; (5) It could generate violence among fishers; and (6) Community features complicate the process.

When asked what their community needs to do to manage fisheries, the fishermen believed that: (1) Fishing communities need to get together and agree about implementing management; (2) All fishermen should organize and belong to the fishermen's associations; (3) Fishermen need to face the problems that threaten their livelihood and get specific training to overcome them; (4) Some fishing communities have shown that successful fisheries management is possible, and their experiences can be useful examples; and (5) Fishermen should learn to compromise and explore other job possibilities. When asked what government needs to do to manage fisheries, the fishermen stated that existing regulations needed to be effectively implemented; they should stop foreign fishermen; and training should be provided.

Local leaders

Similar to the fishermen, local leaders knew nothing about fisheries management. When asked if they believed their community was ready for implementing the bottom-up process, 59 % answered positively (63 % Caribbean vs. 55 % Pacific) and 41 % answered negatively (38 % Caribbean vs. 45 % Pacific). A majority of leaders believed they were ready to implement fishery management by themselves because of:

(1) Rules already established by some communities, (2) Strong fishermen's associations, and (3) Community reliance on fishing for food. Leaders believed that bottom-up management is possible, but only if government and community work together. They reported less confidence in efforts developed solely by communities or imposed by the government.

Leaders who were pessimistic about establishing bottom-up fishery management cited fishermen's attitude - according to which the government must give them everything, but they do not make any effort (7 %). Other difficulties include: (1) Low individual representation; (2) Difficulties in getting many fishermen on the same page; (3) Lack of organization or association among the majority; (4) Prevalence of illegal gear that undercuts participation; (5) Lack of understanding of the need for fisheries management; and (6) No institutional authority is in charge.

When asked what they think that the community needs to do in order to manage its fishing resources; local leaders reported that the entire community must get together and agree to actively participate in implementing fisheries management (the process must include fish sellers). They also stated that fishermen must practice responsible fishing, they must develop a sense of belonging, they must belong to a fishery organization, and they must stop using unfriendly gears. When asked what they thought that the government needs to do in order to manage fisheries resources, local leaders reported that government should: (1) Invest in establishing regulations, including making subsidies available while restrictions apply; (2) Establishing exclusive fishing zones for each community; and (3) Closed seasons. At the same time, leaders believe that it is fundamental that any restriction or regulation be discussed with the community before it is implemented and the government should accord the small-scale fishery sector the importance it deserves.

Experts

Fifty-seven percent of the experts believed that it was possible to establish bottom-up management, but 20 % were skeptical. Experts who believed that bottom-up management should work knew about recent, closely related attempts. These included projects in marine protected areas such as *Guapi* and *Sanquianga* Parks on the Pacific coast, led by WWF; San Andres Island on the Caribbean, led by CORALINA; the Special District in *San Antero* that is zoned and planned to preserve mangrove forests; and a similar effort on the Urabá Gulf. Experts described regional or local committees that could be the basis for coastal zoning or for implementing fisheries management, such as NODOS (Regional Institutional Fishery Councils), INVEMAR (regional Committees for Coastal Management), and local Community Councils (some with their own Natural Resources Code). They also listed fishermen's associations that could be local foundations of bottom-up fisheries management, such as associations in *Las Flores* on the Caribbean, and *Juanchaco* and *Bahía Solano* on the Pacific coast. Experts who were skeptical of bottom- up fisheries management doubted that communities would participate. Other concerns were the poor track records of external institutions going in and out of communities makes implementing fisheries management through external agencies difficult; fishermen's attitude and need for constant supervision and their inability to work alone; and cultural aspects of Indigenous fishing communities.

When asked what they think fishing communities need to do in order to manage their fishing resources, the experts believed that a consensus by the majority of the population in each community was necessary in order to obtain high participation. Additionally, the community should discuss and agree upon their own rules and take responsibility for ensuring compliance; fishermen must belong to a fishermen association, and that the community must have a long term vision.

When asked what they thought that the government needs to do in order to manage fisheries resources, the experts felt that the government needs to be aware that bottom-up management requires a long term vision. Additionally, investing in education in order to increase awareness of fishery management; provide active and consistent inter-institutional support in each community; promote and strengthen fishermen's associations; and adapt to community dynamics.

Public meetings

A series of public meetings to discuss fisheries management were held in selected communities – five communities on the Caribbean coast and four on the Pacific coast. The meetings introduced fishermen to the basics of fishery management, why it is important, how they can be part of this process and start to work from their own communities. After presentations, open discussions allowed the participants to express their individual opinions about this fisheries management and bottom-up management and whether or not it could be useful and applied in their situation.

Even though all meetings brought up internal weaknesses, most participants believed that bottom-up management was possible if they worked hand in hand with government. The fact that some communities had already established rules resulting from their own deliberations suggested that widespread bottom-up fisheries management may be possible. Most weaknesses identified in the meetings matched those identified in the fishermen survey described above. These include disunity among fishermen, lack of participation, the presence of foreign fishermen, and weak fishermen's associations due to poor leadership. Additional weaknesses include authorities' distrust of fishermen, dislike of authority, distrust among fishermen, lack of education, lack of a sense of belonging, the involvement of fishermen in drug trafficking.

The main strengths each community identified were related to human capital such as strong knowledge and experience in fisheries issues; leaders and traditional authorities who believe that bottom-up management is the only solution; and community experience in establishing their own rules. Some believe that the presence of old, established fishermen associations are key to success. Communities cautioned that government must support local decision making processes and impose agreed-upon rules (*Juanchaco*), otherwise bottom-up management will not work. Communities stated that fisheries management needs to be designed for the long run and applied consistently over time in each community that adopts the process. There was a need to integrate non-traditional fisheries stakeholders, such as fishermen's wives, local schools and universities, and seafood supply chains into fisheries management. This shows the extent to which small- scale fishing is importantly integrated into local and national life.

Although none of the communities expressed that they enjoy optimum conditions for implementing bottom-up management, all realized its importance and expressed an urgent need to start the process. Particular characteristics of each community affected their readiness to put bottom-up fisheries management into practice.

DISCUSSION

The introduction to this paper noted that weak governance has been a major factor contributing to the overexploitation and consequent poor condition of small-scale fisheries and many of the communities that depend upon them around the world. This is no less true and Colombia and the data presented here and in Saavedra-Díaz et al. (2015b) contribute to the global comparison of this important component of the fisheries sector. The latter paper identified a number of different problems facing the Colombia small scale fishery, as perceived by different stakeholders – fishermen, local leaders, and fishing experts. While there were some difference among the stakeholders and between the two regions, several key problems emerged including destructive fishing methods, lack of regulation enforcement, pollution and industrial contamination, lack of public services, weakness of fishermen's organizations and leadership. The majority of fisheries experts believed that the small scale fishery sector has been largely ignored by government actions. That analysis concluded that these problems require establishing a fisheries management strategy that can prioritize solutions at different levels: national, coastal, and local.

Identification of problems in any given fishery is necessary but ultimately insufficient. A path forward to solving those problems must be developed. Our approach here is to draw on expertise of fisheries, community leaders and in-country experts to highlight that path.

More than 50 % of each stakeholder group believed that implementing a bottom-up fisheries management approach, such as co-management, is possible. While it is clear that what the details of "co-management" means is still not fully shared, it is still notable that there is significant interest in attempting to take action at a local level to address the problems each community faces, and they all face in common. This approach to using co-management is not only seen to have potential by Colombian stakeholders, but also in other Latin American countries that face similar issues due to fisheries crises, such as Mexico (Salas et al. 2015), Costa Rica (Solís et al. 2015), Brazil (Futemma and Seixas 2008), Uruguay (Trimble and Berkes 2015), and Chile (Marin and Berkes 2010). Even in Colombia, there are examples of cases that have tried participatory management with varying degrees of success (Beardon 2008; Delgado et al. 2010; Navia et al. 2010).

It was notable that all of the stakeholder groups recognize two over-riding needs, 1) that governance of the fishery is urgently needed or conditions will continue to decline, and 2) that whether management is top-down or bottom up, no one group can implement effective management. This is the essence of co-management, shared governance, but shared responsibility. Clearly, fishers perceived the need for the government to help them manage some of the biggest challenges including external challenges (e.g., pollution, criminal elements), while government looked to support from fishers in making real change. Local leaders too, needed a greater willingness of fishers to come together and compromise as needed to make changes, and in dealing with the government.

Surprisingly, Colombian fishermen were the most optimistic about co-management among the three stakeholder groups; however, a majority qualified their answers and listed changes necessary for success. These conditions closely relate to changes mentioned by local leaders and government experts answering similar questions about conditions for success. Optimists believed that existing examples of bottom-up fishery management provide evidence that success is possible. However, pessimists highlighted examples of community or external management efforts that failed and made them concerned about the future of bottom-up management.

From different perspectives, concurrence was also reached on the need to train fishermen in fishery management. Fishermen would share local knowledge, communities would participate in management with greater confidence and the government would provide training that gives everyone a vested interest in the process. Basic changes will be necessary in fishing communities, foremost is increasing membership in fishermen's associations. Experts and leaders agreed on the need to promote participatory research in which fishermen-researchers work closely with teams of biologists.

Each group identified changes in government necessary for insuring successful bottom-up management. Both fishermen and local leaders urgently stressed the need to establish a fisheries agency in order to constantly and consistently oversee the restrictions and regulations agreed to in deliberation. Fishermen and leaders believed that such restrictions would fail without strong subsidies to encourage the compliance of fishermen.

There were some real differences that will need to be resolved in order to make a management system viable. Clearly one of these is the designation of "real" fishermen versus opportunists. In other words, there is a need to develop a shared vision of who is a member of the community and in what capacity, recognized by the community itself, its leaders, and government. Another real difference in perspective, of a different character is the concern fishers have about gear types, particularly gillnets, versus government and leaders support for closed areas. These are not mutually exclusive approaches but a middle ground needs to be developed since it is clearly a major visible issue of concern for many.

Three of the nine study communities were found to have specific features that encourage them to start thinking about fishery management. These are *Ahuyama* and *El Roto* on the Caribbean, and *Bahía Solano* on the Pacific. These are communities with a small fishing population; few fishing methods; small, close- knit fishing neighborhoods; and fishermen's association could easily start the process of fisheries management. These communities were found to have attitudes and activities which support their commitment and awareness of the need for management. It is felt that these three communities could serve as pilot sites for implementing fisheries co-management in the country.

As noted in the introduction, there is a broad theoretical framework for the governance of small-scale fisheries that has been developed by Berkes, Ostrom, Pomeroy and others (Berkes et al. 2001; Ostrom 1990, 2009; Pomeroy et al. 2004, 2011). Colombian small-scale fisheries management does not have all the elements of this framework. However the present paper and previous work in identifying problems and solutions suggests that implementing successful fisheries co-management may now be possible. Long-term direction is needed for fisheries policy in Colombia and it should include a strong governmental framework that enables local bottom-up co-management. Clear a "transition" process, is needed which can be based on these research results, that moves from the current unregulated situation towards sustainable fisheries. Lessons learned as a result of the present study suggest that work during the next five years needs to focus on building strong foundations of social infrastructure during the "transition" stage that will support and maintain viable and resilient fisheries management plans. Particular characteristics of each community affect their readiness to put bottom-up fisheries management into practice. The national framework should be flexible and adaptable to local community needs and priorities. This may be achieved through four activities: (A) small-scale fisheries program of research; (B) social marketing and awareness raising in government and community; (C) restructure governance for co- management; and (D) community organization.

CONCLUSIONS

Understanding small-scale fisheries is key to protecting the health of Colombia's coastal ecosystems and improving the quality of life for coastal fishing communities. All stakeholder groups included in this study generally agreed that both old and new problems could be reduced by implementing co-management strategies at a community level with governmental support. Consistent, long-term policy direction is needed for fisheries in Colombia. If the Colombian administration continues centralized administration of the fishery sector, without adequate support and involvement at the community level, future conditions could well be worse than at present. Socio-ecological systems related to marine and coastal fisheries are fragile, such that fishing communities are vulnerable, food security is at risk, and the health of marine ecosystems is endangered.

All stakeholders engaged in this study believed that bottom-up management is possible in their communities. This concurrence of opinion and the fact that some communities have already established rules resulting from their own deliberations are evidence in favor of co-management. Most communities believed that co-management is possible working hand in hand with government. Communities that successfully established internal rules, which then failed due to the lack of enforcement, prove the need for a partner authority to assist in implementing fisheries management through co-management. This role should fall to national regulatory agencies; however, they often cannot fulfill this role due to frequent changes in policy directions resulting from changes in national administrations.

Effective fisheries management requires a mixture of national and local authority in order to work well, and steps toward a preliminary framework for a two-tiered system are suggested. Hopefully in the future communities will be able to take the lead in local management, and government agencies will grow into trusted partners by coordinating the consistent application of national fisheries policies and protecting national resources against foreign interlopers.

ACKNOWLEDGMENTS

This research was possible because of the participation and collaboration with Colombian marine fishers and local leaders from the communities of Ahuyama, Taganga, Las Flores, San Antero, El Roto, Bahía Solano, Pizarro, Juanchaco, and Tumaco. We also thank the national and regional fisheries experts and fisheries researchers for their open participation and for giving their life's work to make Colombian fisheries better. The main part of the data analysis was done thanks to the help and support of Berta Martín-López. We also thank Karen Alexander for her editorial assistance. This study was supported by the Conservation International Foundation, Universidad del Magdalena, University of New Hampshire, fellowship programs from COLCIENCIAS and Young Women in Science - 2008 (UNESCO – L´Oreal), and Connecticut Sea Grant.

REFERENCES

1. *Bavinck, M, R. Chuenpagdee, M. Diallo, P. van der Heijden, J. Kooiman, R. Mahon, and S. Williams. 2005. Interactive fisheries governance: a guide to better practice. Centre for*

Maritime Research, Amsterdam. Eburon Publishers. http://www.marecentre.nl/fishgovfood/documents/bavinck_interactive.pdf.

2. *Beardon, H. 2008. Del Caos a la Esperanza: cultura, política y protección de los manglares en el Pacífico colombiano. Cali: WWF – World Wildlife Foundation. WWF-UK. DFID y WWF Reino Unido*
3. *Begossi, A, and D Brown. 2003. Experiences with Fisheries co-management in Latin America and the Caribbean. Chapter 8. In: The Fisheries Co-management Experience. Accomplishment, Challenges and Prospects, ed. Douglas Wilson, Jesper Nielsen, and Poul Degnbol, 135 – 152. Fish and Fisheries Series, Volume 6. Dordrecht/Boston/London: Kluwer Academic Publishers*
4. *Berkes, F. 2004. Re-thinking community-based conservation. Conservation Biology 18(3): 621–630. doi:*10.1111/j.1523-1739.2004.00077.x.
5. *Berkes, F. 2010. Shifting perspectives on resource management: resilience and the reconceptualization of 'natural resource' and 'management'. MAST 9(1): 13–40.* http://www.marecentre.nl/mast/documents/Mastvol9no1_Berkes.pdf
6. *Berkes, Fikret, Robin Mahon, Patrick McConney, Richard Pollnac, and Robert Pomeroy. 2001. Managing Small-Scale Fisheries-Alternative Directions and Methods. Ottawa: International Development Research Centre*
7. *Brown, D.N., and R. Pomeroy. 1999. Co-management of Caribbean Community (CARICOM) fisheries. Marine Policy 23(6): 549–570. doi:*10.1016/S0308-597X(98)00040-2
8. *Christy, F. 1997. The development and management of marine fisheries in Latin America and the Caribbean. Washington, D.C.: Inter-American Development Bank.* http://idbdocs.iadb.org/wsdocs/getdocument.aspx?docnum=355303
9. *Chuenpagdaee, R., and S. Jentoft. 2007. Step zero for fisheries management: what precedes implementation. Marine Policy 31(6): 657–668. doi:*10.1016/j.marpol.2007.03.013
10. *Cinner, J., T. McClanahan, M. MacNeilc, N. Grahama, T. Dawd, M. Mukmininf, D. Fearyg, A. Rabearisoah, A. Wamukotai, N. Jiddawik, S. Campbellf, A. Bairda, F. Januchowski-Hartleya, S. Hamedk, R. Laharil, T. Morovel, and J. Kuangel. 2012. Co-management of coral reef socio-ecological systems. PNAS 109(14): 5219–5222. doi:*10.1073/pnas.1121215109
11. *CRC, URI and FIU. 2006. Coastal Resources Center, University of Rhode Island, and Florida International University. Fisheries Opportunities Assessment.* http://www.crc.uri.edu/download/Fish_Opp_Assess_Final_012607_508.pdf.
12. *Cuello, F., and L.O. Duarte. 2010. El Pescador Artesanal, Fuente de Información Ecológica para la Ordenación Pesquera en el Mar Caribe de Colombia. Proceedings of the Gulf and Caribbean Fisheries Institute 62: 463–470.* http://nsgl.gso.uri.edu/flsgp/flsgpc09001/data/papers/073.pdf
13. *Delgado, M.F., W. Gualteros, S. Espinosa, C. Lucero, A.M. Roldan, L.A. Zapata, J.R. Cantera, C. Candelo, C. Palacio, O. Muñoz, G. Mayor, and D.L. Gil-Agudelo. 2010. "Pianguando - Estrategias para el manejo de la piangua" (Cartilla), INVEMAR, ASCONAR, WWF Colombia, Universidad del Valle, UAESPNN – PNN Sanquianga; Co-financiado por el Ministerio de Agricultura, 20. Cali: Serie de publicaciones generales INVEMAR No. 45.* http://issuu.com/natucreativa/docs/piangua
14. *ECOVERSA. 2007. Justificación sobre la necesidad de una nueva estructura para el manejo y ordenación de la pesca y acuicultura a nivel nacional. Corporación ECOVERSA. 30p. Bogotá, Colombia*
15. *FAO. 2006. Food and Agriculture Organization of the United Nations. Report of the Expert Consultation on the Economic, Social and Institutional Considerations of Applying the Ecosystem Approach to Fisheries Management.* ftp:// ftp.fao.org/docrep/fao/009/a0673e/a0673e00.pdf.
16. *Francis, R.C., M.A. Hixon, M.E. Clarke, S.A. Murawski, and S. Ralston. 2007. Ten commandments for ecosystem-based fishery scientists. Fisheries 32(5): 217–233. doi:*10.1577/1548-8446(2007)32[217:TCFBFS]2.0.CO;2

17. *Futemma, C.R.T., and C.S. Seixas. 2008. Is there artisanal fishing territoriality in the Ubatumirim Bay (Ubatuba, SP)? Scale issues in community relations. Biotemas 21(1): 125–138.* http://dx.doi.org/10.5007/2175-7925.2008v21n1p125
18. *García, C. 2010. Conocimiento tradicional: lo que los pescadores artesanales del Caribe colombiano tiene para decirnos. Pan-American Journal of Aquatic Sciences 5(1): 77–89.* http://www.panamjas.org/pdf_artigos/PANAMJAS_5(1)_78-90.pdf
19. *Garcia, S.M., and K.L. Cochrane. 2005. Ecosystem approach to fisheries: a review of implementing guidelines. ICES Journal of Marine Science 62(3): 311–318. doi:*10.1016/j.icesjms.2004.12.003
20. *García-Horta, J.B., and M.T. Guerra-Romos. 2009. The use of CAQDAS in educational research: Some advantages, limitations and potential risks. International Journal of Research & Method in Education 32: 151–165. doi:*10.1080/17437270902946686.
21. *Hart, P. 2003. The fisheries co-management experience. Accomplishment, challenges and prospects. Fish and Fisheries 5(1): 95–96. doi:*10.1111/j.1467-2979.2004.00145_4.x
22. *Kruskal, W., and W. Wallis. 1952. Use of Ranks in One-Criterion Variance Analysis. Journal of the American Statistical Association 47(260): 583–621*
23. *Lam, M., and D. Pauly. 2010. Who is right to fish? Evolving a social contract for ethical fisheries. Ecology and Society 15(3): 16.* http://www.ecologyandsociety.org/vol15/iss3/art16/.
24. *Lebel, L., J. Anderies, B. Campbell, C. Folke, S. Hatfield-Dodds, T. Hugues, and J. Wilson. 2006. Governance and the capacity to manage resilience in regional social-ecological systems. Ecology and Society 11(1): 19.* http://www.ecologyandsociety.org/vol11/iss1/art19/
25. *Marin, A., and F. Berkes. 2010. Network approach for understanding small-scale fisheries governance: the case of the Chilean coastal co-management system. Marine Policy 34(5): 851–858. doi:*10.1016/j.marpol.2010.01.007
26. *McCay, B.J., and S. Jentoft. 1996. From the bottom up: participatory issues in fisheries management. Society & Natural Resources 9(3): 237–250.*
27. *Miles, Matthew B., and Michael Huberman. 1994. Qualitative data analysis: an expanded sourcebook, 2nd ed. Thousand Oaks: SAGE Publications*
28. *Navia, AF, PA Mejía-Falla, J López-García, LA Muñoz y, V Ramírez-Luna. 2010. Pesquería artesanal de la zona norte del Pacífico colombiano: aportando herramientas para su administración, Fase II. Documento técnico Fundación SQUALUS No. FS0110. 100 p. Cali, Colombia*
29. *Olsson, P., C. Folke, and F. Berkes. 2004. Adaptive comanagement for building resilience in social- ecological systems. Environmental Management 34(1): 75–90. doi:*10.1007/s00267-003-0101-7
30. *Olsson, P., L.H. Gudenson, S. Carpenter, P. Ryan, L. Lebel, C. Folke, and C.S. Holling. 2006. Shooting the rapids: navigating transitions to adaptative governance of social-ecological systems. Ecology and Society 11(1): 18.* http://www.ecologyandsociety.org/vol11/iss1/art18/
31. *Ostrom, Elionor. 1990. Governing the commons: the evolution of institutions for collective action. Cambridge: Cambridge University Press*
32. *Ostrom, E. 2009. A General framework for analyzing sustainability of social – ecological systems. Science 325: 419–422. doi:*10.1126/science.1172133
33. *Ozkan, Betul. 2004. Using NVivo to analyze qualitative classroom data on constructivist learning environments. The Qualitative Report 9(4): 589–603.* http://nsuworks.nova.edu/tqr/vol9/iss4/2/
34. *Pomeroy, R., P. McConney, and R. Mahon. 2004. Comparative analysis of coastal resources co- management in the Caribbean. Ocean & Coastal Management 47(9): 429–447. doi:*10.1016/j.ocecoaman.2004.09.005
35. *Pomeroy, Robert, Joshua E Cinner, and Jesper Raakjaer Nielsen. 2011. Conditions for Sucessful Co- management: Lessons Learned in Asia, Africa, the Pacific and the Wider Caribbean. In Small-scale Fisheries Management. Frameworks and Approaches for the*

Developing World, eds. Robert Pomeroy and Neil Andrew. 115–131. UK: CAB International

36. *Posada, Blanca O., and Daniel M. Rozo. 2005. Marco Geográfico. In Informe del estado de los ambientes marinos y costeros en Colombia, Series de publicaciones periódicas No.8, 3–9. Santa Marta: INVEMAR.* http://www.invemar.org.co /redcostera1 /invemar/docs/3801IER_2005_completo.pdf
37. *Preikshot, Dave B, and Daniel Pauly. 1998. A multivariate interdisciplinary assessment of small-scale tropical fisheries. In Proceedings of the International Symposium on Fishery Stock Assessment Models, eds. T.J. Quinn II, F. Funk. J. Heifetz, J.N. Ianelli, J.E. Powers, J.F. Schweigert, P.J. Sullivan, C.-I. Zhang, 803–814. Alaska, USA: Alaska Sea Grant College Program Report.* https://s3-us-west-2.amazonaws.com/legacy.seaaroundus/doc/Researcher+Publications/dpauly/PDF/1998/Books+and+Chapters/MultivariateInterdisciplinaryAssessmentSmallScaleTropical.pdf
38. *Saavedra-Díaz, L, AA Rosenberg, and R Pomeroy. 2015a. Why Colombian marine fisher´s knowledge is a fundamental tool for marine resource management and assessment. In Fishers´ knowledge and the ecosystem approach to fisheries: applications, experiences and lesson in Latin America, eds. Johanne Fishers, John Jogersen, Helga Josupeit, Daniela Kalikoski, and Cristine M. Lucas. 89 – 106. Rome: FAO Fisheries and Aquaculture Technical Paper No. 591.* http://www.fao.org/3/a-i4664e.pdf
39. *Saavedra-Díaz, L, AA Rosenberg, and B Martín-López. 2015b. Social perceptions of Colombian small marine fisheries conflicts: insights for management. Marine Policy 56: 61–70. doi:*10.1016/j.marpol.2014.11.026
40. *Salas, S., R. Chuenpagdee, A. Charles, and J.C. Seijo (eds.). 2011. Coastal fisheries of Latin America and the Caribbean, FAO Fisheries and Aquaculture Technical Paper. No. 544, 430. Rome: FAO.* http://www.fao.org/docrep/014/i1926e/i1926e.pdf
41. *Salas, S., R. Regist, C. Zapata, M. Cabrera, and J. Euán. 2015. How much we can learn from fishers about ecology and fisheries management: case studies on spiny lobster fishery in Mexico. In Fischers´ knowledge and the ecosystem approach to fisheries: applications, experiences and lesson in Latin America, ed. Fishers Johanne, Jogersen John, Josupeit Helga, Kalikoski Daniela, and Cristine M. Lucas, 107–124. Rome: FAO Fisheries and Aquaculture Technical Paper No. 591.* http://www.fao.org/3/a-i4664e.pdf
42. *Saldaña, Johnny. 2009. The Coding Manual for Qualitative Researchers, 1st ed. London: SAGE Publications.*
43. *Schumman, S. 2007. Co-management and "consciousness": fishers' assimilation of management principles in Chile. Marine Policy 31(2): 101–111. doi:*10.1016/j.marpol.2006.05.008
44. *Solís, V., A. Muñoz, and M. Fonseca. 2015. Integrating traditional and scientific knowledge for the management of small scale fisheries: an example from Costa Rica. In Fischers´ knowledge and the ecosystem approach to fisheries: applications, experiences and lesson in Latin America, ed. Fishers Johanne, Jogersen John, Josupeit Helga, Kalikoski Daniela, and Cristine M. Lucas, 179–190. Rome: FAO Fisheries and Aquaculture Technical Paper No. 591.* http://www.fao.org/3/a-i4664e.pdf
45. *St. Martin, K., B.J. McCay, G. Murray, T. Johnson, and B. Oles. 2007. Communities, knowledge and fisheries of the future. International Journal of Global Environmental Issues 7(2/3): 221–239.* http://umsms.siteturbine.com/faculty/faculty_files /publications/912/St.%20Martin%20et%20al.pdf
46. *Thayer, A., M. Evans, A. McBride, M. Queen, and J. Spyridakis. 2007. Content analysis as a best practice in Technical Communication Research. Journal Technical Writing and Communication 37: 267–279. doi:*10.2190/TW.37.3.c
47. *Trimble, M., and F. Berkes. 2015. Towards adaptive co-management of small-scale fisheries in Uruguay and Brazil: lessons from using Ostrom´s design principles. Maritime Studies 14: 14. doi:*10.1186/s40152-015-0032-y

48. *Tyler, Stephen. 2006. Comanagement of Natural Resources. Local Learning for Poverty Reduction. Otawa: Centro Internacional de Investigación para el Desarrollo.* http ://www.idrc.ca /EN/Resources/Publications/openebooks/346-1/index.html
49. *Yin, Robert. 2003. Case study research: design and methods. Applied social research methods series. New York, USA: SAGE Publications*

CHAPTER 7

An Assessment of The Economic Contribution of EU Aquaculture Production and The Influence of Policies for Its Sustainable Development

John Bostock[1], Alistair Lane[2], Courtney Hough[3], Koji Yamamoto[1]

[1]*Institute of Aquaculture, University of Stirling, Stirling FK9 4LA, UK*

[2]*European Aquaculture Society, Slijkensesteenweg 4, 8400 Ostend, Belgium*

[3]*Federation of European Aquaculture Producers, rue de Paris 9, 4020 Liege, Belgium*

ABSTRACT

This review paper examines the structure of the EU aquaculture sector, the contribution it makes to the EU economy and the policy environment for past and future development. The primary analysis uses statistical data from the Food and Agriculture Organization of the United Nations which has been re-categorized according to species groups established by the European Aquaculture Technology and Innovation Platform (EATiP) and by culture system type using expert knowledge. Additional data sources for the analysis include the European Market Observatory for Fisheries and Aquaculture Products (EUMOFA) and the European Commission Scientific, Technical and Economic Committee for Fisheries. EU aquaculture production was 1.34 million tonnes in 2012 with a first sale value of €4.76 billion. Shellfish comprised 45 % by volume and 28 % by value; marine fish

30 % by volume and 53 % by value; and freshwater fish 25 % by volume and 19 % by value. The total production volume has actually fallen slightly from 1.4 million tonnes in 2000, whilst the value has increased significantly from 2.79 billion in 2000, mainly due to a growth in Atlantic salmon production. Five countries accounted for around 78 % of the direct output value of EU aquaculture in 2012, the UK, France, Greece, Italy and Spain. Around 50 % of the direct output value was generated using marine cage systems (28 % by volume), whilst less than 3 % of value was generated in recirculated aquaculture systems (<1.5 % by volume). Around 5 % of value was contributed by extensive to semi-intensive inland and coastal pond systems. STECF (2014) estimates there are between 14,000 and 15,000 aquaculture enterprises in the EU employing around 80,000 people, approximately 40,000 full-time equivalent (FTE). The greatest number of jobs (FTE) is provided by the freshwater pond and suspended shellfish sectors due to much lower productivity figures. This could be seen as a social benefit in rural and coastal regions, but corresponding low wages could also discourage young entrants to the industry and lead to dependency on migrant workers. Where efficiencies can be improved through capital investment there is likely to be significant scope for consolidation of ownership as can be observed in the marine fish sector. The output from aquaculture has to find a place within the wider fish and seafood market where volumes are generally inversely related to price. The potential growth of the sector is therefore constrained both in relation to the overall market and with respect to competition from substitute products. These include product from EU capture fisheries as well as imports from third countries (sourced from aquaculture and capture fisheries). Whilst interactions between individual products can be hard to demonstrate, any increase in production costs is likely to lead to lower output volumes, whilst improvements in production efficiency can lead to increased output volumes. With around 60 % of EU fish and seafood supply obtained through imports, and little prospect of increasing outputs from capture fisheries, EU policy is generally supportive of sustainable aquaculture development for reasons of food security and economic development. The underlying basis for this is maximizing the quality and health benefits of farmed products, whilst improving resource efficiency and minimizing impacts. This is expressed through funding support for research and technological development and structural funds to the fisheries and aquaculture industries. However, constraints to growth also exist in the form of regulatory barriers and costs that reduce industry competitiveness. Changing market requirements are also a factor. Prospects for growth have been assessed using the results of EATiP stakeholder workshops combined with the analysis of the sector by system type. These suggest an overall increase in production by 55 % is possible by 2030 based mainly on expansion of marine cage-based farming using larger systems in more exposed sites and similarly shellfish farming using larger-scale

suspended systems. Expansion of recirculated aquaculture systems appears likely based on entrepreneurial and European policy for research and technological development activity, although constrained by currently low competitiveness.

Keywords: Aquaculture EU Europe Economics Policy Systems Competitiveness

INTRODUCTION

Fish and shellfish have been produced using aquaculture techniques for hundreds of years. However, a new phase of growth commenced in the 1970s–1980s as production from capture fisheries reached a plateau, whilst demand for aquatic food products continued to rise. Total EU output increased throughout the 1990s but stabilized in the early 2000s although consumption and imports maintained an upward trend. This review paper examines the structure of the EU aquaculture sector, the contribution it makes to the EU economy and the policy environment for past and future development. The paper is based on work carried out by the authors for the European Parliament in 2014 (Lane et al. 2014) which in particular featured a preliminary (i.e. based on expert knowledge rather than survey-based farm data) disaggregation of production and value data by system type. Whilst this cannot be taken as definitive, it provides another view on the industry which should help inform future analysis and policy development.

The scale and structure of the European aquaculture industry

This section provides a short overview of production of the key European species. The authors have identified the main technologies used and have developed a new classification of 'production by technology' based on expert knowledge of the sector.

Aquaculture development in European countries

Whilst global aquaculture production expanded at an average annual rate of 6.2 % in the period 2000–2012 (9.5 % in 1990–2000) (FAO 2014a) to reach 90.4 million tonnes (live weight equivalent), growth in Europe was well below this at an average of 2.9 % and total production is only 4.3 % of the global total at 2.88 million tonnes. Most of the positive growth was due to Norway, whilst production in the EU actually declined after 1999 (See Fig. 1).

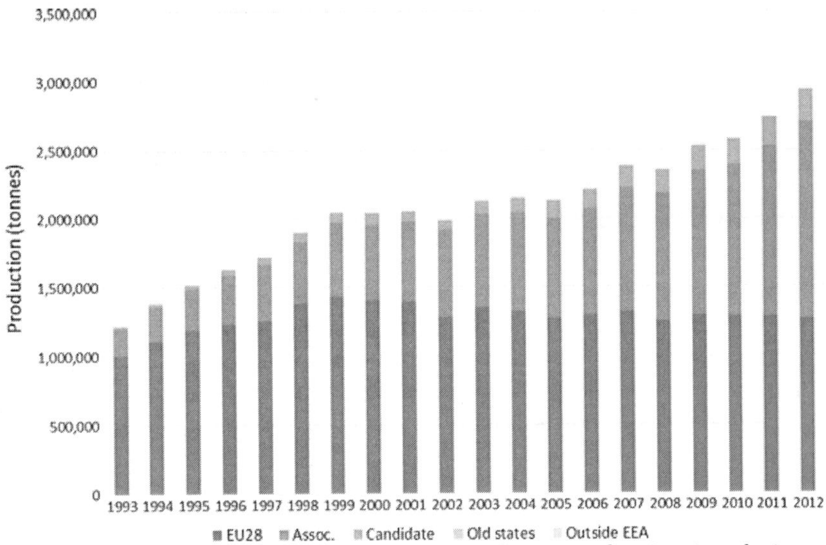

Figure. 1 Development of European aquaculture production in relation to EU membership (*Source:* developed from FAO FishStat database)

Aquaculture is practised in almost every European country with the largest producers being Spain, France, UK, Greece and Italy (Fig. 2).

Production by species of the principal sub-sectors of European aquaculture

In volume terms, the largest segment of EU aquaculture is shellfish—mainly of mussels and oysters. In value terms, however, fish production is more significant. More than 70 different fish species are cited for aquaculture in the EU (FAO 2014b), but production (in 2012) was dominated by rainbow trout, Atlantic salmon, gilthead sea bream, European sea bass and common carp. These five species make up 90 % of all fish production in the region.

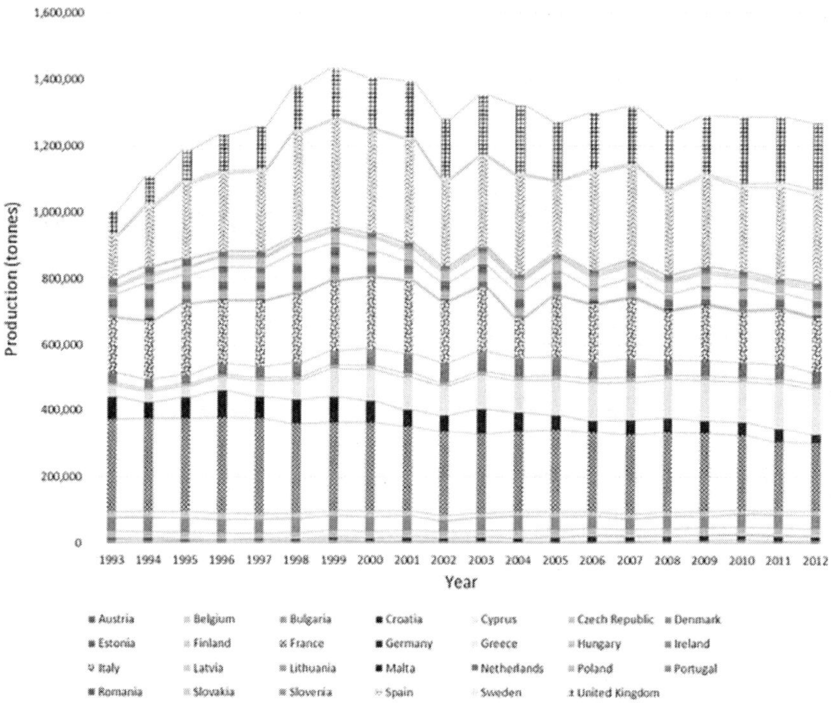

Figure. 2 Geographic distribution of EU production (*Source:* developed from FAO FishStat database)

For the purpose of this study, we consider the EU aquaculture industry as comprising five separate segments, namely: coldwater marine fish, warmwater marine fish, freshwater fish, shellfish and algae and other aquatic products (Fig. 3).

Coldwater marine fish

The cultivation of Atlantic salmon dominates this segment and is seen as the major success story in this sector. Production rose from only 900 tonnes in 1980 (FAO 2014b) to 168,000 in 2013 (FEAP 2014), reared mainly in Scotland (UK) with 11,500 coming from Ireland. The EU production is, however, significantly overshadowed by Norway (1.2 million tonnes) in 2013 (FEAP 2014). This is supplemented by the rearing of large rainbow trout in marine/brackish water conditions (Scotland, Denmark, Finland and Sweden) of 24,000 tonnes (FEAP 2014) (Fig. 4).

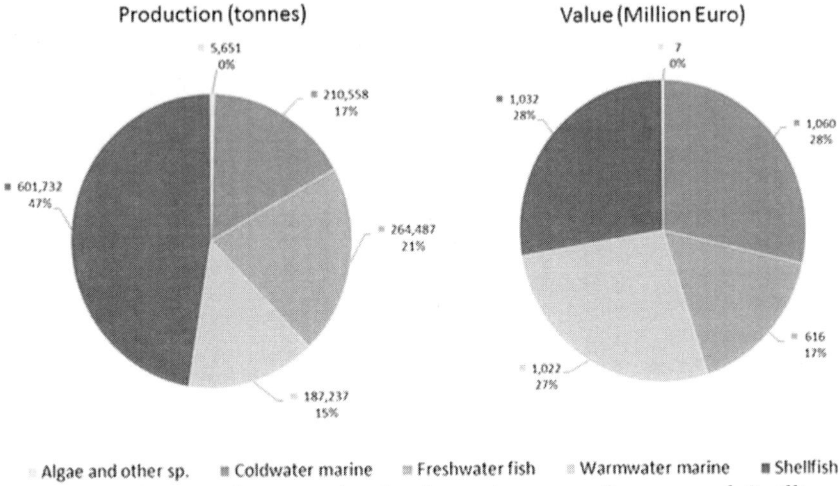

Figure. 3 EU aquaculture production by major group (tonnes and €million in 2012) (*Source:* FAO FishStat data)

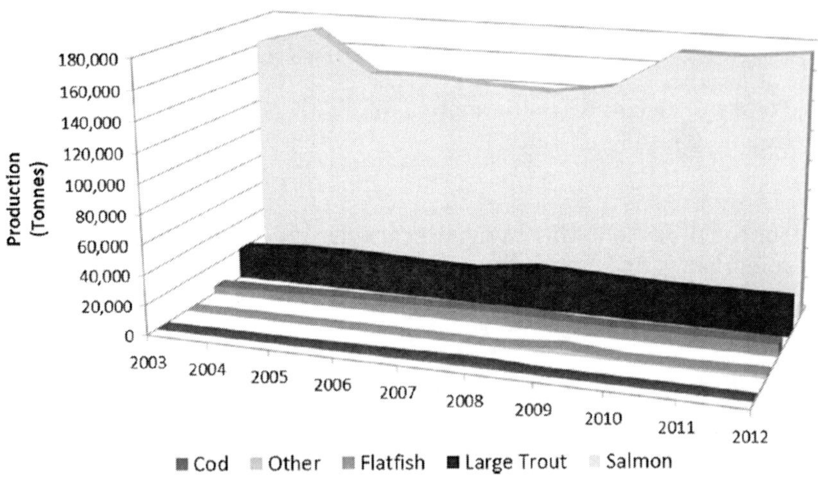

Figure. 4 EU production of marine coldwater fish (*Source:* FAO FishStat data)

Efforts on diversification have focused on cod, sole and halibut—but so far with a small number of operations servicing mostly niche markets. Norway invested heavily in cod production, resolving technical challenges and reaching annual production levels exceeding 20,000 tonnes. However, resurgent wild stocks and market/price competition led to a crash and suspension of this activity (Jensen et al. 2014; Norwegian Directorate of Fisheries 2015).

Warmwater marine fish

This sector covers fish farming in warmer marine conditions, notably in the Mediterranean and Southern Europe (France, Portugal, Spain). The dominant species are gilthead sea bream and European sea bass, with turbot and meagre leading the remainder. EU production totalled 218,000 tonnes in 2013 (FAO 2014b). As with salmon farming, the total production of these 3 species in 1980 was less than 500 tonnes. Growth started in the 1990s, when hatcheries were able to produce juveniles more readily for stocking in cages (Fig. 5).

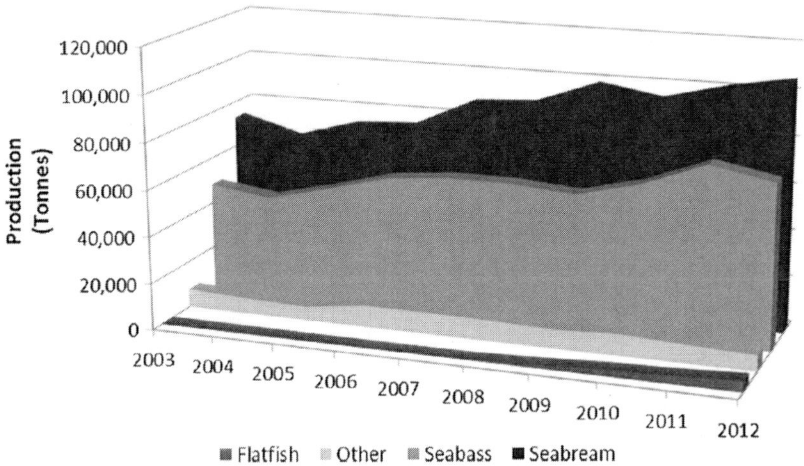

Figure. 5 EU production of warmwater marine fish (*Source:* FAO FishStat data)

The growth in Turkish aquaculture (of sea bass and sea bream) has also been significant, especially as it exports mainly to the EU. From 25,000 tonnes of sea bass/sea bream in the late 1990s, Turkey produced 78,000 tonnes in 2010 and is estimated to be over 85,000 tonnes in 2013 (FEAP 2014).

Freshwater fish

Whilst the freshwater sector is by far the oldest in European fish farming, it is divided into 2 main components, those of rainbow trout and common carp. This distinction principally reflects climatic conditions where rainbow trout is more suitable for temperate environments with carp being better in the more extreme continental conditions seen in Central Europe (hot summers, cold winters) (Fig. 6).

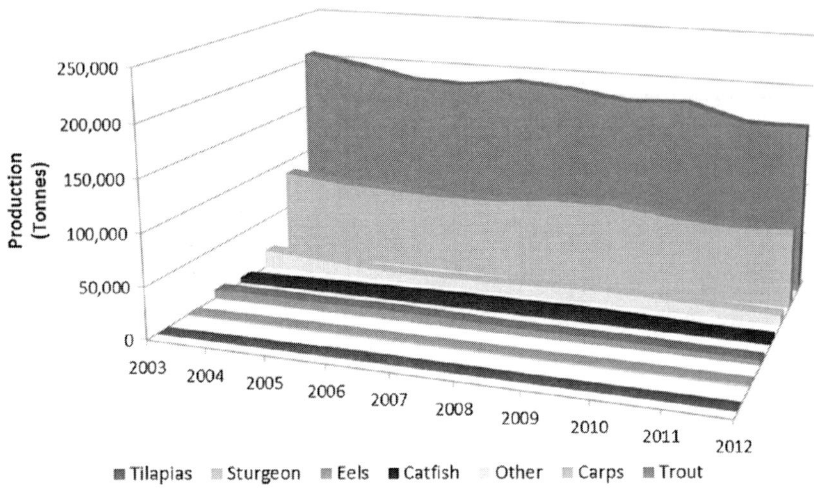

Figure. 6 EU production of freshwater fish (*Source:* FAO FishStat data)

Rainbow trout production expanded from the 1980s, moving from 100,000 tonnes to a peak of approximately 250,000 tonnes in 2000–2001. This has since fallen to around 165,000 tonnes in 2012, although combined with the large trout reported under coldwater marine fish, production still exceeded that of salmon in 2012. It is produced not only for consumption (principally as a portion-size fish (±250–350 g.) that is produced in one growing season) but also for stocking lakes and rivers for sport fishing. The decline in production is attributed to operational and licensing difficulties, in relation to environmental legislation, and the difficulties faced by smaller companies to deal with evolving market conditions and competition, notably with salmon. Significant reductions have been seen in Italy, Germany, Denmark, France and Spain (FEAP 2014).

Common carp attained 90,000 tonnes of production in the 1970s; major producers were Poland, Romania, Hungary, Germany, Czech Republic and Poland (each >10,000 tonnes). Smaller production levels were reported for France, Croatia and Lithuania. Carp is seen as a 'traditional' species, being a cultural fish dish for many inland countries that did not have access to marine seafood. EU carp production is now around 60,000 tonnes (FEAP 2014). This reduction is attributed to different causes; predation from wild birds (cormorants, herons) is a major factor and disease (koi herpes virus) is an additional issue. In addition, with adhesion of Central European States to the EU, supermarkets have replaced traditional markets and access to other seafood and fish products has ensued (e.g. marine fish, salmon, pangasius catfish). Carp markets have remained traditional, and added-value products are relatively few (STECF 2013a, b).

Eel farming reached a maximum of 11,000 tonnes in 2000 but is entirely dependent on the wild catch of young eels (glass eels) for its initial stock; the reductions in wild catch availability and high price competition for the stock with Asia have led to a fall in production, which was around 6000 tonnes in 2012 (FAO 2014b).

Many other freshwater species have been cultured at smaller scale in the EU, including Arctic char and other trout species, sturgeon—mainly for caviar, perch, pike-perch, roach and tench, African catfish—reared in warmwater, recirculation systems (mainly in the Netherlands) and tilapia and barramundi—tropical species reared in warmwater and recirculation systems. With the exception of African catfish and sturgeon, these all remain very minor components of EU aquaculture due to technical and/or marketing issues. Sturgeon rearing for caviar has expanded significantly in recent years, assisted by restrictions on the exploitation of wild sturgeon and trade in wild caviar under CITES advice and agreements (Bronzi et al. 2011; Doukakis et al. 2012).

Shellfish

European shellfish production is mainly of oysters (the Pacific cupped oyster, *Crassostrea gigas*, and the European flat oyster, *Ostrea edulis*) and mussels (the blue mussel, *Mytilus edulis*, and the Mediterranean mussel, *Mytilus galloprovincialis*). Together, oyster and mussel culture represent 93 % of the total European cultivated mollusc production (Eurostat 2011). The third element of shellfish production is of clams, cockles and arkshells (Fig. 7).

France is by far the leading producer of oysters (± 85,000 tonnes in 2011), Spain of mussels (±209,000 tonnes in the same year) and Italy of clams (± 32,000 tonnes in the same year).

Over recent years, the production of oysters has declined with abnormal mortality events of *Crassostrea gigas* adults reported in most French oyster production areas, generally linked with oyster herpes virus (Martenot et al. 2011). It is now thought that the bacterial pathogens *Vibrio splendidus* and *V. aestuarianus* may play a more significant role in the mortality events, with interactions between pathogens and the development of some resistance by previously infected stocks playing a role in the severity of losses (Vezzulli et al. 2015; Petton et al. 2015). The UK, Jersey, Ireland and the Netherlands have all suffered recent similar mortalities (Renault 2011).

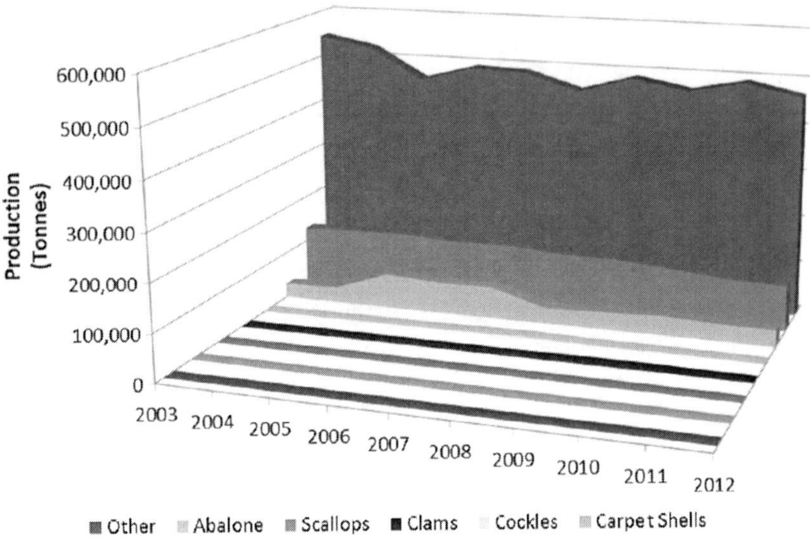

Figure. 7 EU production of shellfish (*Source:* FAO FishStat data)

Algae and other aquatic products

European production of algae and other species has only really started to evolve since 2007. Of these, aquatic plants (including brown seaweeds of the genus *Phaeophyceae*) have seen the most growth and this has been reported for Denmark (FAO 2014b).

Other species groups include crustaceans (mostly freshwater crayfish species in Central and Eastern European countries, but also several prawn species including the kuruma prawn, *Penaeus japonicus* in Southern Europe).

Mollusc species include mainly octopus and the miscellaneous aquatic animal group such as sea urchins and non-specified aquatic animals from inland waters.

European aquaculture technologies and estimation of production by technology

Classification of systems and overview of production

European aquaculture is a very diverse activity. Not just in terms of species produced, but also in the technologies that are used for rearing them.

Aquaculture can be classified according to various primary characteristics such as scale (Lazard et al. 1991), intensity of production and feeds used (e.g. Edwards 1993), and finally by a matrix of environment and/or species (or species group).

A technology-based classification primarily considers the nature of the containment system, and a primary distinction is made between land-based and water-based systems. For the former, there are potential benefits both to the farm and to the environment if water is recycled and treated on the farm. 'Recirculated aquaculture systems' (RAS) are normally defined by replenishing less than 5 % of the system volume per day. This approach allows systems to be effectively isolated from the environment and water quality conditions fully controlled. Water-based containment systems are either floating structures that are moored in position or static constructions usually placed in the intertidal zone or shallow water and fixed into the substrate.

A recent study (Lane et al. 2014) performed for the Committee on Fisheries of the European Parliament selected 11 production technologies (see Table 1) as being the most dominant technologies for EU production and, using data from FAO, characterized by country and by species by assigning the 2012 production to the most dominant technology used for (ongrowing) production (Fig. 8). Hatchery and nursery technologies were therefore not included as their impacts are small and included in overall production calculations.

Table 1 Key to system types used in analysis

EP code	System type
EP01	Freshwater pond aquaculture (extensive to semi-intensive)
EP02	Coastal pond aquaculture (mostly semi-intensive)
EP03	Intensive freshwater flow-through and partial recirculation systems (mostly tanks, raceways and small ponds)
EP04	Intensive marine flow-through and partial recirculation systems (mostly large tanks)
EP05	Indoor land-based recirculated aquaculture systems (freshwater)
EP06	Indoor land-based recirculated aquaculture systems (marine)
EP07	Small cage systems—freshwater
EP08	Small cage systems—sheltered marine

EP09	Large cage systems—marine in exposed sites, using mechanized systems
EP10	Marine bottom culture (non-fed sedentary and attached animals and plants)
EP11	Marine supported and suspended culture (non-fed sedentary and attached animals and plants)

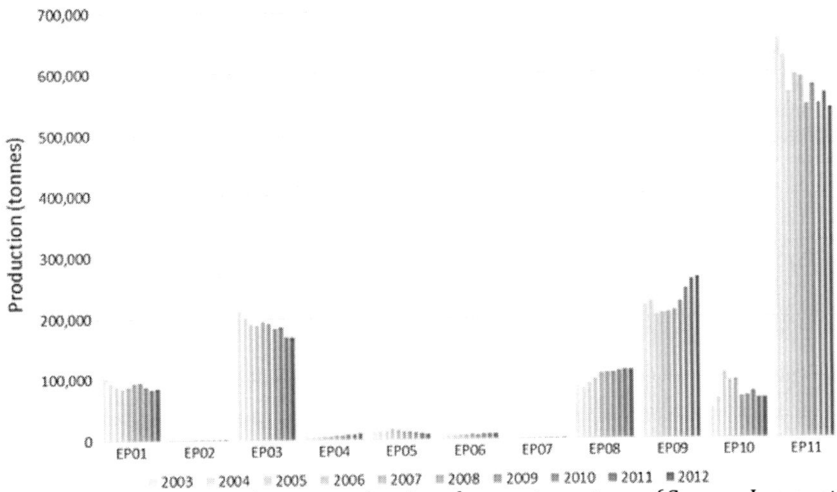

Figure. 8 EU aquaculture production by system type (*Source:* Lane et al. 2014)

Analysis of main segments by production system

Coldwater marine fish

Cage/net production is the primary ongrowing technology (Table 2), and there has been a gradual expansion of unit size accompanied by automatically controlled feed distribution, remote-monitoring and mechanized movement/harvesting of fish. This has significantly reduced labour costs as well as providing safer working conditions in the marine environment (Bostock et al. 2010).

Table 2 Coldwater marine fish—estimated production quantity by system type

Technology	Estimated 2012	
	Production (tonnes)	**Value (million euro)**
Large cage systems—marine in exposed sites, using mechanized (automated) systems	190,090	870.21
Small cage systems—sheltered marine	11,000	37.46
Indoor land-based recirculated aquaculture systems (marine)	9180	51.85
Coastal pond aquaculture	180	0.75
Intensive marine flow-through and partial recirculation systems (mostly large tanks)	100	0.87
Total	210,550	961.14

Source: Lane et al. (2014) and present study

The vast majority of cage-based marine fish farming is conducted in relatively sheltered coastal areas that provide protection from storms and ease of access by personnel (for maintenance work, stocking, feeding and harvesting, etc.). However, the number of such sites that have been licenced for development over the past 15 years has been limited due to planning authorities responding to expressed concerns over the potential and perceived impacts of the waste produced by farms, influence of escapees (e.g. from storm damage) on wild stocks, visual impact (e.g. to coastal residences and hotel businesses) or interactions with wildlife and conservation interests. For these reasons both 'offshore' farming (further away from the coastal zone) and closed containment system (e.g. onshore RAS) have been promoted as being a panacea for the future (Olsen et al. 2008; Klinger and Naylor 2012).

Warmwater marine fish

As for the coldwater marine segment, ongrowing is usually done in cages, although generally smaller than used for salmon (Table 3). It is anticipated, however, that production in larger cage sizes will be prevalent in the future and that production will develop more in exposed or high energy sites.

Table 3 Warmwater marine fish—estimated production quantity by system type

Technology	**Estimated 2012**	
	Production (tonnes)	**Value (million euro)**
Small cage systems—sheltered marine	102,420	558.69
Large cage systems—marine in exposed sites, using mechanized (automated) systems	74,330	369.88
Intensive marine flow-through and partial recirculation systems (mostly large tanks)	9910	58.47
Coastal pond aquaculture and valliculture	900	4.18
Indoor land-based recirculated aquaculture systems (marine)	340	3.57
Total	187,900	1021.80

Source: Lane et al. (2014) and present study

There are a few specialized pump-ashore marine tank systems mostly for flatfish, and some use of coastal ponds, particularly in Italy and Portugal, but the contribution to total marine fish production is low.

Freshwater fish

Freshwater aquaculture—compared to marine aquaculture—is fixed to the site where the installations are made and is largely dependent on adequate access to water for its production potential (Table 4). Traditionally, incoming water provides the environment for growing the fish and site selection is usually based on a combination of clean water and temperatures appropriate for the species.

Table 4 Freshwater fish—estimated production quantity by system type (tonnes)

Species	Freshwater pond	Intensive freshwater flow-through	Indoor land-based recirculated	Small cage	Total production
Trout	310	164,390	0	250	164,950
Carps	77,860	0	0	0	77,860
FW other	6640	2580	290	0	9510
Catfish	1140	400	3790	0	5330
Eels	0	0	4960	0	4690
Sturgeon	400	1910	0	0	2310
Tilapias	0	0	450	0	450
Salmon	0	5	0	0	5
Total (*t*)	86,350	169,285	9220	250	265,105
Total (€M)	184.22	464.63	60.66	2.07	711.58

Source: Lane et al. (2014) and present study

Both trout and carp were originally reared in earthen ponds that require considerable maintenance (banks, floor, inlets/drains, etc.). Trout farming expanded as a result of increased productivity, due to the development of compound feeds adapted to nutritional requirements and the use of concrete tanks/raceways and latterly oxygenation allowing higher stocking levels and greater mechanization and automation for feeding, grading and harvesting fish.

Carp and coarse fish are generally reared in large freshwater ponds with minimal extra feeding (extensive production); the use of specific compound feeds is much rarer. This means that the aquaculture process to provide fish of market size is longer (2–3 years) and less direct stock management (e.g. grading and vaccination) is practised. Nonetheless, the infrastructure of extensive pond farming is recognized as providing environmental services—providing flood protection, habitats for wild birds and animals, and water treatment (outlet water is often cleaner than inlet) (Szücs et al. 2008).

Sturgeon production for caviar often uses water recirculation systems—particularly for younger fish—and uses advanced monitoring procedures of individual fish, so as to follow ovary development for caviar harvesting.

Recirculated aquaculture systems (RAS) have been in use since the 1980s as a means of reducing water usage (and thereby increasing potential production for any given site/water supply), maintaining stable and more biosecure environments, and in some cases to assist the conservation of heat energy. Installations and monitoring equipment have evolved considerably, and large-scale units have been installed for different freshwater species (notably for salmon juvenile production, eels, African catfish, barramundi and tilapia). State-of-the-art systems use approximately 500 litres of water per kg production, but with increased capital and energy costs (Verdegem et al. 2006; Martins et al. 2010).

Shellfish production

European oyster production is usually a three-year process that starts with the collection of small oysters on a support from which they can be easily removed (spat collector, e.g. white-painted roofing tiles, as practised in the Bassin d'Arcachon) or on a substrate such as mussel shells. The newly settled oyster juveniles are then transferred to intertidal or sub-tidal bottom culture plots, or in intertidal bags on trestles, or suspended from frames.

There are three different culture techniques for mussel culture—using poles ('bouchot'), suspended ropes or bottom culture. A 'bouchot' is a wooden pole, placed upright into the sand. Mussel seed, collected (usually around March) either on poles (placed further out to sea) or on ropes, is transplanted onto the growing poles and harvested after 12–15 months. Suspended rope culture: Ropes covered with mussel seed are suspended from either frames, floating structures (rafts), or longlines with surface buoys. Frames are built from metallic poles, placed upright into the ground, at water depths ranging between three and nine metres. Bottom culture: Based on the harvesting of naturally producing mussel beds—a technique that is widely practised in the Netherlands and Germany and to some extent in UK and Ireland. Since the 2000s mussel seed is also derived from suspended seed collectors and then relayed on bottom culture plots, particularly in the Netherlands and Germany.

Scallop, clam and cockle species are generally produced using bottom culture techniques, with juveniles from the wild or in some cases, hatcheries (Table 5).

Table 5 Shellfish—estimated production quantity by system type

Technology	**Estimated 2012**	
	Production (tonnes)	**Value (million euro)**
Marine supported and suspended culture (non-fed sedentary and attached animals and plants)	537,050	837.21
Marine bottom culture (non-fed sedentary and attached animals and plants)	65,440	204.01
Total	602,490	1041.23

Source: Lane et al. (2014) and present study

Algae and other aquatic products

The majority of reported production volume is of brown seaweeds using suspended culture techniques. Other species groups, such as crayfish and shrimps, are generally produced in ponds, whether freshwater or brackish water (Table 6).

Table 6 Algae and other aquatic products—estimated production quantity by system type

Technology	**Estimated 2012**	
	Production (tonnes)	**Value (million euro)**
Marine supported and suspended culture (non-fed sedentary and attached animals and plants)	5352	0.01
Coastal pond aquaculture	247	2.39
Freshwater pond aquaculture	38	0.31
Marine bottom culture (non-fed sedentary and attached animals and plants)	9	2.95
Small cage systems—sheltered marine	5	0.02
Intensive freshwater flow-through and partial	1	0.007

recirculation systems		
Indoor land-based recirculated aquaculture systems (marine)	1	0.01
Total	5653	5.70

Source: Lane et al. (2014) and present study

Small cage marine systems are used for the culture of cephalopods, mainly octopus species.

The economic value of EU aquaculture

Overview of value chain structures

The EU Fish Processors and Traders Association estimated the total turnover of the seafood sector in the EU to be around €27 billion (AIPEC-CEP 2015). The contribution of EU aquaculture producers and capture fisheries to this was, respectively, €4 billion and €7 billion (STECF 2013a, b). The remainder of the €16 billion mostly comprises the value of imports. This value is doubled, however, once retailing and food service are taken into account. The European Market Observatory for Fisheries and Aquaculture Products estimated total expenditure on fish and seafood products in the EU in 2011 to be €52.2 billion (EUMOFA 2014). Their figures also suggest the value of imports to be €19.2 billion in 2012. These figures can best be understood through value chain analysis.

The value of any product at the point of consumption can be considered to be the sum of the value added through many intermediate activities needed to produce it and deliver it to the consumer. Such value chains can be quite complex, but with respect to aquaculture production can be considered to consist of farming itself and associated upstream (suppliers) and downstream (processing and distribution) activities. The price of a product at any point is a combination of the cost of all the inputs required to reach that point in the value chain plus a contribution to company profits. Where each step is carried out by separate companies, there will be an element of profit taken at each step. If all the activities are carried out by a single company (vertical integration), the profit margin may only be applied at the final stage (Fig. 9).

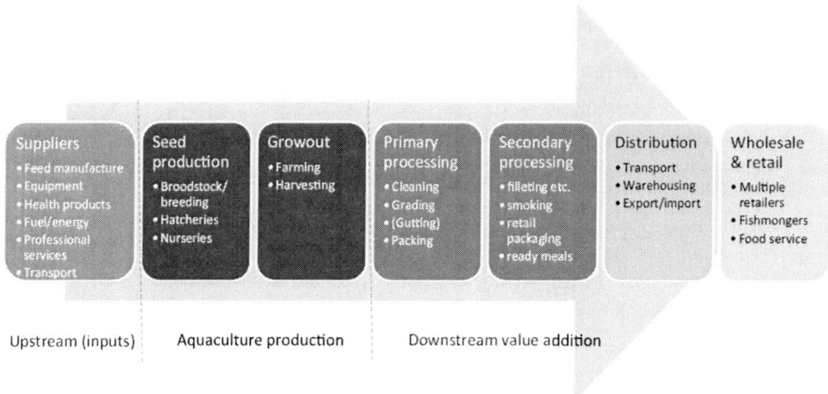

Figure. 9 Generic value chain (simplified)

The value chain for aquaculture products varies regionally depending on the nature of the production process and the degree of processing prior to market. The simplest will be for shellfish that are marketed without substantive processing and packaging, whilst the more complex will be for salmon or other species that have substantial upstream inputs and downstream processing and distribution. Based on limited studies of seafood value chains (e.g. Gudmundsson et al. 2006; EUMOFA 2013) raw material (e.g. output from fishing or farming) accounts for between 15 and 40 % of the final retail price. Processing and distribution adds between 10 and 30 % to the final price, whilst retail costs and margins adds between 25 and 40 % to the final price. A specific example for farmed fish is a study by EUMOFA (2014) on fresh sea bream in Italy (i.e. no secondary processing) which found ex farm values to be around 40 % of the final retail price. Care needs to be taken in such an analysis to take account of wastage (e.g. in gutting and filleting) although where value is obtained from by-product that should also be considered. In the case of food service (e.g. institutional catering and restaurants), the ingredient cost will only represent 30–35 % of the price charged per meal,[1] so the cost of primary production (e.g. a farmed fish) may only represent 10–15 % of the final value of the product as sold to a consumer. For this reason, it is important to consider the whole value chain as changes in output from the production sector, for instance, can have a larger economic impact in the downstream activities than in the production sector itself. For instance, an analysis of the Scottish salmon farming industry (IMANI/SRSL 2014) estimated the turnover from primary production to be £537 (€660) million in 2012 with downstream processing, distribution, retail and catering adding a further £270 (€332) million within Scotland and over £860 million (€1.06 billion) across the whole UK value chain (total £1.4 (€1.72) billion). On the upstream side, feed is generally the most significant input for aquaculture. In the case of the Scottish salmon industry, this was estimated to account for between 50 and 60 % of production costs (IMANI/SRSL 2014). Other activities supplying aquaculture producers include engineering, transport and equipment supply companies, fuel supply and professional services.

Economic assessment at the EU level

Since many companies involved in both the upstream and downstream activities are also servicing other sectors, they are generally not fully reliant on aquaculture production for their turnover. In some cases, they could (and often do in the case of secondary processor, retailers and food service) substitute local aquaculture products with fisheries products or imported products. For this reason, a formal analysis of economic impact of an activity (e.g. aquaculture production) is considered in terms of direct impact (e.g. hatcheries, farms, primary processing and sales operations), indirect impact (suppliers and wider business transactions) and induced (expenditure of company employees in the broader economy, etc.). Economic input–output models are used to quantify these effects, usually within defined geographic boundaries. Few examples of this approach are available, but a study by Fisheries and Oceans Canada (2010) quotes a direct output value of the Canadian aquaculture industry in 2007 to be C$1.026 billion, whilst the gross output (including calculated indirect and induced impacts) was C$2.1 billion (i.e. over double). This multiplier effect is industry and region dependent and can be applied to other measures of economic value including employment and value added (discussed below). In the case of the aquaculture sector in Scotland for instance, the output multiplier for indirect effects is 1.7, whilst the multiplier for both indirect and induced effects is 1.9 (Scottish Government 2015—data for 2012).

Whilst a full value chain approach or the use of input–output models provides an important overall perspective on the value of aquaculture activities, obtaining the appropriate economic data can be difficult or expensive. Many studies (e.g. STECF 2014) focus only on direct impacts for which data are more readily available. The main economic indicator used so far is output which is the total value of the products from aquaculture—essentially the same as turnover. However, it is more meaningful in economic terms to consider the value that is added by a particular activity (e.g. the farming process) as gross value added (GVA). This is the value of the output minus the cost of purchased inputs (e.g. feed, medicines, fuel, services, etc.). The value that is added is therefore comprises employee costs, the cost of capital employed and the profit that is made. GVA values are therefore substantially lower than output values. Hence, the total GVA for the EU aquaculture sector in 2012 was reported to be €1.294 billion, whilst total turnover (output) was €4.365 billion (i.e. the multiplier from turnover to GVA is around 0.3) (STECF 2014). However, this varies by sector with shellfish averaging 0.54 and sea bass and sea bream less than 0.1. Shellfish provides a greater GVA in relation to output because there are no costs for feed. The GVA for sea bass and sea bream is low due to negative profitability in the sector at this time (2012).

A further economic indicator is profitability (measured as earnings before interest and taxes (EBIT) albeit with some variation in definition) and ratios of EBIT to turnover and capital employed. These are particularly useful for assessing the economic health of the sector and potential for reinvestment or tax revenues. STECF (2014) using data from 18 countries (excluding Greece and Poland) found the overall ratio of EBIT to turnover was around 8 % in 2012 and return on investment (EBIT divided by total assets) was 7 % (down from 9 % in 2011). Further analysis and perspectives on EU aquaculture sector profitability are provided by Guillen et al. (2015) using data from submitted company accounts and somewhat different means of calculation. This analysis by country and segment illustrates significant variability over time suggesting the need for relatively long-term financial planning on the part of companies.

The social economic value of the sector is generally measured in terms of the number of people employed and the normalized measure of FTE (full-time equivalent). The employment generated by aquaculture is mostly in rural and coastal regions with fewer economic opportunities and, hence, may have a greater social impact than an equivalent number of jobs in cities. A more precise measure of economic value is labour income, which is a measure of earnings by the people employed in the industry. STECF (2014) estimates total employment in EU aquaculture to be around 80,000 people and around 40,000 FTE (indicating a fairly high degree of part-time and seasonal work). Average wages per FTE (based on data from 19 countries) were €22,100 per annum in 2012, but this varied from €3100 in Bulgaria to €70,000 in Denmark. Total labour earnings in EU aquaculture were therefore around €884 million.

Economic value by sub-sector

A detailed analysis of the economic performance of the EU aquaculture sector by species/system and country has been carried out by the EU Scientific, Technical and Economic Committee for Fisheries (STECF 2014) as far as data are available. Data from that report have been combined with the analysis of FAO FishStat data in order to provide a perspective using the species groupings and system types defined earlier.

Five countries accounted for around 78 % of the direct output value of EU aquaculture in 2012, the UK, France, Greece, Italy and Spain. For the UK, the main component is Atlantic salmon culture, whilst for France the output is dominated by shellfish production. Greece is predominantly sea bass and sea bream, whilst both Italy and Spain have more diverse aquaculture sectors involving shellfish and both marine and freshwater finfish (Fig. 10).

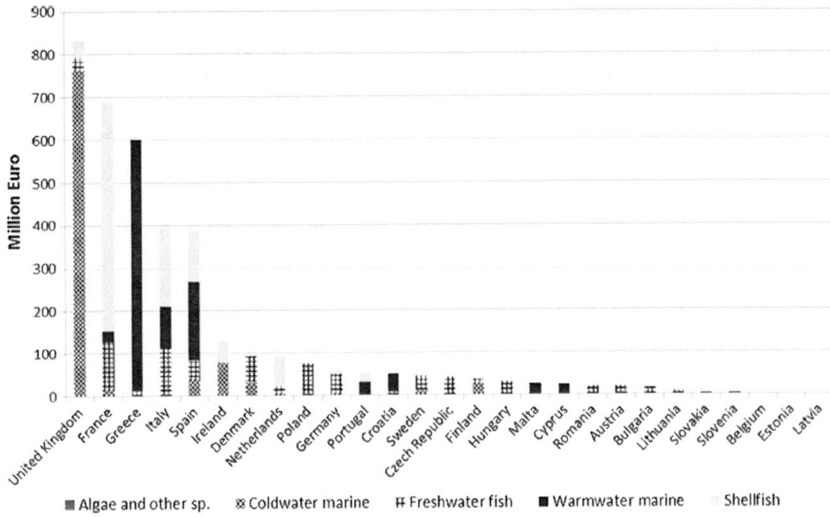

Figure. 10 Direct output value of EU aquaculture production (2012) [*Source:* analysis of FishStat (FAO 2014b) data]

The importance of the UK (Scottish) salmon production by output value carries through to an analysis by system type with large marine cage systems (supplemented by some production of sea bass and sea bream in several Mediterranean countries) providing the greatest direct output value (€1.26 billion). This is followed by the marine shellfish sector and in particular suspended systems (e.g. mussel culture in Spain) (€0.84 billion). Interestingly, this achieves a higher GVA (€436 million) than large cage culture (€360 million) due to the lower purchased inputs for shellfish farming. Conversely, smaller marine cage systems for sea bass and bream and intensive systems for trout had lower GVAs (around 20 %) in relation to output value due to poor profitability in 2012 (Fig. 11).

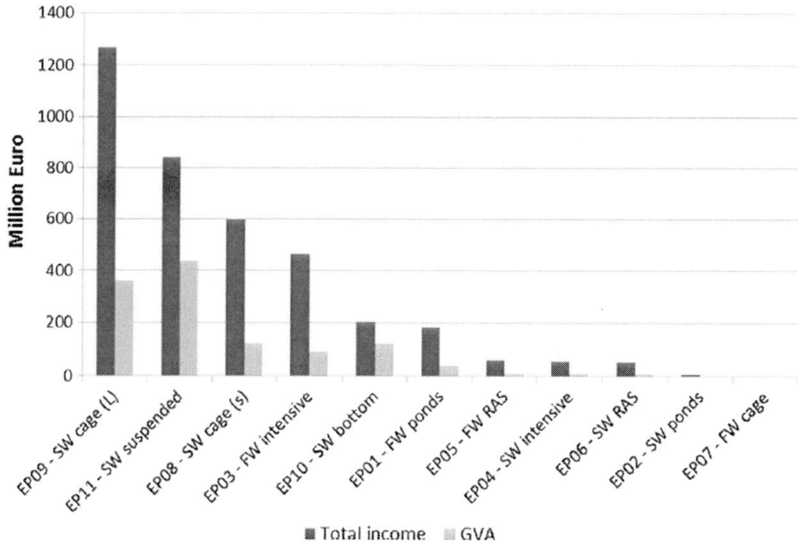

Figure. 11 Output value and estimated GVA of EU aquaculture production by production system type (2012) [*Source:* based on FishStat (FAO 2014b) and STECF (2014) data]

Some analysis of employment patterns is possible using data from STECF, although data are not available for all countries or species/system combinations. Dividing output value by FTE employment provides an indication of labour productivity and can also be used to estimate the FTE employment by sub-sector. The highest productivities are seen in the Scottish salmon industry at around €490,000 output per FTE employee. On the basis of available data, the lowest productivity can be found in the freshwater pond sector with an output of only €14,000 per FTE employee. The shellfish sector is second lowest at around €70,000 per FTE. This means that these sectors provide more employment in relation to output and hence will have greater social impact. However, in practice it can mean low wages which can prompt the local population to seek higher returns from alternative occupations and increased dependency on immigrant labour from economically weaker countries (Fig. 12).

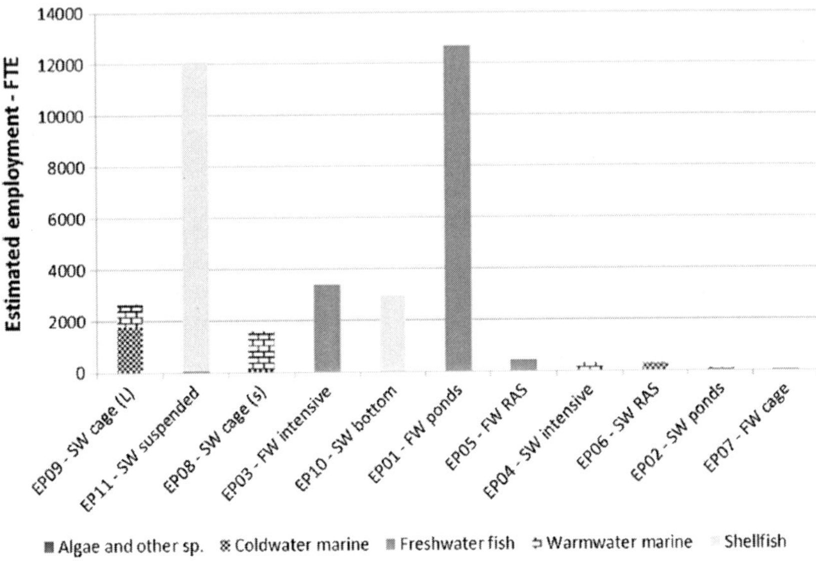

Figure. 12 Estimated employment (FTE) in EU aquaculture by system type (2012) [*Source:* calculated using FishStat (FAO 2014b) value data and STECF data on employment. Note that limited data mean that the same employment ratios are applied to all shellfish systems and to minor fish systems]

The labour productivity figures are best in the sectors which are most invested in technology and larger-scale farming operations. Low productivity is associated with smaller-scale enterprises with relatively low mechanization and use of capital. Analysis carried out by STECF (2014) confirmed the EU production sector in general to be highly fragmented with 12,466 companies recorded (14–15,000 estimated) with aquaculture as their main activity in 2012. These have an average of 5.5 employees with only 9 % of companies having more than ten. In contrast, the salmon industry is increasingly consolidated. The number of active companies in Scotland has fallen from 132 in 1993 to 15 in 2013. These employ an average of 72 people (Marine Scotland Science 2014). The sea bass and bream sector in Southern Europe are following a similar trend.

Competitiveness of aquaculture

The outputs from European aquaculture enterprises find a place on the market alongside products from capture fisheries and imported seafood (both captured and farmed) from around the world. This implies competition on price, quality and other attributes determined by consumer perceptions. Price is the dominant factor as conventional supply and demand economics predicts that demand falls as prices rise, whilst the incentive and then ability of producers to supply the market fall as prices fall. Price is the mechanism that balances supply and demand. As consumers have choice over their expenditure, and particularly in a commodity market such as unprocessed fish and seafood, potential market volumes for aquaculture producers are increasingly limited as production costs rise. This is illustrated in Fig. 13 which shows sales volumes of EU capture fish and aquaculture production in relation to wholesale price band (2012). It must be noted that the capture fisheries data include industrial fisheries for reduction to fishmeal (approximately 50 % of the €0–€1/kg price band) and that sales values recorded for aquaculture products will be higher than capture fisheries products due to most aquaculture prices including primary processing and distribution, whilst capture fisheries prices are generally based on landed auction prices. Note that higher price categories are included in the following charts as there are species falling into most of these categories, although production is sometimes too small to register on the chart.

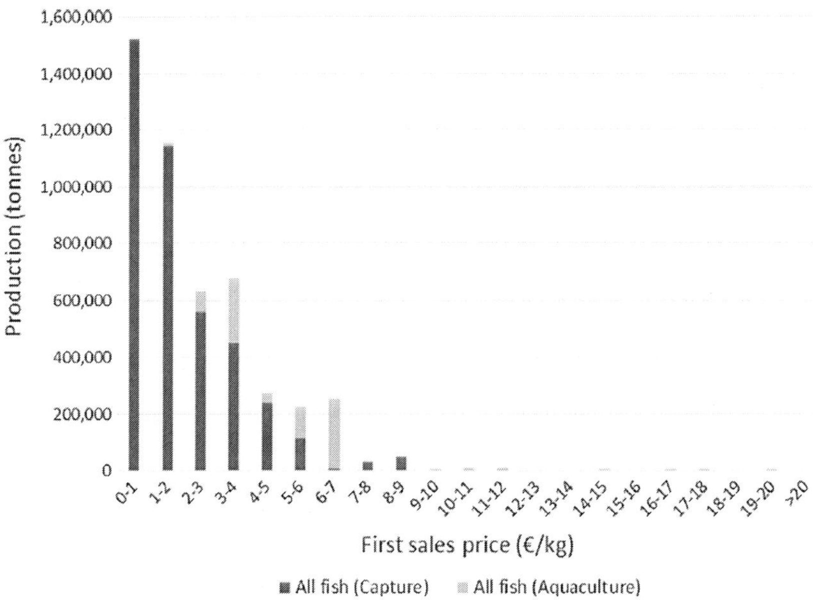

Figure. 13 EU capture finfish landings and EU aquaculture fish production (2012) by price band [*Source:* calculated using FishStat production data and value data for aquaculture (FAO 2014b) and capture fisheries prices from EUMOFA (http://www.eumofa.eu)]

Although the prices are not strictly comparable, it is interesting to note that over 50 % of EU finfish production has a first sale price below €2/kg, which is below the cost of production for most European fish species using existing aquaculture systems. Aquaculture starts to make a contribution to supplies as prices rise above €2/kg and particularly over €3/kg. It is dominant in the €6–7/kg category. This picture represents around 40 % of total EU supplies (due to imports), but gives some indication of overall market structure.

Breaking this down by sub-sector shows that freshwater aquaculture and shellfish farming are more competitive with capture fisheries products than finfish (Figs. 14, 15, 16, 17).

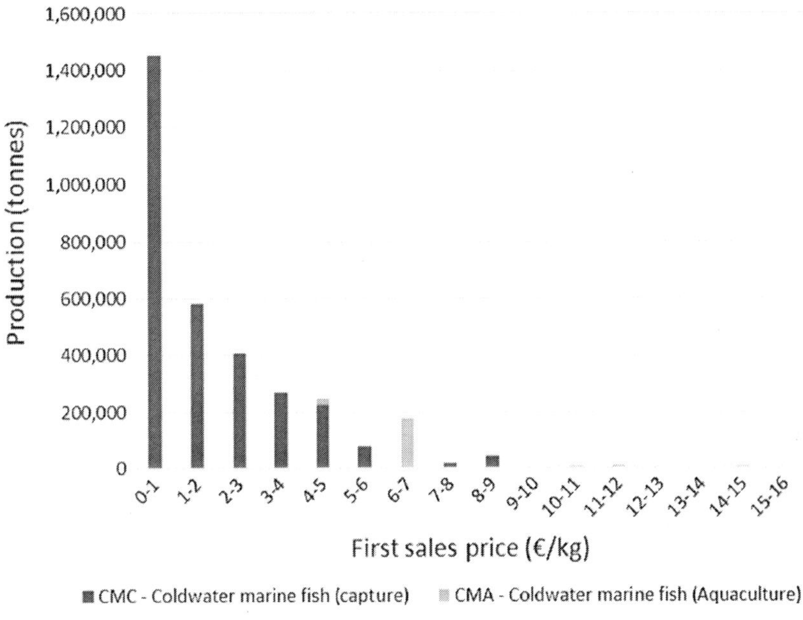

Figure. 14 Price structure of EU capture fisheries and aquaculture production of coldwater marine fish species (2012). [*Source:* calculated using FishStat production data and calculated prices for aquaculture (FAO 2014b) and capture fisheries prices from EUMOFA (http://www.eumofa.eu)]

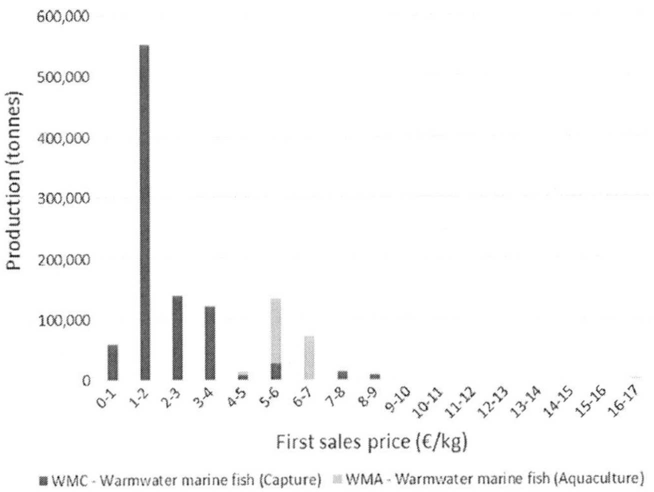

Figure. 15 Price structure of EU capture fisheries and aquaculture production of warmwater marine fish species at first sale (2012) [*Source:* calculated using FishStat production data and calculated prices for aquaculture (FAO 2014b) and capture fisheries prices from EUMOFA (http://www.eumofa.eu)]

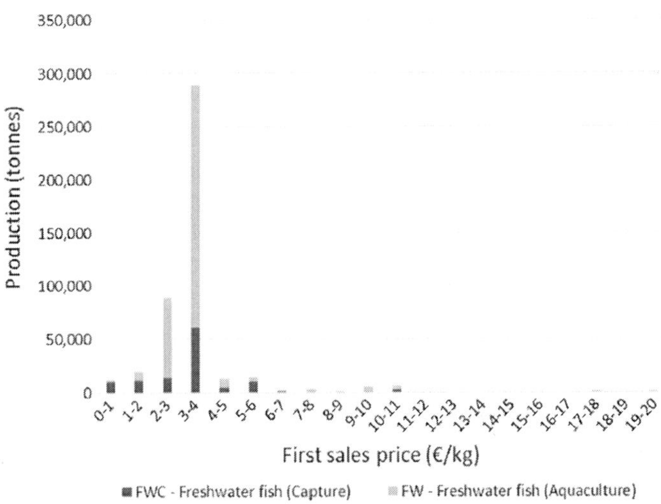

Figure. 16 Price structure of EU capture fisheries and aquaculture production of freshwater fish species at first sale (2012) [*Source:* calculated using FishStat production data and calculated prices for aquaculture (FAO 2014b) and capture fisheries prices from EUMOFA (http://www.eumofa.eu)]

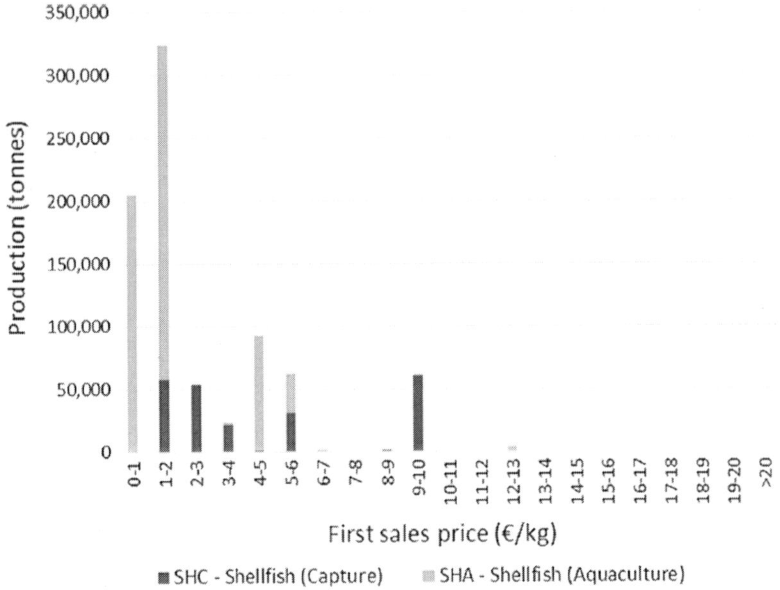

Figure. 17 Price structure of EU capture fisheries and aquaculture production of shellfish species at first sale (2012) [*Source:* calculated using FishStat production data and calculated prices for aquaculture (FAO 2014b) and capture fisheries prices from EUMOFA (http://www.eumofa.eu)]

If products from aquaculture and capture fisheries readily substitute for each other, the volume of aquaculture production within the EU can be expected to be strongly influenced by its competitiveness on price in relation to other sources (capture fisheries and imports). Technical and management innovations that reduce the cost of production would be expected to increase competitiveness and increase the potential for EU production, whilst higher input and regulatory compliance costs should reduce the scope for EU production (Bostock et al. 2009). On this basis, there is the potential for aquaculture to be more competitive and eventually replace some of the capture fisheries supplies. This hypothesis was examined by Asche et al. (2001) which found little evidence of market interactions between species other than where both farmed and wild species were available together. Overall, it was considered that aquaculture production expanded the market for seafood. This conclusion was further strengthened in a later study (Asche et al. 2009) on the European whitefish market. However, stronger market interactions have been found in some cases. For instance, whilst Norman-Lopez (2009) found no interactions between farmed catfish and tilapia in North America, fresh farmed tilapia fillets were found to compete with wild whole red snapper wild fresh fillets of sea dab, and blackback flounder. Regnier and Bayramoglu (2014) found greater evidence of differentiation in the French fresh fish market between farmed and wild sea bass than between farmed and wild sea bream. They attributed this to sea bass being a higher value product with consumers being more willing to pay higher prices (in this case for wild caught fish) to make ethical choices with higher priced products. As aquaculture production overtakes capture fisheries for human consumption, a greater diversity of market segments and interactions can be anticipated particularly with respect to ethical issues relating to production methods (Natale et al. 2013).

Policy support measures and mechanisms for European aquaculture and their influence

Studies and policy development

Identifying challenges for sustainable development of aquaculture, and hence growth, has to refer to an extensive combination of economic, social and technical issues. Whilst some of these apply specifically to aquaculture, many are more general in nature, covering livestock rearing conditions, water use, environmental considerations and food supply in general. The complex interweaving of legislative and societal approaches that cover these issues tends to show that, until recently, aquaculture has been an 'add-on' that remained outside core policy considerations.

This is most clearly demonstrated by the history of the Common Fisheries Policy (CFP) where aquaculture was included primarily because its products are in the same market(s) as those of fisheries. This is reflected in the requirements of the oldest component of the CFP, the Common Organisation of the Markets for Fisheries and Aquaculture Products, which has existed since 1970. As aquaculture grew in the 1980s–1990s, both in Europe and globally, its higher level of visibility meant that both markets and legislation had to adapt to this new developing activity.

After different reforms, the CFP[2] in 2014 recognized aquaculture to be a key component of its scope, alongside traditional fisheries and seafood processing.

Nonetheless, whilst the CFP is the legislative instrument used for the management of European fish stocks that are a common resource/good, **aquaculture is predominantly an economic activity of the private sector, where the stock belongs to the operator**. This gives different responsibilities to both the operator and suppliers, which include:

1. Environmental respect and, in many cases, management, covering:
    a. Water use and waste management
    b. Adaptation to: for freshwater environments the Water Framework Directive, River Basin Management principles and for marine environments the Marine Strategy Framework Directive (2008), the Directive for Maritime Spatial Planning (2014) and, more generally, Natura2000 and the Habitats Directive.
2. Health and welfare of livestock (mainly covered by EFSA advice)
    a. Standards for the protection of animals bred or kept for farming purposes (including fish)[3]
    b. Aquatic Animal Health Directive[4]
    c. Live Animal Transport Directive[5] (for stock movements [e.g. from hatcheries])
    d. Stunning and killing
3. Feed composition
    a. Awareness of the finite nature of fishmeal and fish oil sources
    b. Ban on ruminant protein sources (following TSE and dioxin crises) although non-ruminant PAPs were reintroduced in 2013
    c. Promotion of plant-based and alternative protein/oil sources
4. Processing standards
    a. European food processing standards
    b. Adherence to market-oriented quality and/or certification labels

Complementary policies (non-exhaustive) that influence aquaculture include the Integrated Maritime Policy, Blue Growth, the Circular Economy and the Bioeconomy.

Challenges to European aquaculture development are numerous, and the most important ones identified by Lane et al. (2014) are:

- Strong market competition within the EU seafood market, particularly with imported seafood (fisheries and aquaculture), which has kept market prices down for several years.

- Administrative burdens, particularly slow times to licensing (and the very number of licences/permits needed to operate), that restrict investments and expansion potential.
- High labour and employment costs and working conditions (e.g. 35 h week).
- Adequate access to bridging finance for working capital (investment in stocks).

These challenges are at the core of the European Commission's strategic guidelines for the sustainable development of European aquaculture,[6] where they were described in detail and specific targets for each of these were set for compliance by Member States and also for the Commission.

The European aquaculture industry agreed that the main challenges to progress are (EATIP 2012):

- Competition in the marketplace, principally from imports
- Access to and competition for space for coastal and inland aquaculture
- Maintaining health and welfare of livestock
- Improving resource use (husbandry, feeds, farm technology)
- Governance within the Common Fisheries Policy

Support for research and technological development

The EU aquaculture sector has been supported through both national and EU-level research funding. EU funding is mainly through the RTD framework programmes, usually through primary research themes but also included in various cross-cutting programmes which have supported networking, mobility, access to research infrastructures and SME participation, etc. (Table 7).

Table 7 Major EC funding for aquaculture research and technological development

Framework programme	**Main theme including aquaculture**	**Aquaculture priorities**	**Funding allocation**
FP4 (1994–1998)	FAIR	The impact of environmental factors; ecological impact of aquaculture; biology of	103 M ECU for fisheries and aquaculture

		species and optimization of aquaculture; improved methodology for data collection and analysis	
FP5 (1998–2002)	Quality of live (QoL)—key action 5, sustainable agriculture, fisheries and forestry and integrated development in rural areas including mountain areas	Sustainable production systems with the reduction of the impact on ecosystems and diversification of cultivated species (both plant and animals); improvement of production techniques, genetic improvement; disease resistance and control	520 M EUR for action 5 (sustainable agriculture and fisheries and forestry....)
FP6 (2002–2006)	Scientific support to policy (SSP—linked with the Common Fisheries Policy); Priority 5—'Food Quality and Safety' and Priority 6—'Sustainable Development'	Integration of environmental requirements into the CFP; sustainable aquaculture production; synthesis and dissemination of results to end-users; to support a better seafood production, improving the knowledge about diseases, health conditions, processing, etc	159 fisheries and/or aquaculture-related projects were funded with a total budget of 292.9 million euro. Of these, 75 projects (98 million euro) were aquaculture specific
FP7 (2007–2013)	Food, agriculture and biotechnology	Sustainable production and management of biological aquatic resources; fork to farm—seafood, health and well-being	1760 M euro on food, agriculture, fisheries and biotechnology
Horizon 2020 (2014–	Food security, sustainable agriculture and	Domestication of new species; interactions with the environment;	4152 M euro for all food security

2020)	forestry, marine, maritime and inland water research and the bioeconomy (under societal challenges pillar)	health and nutrition of farmed aquatic organisms; reproduction and breeding; social aspects of the seafood sector for the cohesion of the social fabric in the European coastal areas	including funding of the European Institute of Innovation and Technology

Research funding under the main RTD programmes has been increasingly directed by the European Commission to ensure it properly supports strategic and policy objectives. In responding to the needs for consultation on research needs in aquaculture, a multistakeholder technology platform was created in 2008, the European Aquaculture Technology and Innovation Platform (EATiP). Recognized as an official European Technology Platform (ETP)s, EATiP has to have a strategy, mobilization and dissemination function, encompassing.

- The preparation of an industry-focused **strategic research and innovation agendas** including technology roadmaps and implementation plans;
- encouraging **industry participation in Horizon 2020**, the EU's framework programme for research and innovation, and cooperating with networks in Member States;
- fostering **networking opportunities** with other ETPs and other partners along the value chain to address cross-sectoral challenges and promote the move towards more open models of innovation;
- identifying **opportunities for international cooperation;**
- acting as one of the channels of external **advice** for the programming and implementation of Horizon 2020; notably, ETPs have been a key driving force behind the launch of high profile public–private partnerships under the programme.

EATiP provided a detailed Vision document, and a strategic research and innovation agenda (SRIA) was made in 2012, which identified specific sectoral needs—in terms of both legislation and research requirements. Its value for policy was recognized within the EC strategic guidelines for the sustainable development of European aquaculture, and several of the topics identified in its SRIA have become topics for the subject of calls in recent work programmes (Fig. 18).

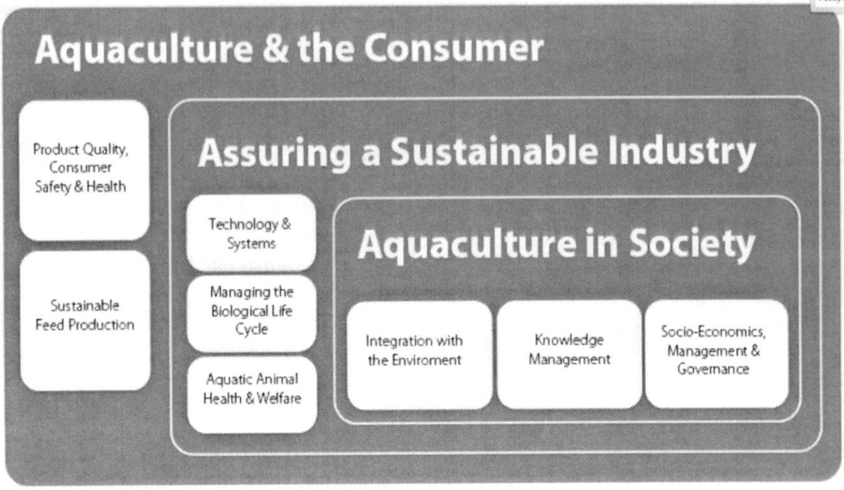

Figure. 18 Thematic priorities identified by EATiP

Further inputs on strategic research priorities are provided by the European Fisheries and Aquaculture Research Organisation (EFARO—http://www.efaro.eu) and the COFASP project (Cooperation in Fisheries, Aquaculture & Seafood Processing—http://www.cofasp.eu).

Support for aquaculture industry development

Direct support for aquaculture industry investment is provided through EU structural funds. The first of these was the Financial Instrument for Fisheries Guidance (FIFG) which ran over the period 1994–1999 and 2000–2006. These provided grants that were part-funded by national governments within a framework of multiannual support programmes with differing levels of contribution depending on country region. The total programmed expenditure under FIFG 1994–1999 was ECU 2.676 billion and for FIFG 2000–2006 €3.746 billion.[7] In the latter programme, €0.317 billion was provided to support a total investment of €1.199 billion in the aquaculture sector (EC 2010).

FIFG was followed by the European Fisheries Fund (EFF), which ran from 2007 to 2013. With a budget of €4.3 billion, its aim was to improve the competitiveness of the sector and help it become environmentally, economically and socially sustainable. Aquaculture, processing, marketing and inland fisheries made up Axis 2 of the fund, with aid being available for diversification into new aquaculture species with good market prospects, environmentally friendly aquaculture, public and animal health measures, processing and marketing and lifelong learning. Special provision existed for inland fishing, reflecting its importance in Central and Eastern Europe.

In its interim evaluation report of the EFF (Ernst and Young 2011), Ernst and Young reported that of a total of €1.24 billion programmed under Axis 2, €518 million (43 %) had been committed to 3556 projects across all Member States. Delays in implementation were (in general) stated to have been due to limited co-financing in an unfavourable economic environment, and the late launch of the programme, mostly due to delays in validation of the operating programmes.

Specifically under Axis 2, measure 2.1 (aquaculture) represented 27 % of the EFF spend, measure 2.2 (inland fishing) less than 1 % and measure 2.3 (fish processing and marketing) the vast majority with 72 %. Projects under measure 2.1 were principally focussed on productive investments, although many were reported as being constrained by environmental impact assessment, requirements. Investments in 'environment measures' were observed to be well implemented, but animal and public health measures were considered by many Member States to be unsuitable and not adapted to the reality in both the fish and shellfish sectors. Under measure 2.3, the majority of investments were in increasing production capacity, improving systems and improving hygiene and working conditions in processing. These reportedly produced positive impacts on employments, but the national evaluations did not enable an assessment on their effects on competitiveness and sustainability.

In its 6th annual report on EFF (COM (2013) 921 final), the European Commission reported that the expenditure in aquaculture measures had leveraged additional national public contribution of €183 million and a further €538 million of private funding. Hence €1 of EFF funding had a leveraging effect of €1.68.

The EFF has been superseded by the European Maritime and Fisheries Fund (EMFF) which will run from 2014 to 2020. This has a total EC budget of €6.4 billion (Table 8).[8] Each country is required to draw up an operational programme[9] which takes account of both the EU fisheries and aquaculture policies and national policies and priorities (in the case of aquaculture expressed through a National Aquaculture Plan). Applicants are required to show compliance with the relevant national operational programme, and as with previous programmes, there will be an element of national co-funding. A new element is Fisheries and Aquaculture Monitoring and Evaluation (FAME) which will be carried out by an expert support unit based in Brussels. This has the remit to (i) assist in monitoring and evaluating the implementation of the EMFF and provide the Commission with regular updates and analysis of the state of play of the implementation of the EMFF, and (ii) build capacity across the Member

States and in the Commission on evaluation and monitoring methodologies, indicators and good practices.[10]

Table 8 Total EU allocations of European Maritime and Fisheries Fund 2014–2020 (unit €, current prices)

MS	2014	2015	2016	2017	2018	2019	2020	Total	%
BE	5,722,130	5,795,229	5,848,204	5,942,991	6,081,279	6,122,861	6,233,357	41,746,051	0.7
BG	12,071,289	12,225,498	12,337,253	12,537,214	12,828,942	12,916,663	13,149,763	88,066,622	1.5
CZ	4,263,975	4,318,446	4,357,922	4,428,555	4,531,602	4,562,588	4,644,927	31,108,015	0.5
DK	28,559,270	28,924,111	29,188,510	29,661,596	30,351,790	30,559,328	31,110,815	208,355,420	3.6
DE	30,100,054	30,484,577	30,763,242	31,261,850	31,989,281	32,208,016	32,789,256	219,596,276	3.8
EE	13,840,012	14,016,816	14,144,946	14,374,205	14,708,679	14,809,253	15,076,507	100,970,418	1.8
IE	20,231,798	20,490,256	20,677,561	21,012,701	21,501,645	21,648,669	22,039,349	147,601,979	2.6
EL	53,289,776	53,970,543	54,463,896	55,346,644	56,634,503	57,021,756	58,050,796	388,777,914	6.8
ES	159,223,336	161,257,387	162,731,468	165,369,007	169,216,972	170,374,037	173,448,682	1,161,620,889	20.2
FR	80,594,423	81,624,003	82,370,140	83,705,190	85,652,923	86,238,597	87,794,897	587,980,173	10.2
HR	34,629,786	35,072,176	35,392,777	35,966,420	36,803,321	37,054,974	37,723,684	252,643,138	4.4
IT	73,642,561	74,583,332	75,265,111	76,485,002	78,264,728	78,799,884	80,221,941	537,262,559	9.3
CY	5,443,762	5,513,306	5,563,703	5,653,880	5,785,440	5,824,999	5,930,119	39,715,209	0.7
LV	19,167,006	19,411,862	19,589,309	19,906,810	20,370,021	20,509,307	20,879,427	139,833,742	2.4
LT	8,694,653	8,805,725	8,886,220	9,030,247	9,240,371	9,303,555	9,471,451	63,432,222	1.1
HU	5,358,928	5,427,387	5,477,000	5,565,770	5,695,280	5,734,223	5,837,705	39,096,293	0.7
MT	3,101,540	3,141,162	3,169,876	3,221,253	3,296,208	3,318,746	3,378,637	22,627,422	0.4
NL	13,915,788	14,093,559	14,222,391	14,452,906	14,789,211	14,890,336	15,159,053	101,523,244	1.8
AT	954,693	966,888	975,727	991,541	1,014,613	1,021,551	1,039,987	6,965,000	0.1
PL	72,814,233	73,744,422	74,418,532	75,624,702	77,384,410	77,913,547	79,319,610	531,219,456	9.2
PT	53,797,969	54,485,229	54,983,288	55,874,453	57,174,593	57,565,539	58,604,393	392,485,464	6.8
RO	23,085,512	23,380,425	23,594,150	23,976,562	24,534,471	24,702,232	25,148,019	168,421,371	2.9
SI	3,400,584	3,444,026	3,475,509	3,531,839	3,614,022	3,638,734	3,704,400	24,809,114	0.4
SK	2,163,649	2,191,290	2,211,321	2,247,162	2,299,451	2,315,174	2,356,953	15,785,000	0.3
FI	10,197,069	10,327,335	10,421,739	10,590,653	10,837,087	10,911,188	11,108,097	74,393,168	1.3
SE	16,469,779	16,680,178	16,832,654	17,105,477	17,503,503	17,623,188	17,941,225	120,156,004	2.1
UK	33,327,114	33,752,863	34,061,403	34,613,468	35,418,887	35,661,073	36,304,629	243,139,437	4.2
EU 27*	788,060,689	798,128,031	805,423,852	818,478,098	837,523,233	843,250,018	858,467,679	5,749,331,600	100

Source: DG MARE

* LU is excluded, because it is not a recipient of EMFF

Future directions

At the present, the principle EU policy for aquaculture development remains the Common Fisheries Policy, supported by the Common Organisation of the Markets for Fisheries and Aquaculture Products and the financial instrument, the EMFF.

The submission of the National Aquaculture Plans foreseen in the strategic guidelines for the sustainable development of European aquaculture has been delayed, principally due to late finalization of the EMFF and the need to combine these multiannual plans with the operational programmes of each Member State. It is anticipated that these will be published in 2016.

From the consultation within EATiP, during the preparation of its Vision document, the following forward production projections were prepared from expert inputs and assessments for individual species and sectors (Table 9).

Table 9 EATiP exercise and forward production projections (a) coldwater marine assumptions, (b) warmwater marine assumptions, (c) fresh water marine assumptions, (d) shellfish assumptions, (e) summarized assumptions

	Forecast growth (tonnes)				Value 2010	Value 2030	Jobs		
	APR (%)	2010	2020	2030	M€	M€		2010	2030
(a) Coldwater marine assumptions									
Cod	4.0	1	1	2	€ 0.00	€ 0.01	Upstream	143	191
Flatfish	9.6	189	473	1,079	€ 1.21	€ 5.39	On-farm	951	1273
Large trout	2.0	30,220	36,838	44,025	€ 75.55	€ 110.06	Added-value (processing)	951	1273
Salmon	4.0	159,912	236,709	336,910	€ 495.73	€ 1044.42	Downstream	95	127
Subtotal		190,322	274,021	382,016	€ 572.49	€ 1159.88	Total jobs	2139	2865
Total increase	101.0			191,694		587.40	Feeds (tonnes)	285,561	458,419
(b) Warmwater marine assumptions									
Flatfish	5.0	10,102	16,455	25,527	€ 89.30	€ 225.66	Upstream	912	1358
Other	4.0	16,453	24,355	34,664	€ 151.70	€ 319.60	On-farm	6080	9051
Sea bass	4.0	74,907	110,881	157,818	€ 346.07	€ 729.12	Added-value (processing)	6080	9051
Sea bream	4.0	111,322	164,784	234,539	€ 445.29	€ 938.15	Downstream	608	905
Subtotal		212,784	316,474	452,548	€ 1,032.36	€ 2,212.54	Total jobs	13,679	20,365
Total increase	112.0			239,764		€ 1,180.18	Feeds (tonnes)	383,011	543,057
(c) Fresh water marine assumptions									
Carps	1.5	67,484	78,318	89,548	€ 131.59	€ 174.62	Upstream	1175	2259
Catfish	6.0	7,279	13,036	22,023	€ 10.41	€ 31.49	On-farm	12,653	15,060
Eels	5.0	6,561	10,687	16,579	€ 99.73	€ 252.01	Added-value (processing)	7,592	10,583
Other	3.0	1,988	2,672	3,486	€ 10.73	€ 18.82	Downstream	1265	1506
Sturgeon	8.0	1,928	4,162	8,320	€ 15.19	€ 65.56	Total jobs	22,686	29,409
Tilapias	25.0	130	1,211	9,021	€ 0.22	€ 15.33			
Trout	1.5	246,498	286,071	327,090	€ 601.45	€ 798.10			
Subtotal		331,868	396,156	476,068	€ 869.33	€ 1355.94	Total jobs	17,473	25,925
Total increase	43.0			144,200		€ 486.61	Feeds (tonnes)	305,658	367,977

Based on our categorization of the sector by production technology (Tables 2, 3, 4, 5, 6) and by the outcomes of the consultations within EATiP and directly with the professional sector, an estimation on the trends (for 2030) of the use of technologies for each of the production sub-sectors was prepared. This is shown in Table 10. The arrow indicates the trends (upwards, stagnant and downward).

Table 10 Trends in prevalence of production technologies

Technology	Estimated 2012 production (tonnes)	Trend in use of technology 2030
Coldwater marine finfish		
Coastal pond aquaculture	180	↘
Indoor land-based recirculated aquaculture systems (marine)	100	↗
Intensive marine flow-through and partial recirculation systems (mostly large tanks)	9800	↗
Small cage systems—sheltered marine	11,000	↘
Large cage systems—marine in exposed sites, using mechanized (automated) systems	190,090	↗
Warmwater marine finfish		
Coastal pond aquaculture and valliculture	900	→
Indoor land-based recirculated aquaculture systems (marine)	9910	↗
Intensive marine flow-through and partial recirculation systems (mostly large tanks)	340	↗
Small cage systems—sheltered marine	102,420	↘
Large cage systems—marine in exposed sites, using mechanized (automated) systems	74,330	↗

Freshwater fish		
Freshwater pond aquaculture (extensive to semi-intensive)	86,350	↗
Intensive freshwater flow-through and partial recirculation systems (mostly tanks, raceways and small ponds)	169,285	↘
Indoor land-based recirculated aquaculture systems (freshwater)	9220	↗
Small cage systems—freshwater	250	→
Shellfish		
Marine bottom culture (non-fed sedentary and attached animals and plants)	65,440	→
Marine supported and suspended culture (non-fed sedentary and attached animals and plants)	537,050	↗

Source: estimated by the authors of this study

The main conclusions of this foresight exercise were as follows:

- **Productivity/competitiveness drive towards larger cages**, particularly in offshore locations, seen as increasing location trend for both Mediterranean and coldwater farming.
- **Coastal pond aquaculture will decline or stagnate**, principally because of lower yields and competition for space (availability and licences)
- **Indoor marine recirculation systems will increase for hatcheries,** but less likely for ongrowing (costs of investment, energy costs) due to cost comparison with cage production (Mediterranean and coldwater production) although may play a role in the development of some higher value species such as sole and shrimp
- **Freshwater pond production, stable or increasing**, dependent on a combination of market demand, diversification activities and recognition of environmental services
- **Intensive flow-through systems for freshwater will probably decline**, dependent on a combination of market demand, water

availability and diversification towards specialized/niche markets (e.g. organic label) where lower intensity demanded

- **Freshwater recirculation systems will increase**, notably for high-value production (sturgeon, pike-perch) and potentially for warmwater species that can be produced at high density (e.g. African catfish, eel, tilapia)
- **Shellfish production will continue to be dominated by supported/suspended** cultivation systems

It is anticipated by many stakeholders that an increasingly proportion of the expected increase in large cage culture for fish species and supported and suspended culture for shellfish species will be done in integrated multitrophic aquaculture (IMTA) systems, where species are combined (e.g. salmon, seaweeds, mussels) within a complementary area so as to best use space and to mitigate environmental impact. At this time, it is not possible to estimate the percentage production from such systems, since practicality, legal frameworks, licensing conditions and economic viability need to be elaborated by region or coastal area.

CONCLUSIONS

EU aquaculture production contributes the equivalent of 2.66 kg (live weight) of fish and seafood per EU citizen with a first sale value of €4.76 billion contributing a GVA of €1.29 billion and employing around 80,000 people (40,000 FTE) (STECF 2014). Whilst the value of EU aquaculture industry increased by 70 % between 2000 and 2012, actual volumes declined slightly. As the overall market for fish and seafood products in the EU is in the region of €27 billion (AIPEC-CEP 2015), the lack of output growth has been widely attributed to competitiveness issues including access to suitable sites (EC 2013). Recognizing this, and based on widespread recognition of the potential contributions that the sector could provide to European seafood supplies, the 2014 reform of the CFP provided a new impetus for the sustainable development of European aquaculture.[11] This was accompanied by a number of different studies and position papers on issues and challenges that affect directly or indirectly the potential for development. Consultations with industry stakeholders presented above indicate potential to responsibly increase output volume by 55 %, value by 77 % and employment by 40 % by 2030 if suitable social and economic conditions prevail. The overall strategy to achieve this sustainable development of (European) aquaculture is by maximizing the quality and health benefits of farmed products, whilst improving resource efficiency and minimizing impacts. Based on this strategy, and within the framework of the CFP, EU Member States are preparing national multiannual plans for aquaculture development. These will, for the first time, provide an overview of Member State intentions to grow aquaculture and the approaches that are prioritized to achieve this.

Monitoring progress is another challenge, and the STECF (2013a, b) study underlined the difficulties of obtaining accurate and up-to-date statistics on the performance of the EU Aquaculture Sector. However, the Data Collection Framework (DCF) continues to be implemented and should be adapted to include all forms of aquaculture, be it marine or freshwater. This was also a clear recommendation to the Commission from the European Court of Auditors (2014) when considering the effectiveness of the EFF support for aquaculture. With improved data it will be easier to understand developments within the sector and develop appropriate policies to support social and economic development.

There is increasing recognition that aquaculture can contribute to additional EU policies and strategies, including Blue and Green Growth and the Bioeconomy, the use of sustainable resources, food security and public health—by providing sustainably produced, high quality and healthy food. Clarification of the position of aquaculture within the existing environmental legislation has been initiated with recommendations existing for Natura2000 and is under development for the EU Water Framework Directive and for the Marine Strategy Framework Directive. This is essential so as to boost the potential integration of the activity and the expansion of site availability. Assessment and agreed quantification of environmental services provided by aquaculture requires to be integrated, alongside potential incentives.

Finally there is a need to better understand market interactions and competitiveness for EU aquaculture products to guide future investment and support policies. The establishment of EUMOFA[12] is an important step in this direction together with the PrimeFish Horizon 2020 project[13] which is studying several key species and developing models that will help understand boom and bust cycles, the success factors for new product development and the interactions between key aquaculture, capture and imported whitefish species.

Taken together, these policy positions, strategies and project initiatives, particularly the requirement for National Aquaculture Plans, will form the basis of the planning required to move towards the next stage of EU aquaculture policy realization and implementation.

FOOTNOTES

1. http://restaurants.about.com/od/menu/a/foodcost.htm.
2. Regulation (EU) No 1380/2013.
3. Directive 98/58/EC.
4. Directive 2006/88/EC.
5. Directive 95/29/EC.
6. COM(2013) 229.
7. http://www.eurocbc.org/page355.html.
8. http://ec.europa.eu/fisheries/documentation/publications/2015-cfp-funding_en.pdf.
9. See http://ec.europa.eu/fisheries/cfp/emff/country-files/index_en.htm.
10. http://ec.europa.eu/fisheries/cfp/emff/fame/index_en.htm.
11. http://ec.europa.eu/fisheries/cfp/index_en.htm.
12. http://www.eumofa.eu.
13. www.primefish.eu.

REFERENCES

1. AIPCE-CEP (2015) Finfish study 2015. EU Fish Processors and Traders Association, Brussels. http://aipcecep.drupalgardens.com/content/white-fish-study
2. Asche F, Bjørndal T, Young JA (2001) Market interactions for aquaculture products. Aquac Econ Manag 5(5–6):303–318. doi:10.1080/13657300109380296
3. Asche F, Roll KH, Trollvik T (2009) New aquaculture species—the whitefish market. Aquac Econ Manag 13(2):76–93. doi:10.1080/13657300902881641
4. Bostock J, Murray F, Muir J, Telfer T, Lane A, Papanikos N, Papergeorgiou P, Alday-Sanz V (2009) European aquaculture competitiveness: Limitations and possible strategies. Report for the European Parliament, Directorate-General for Internal Policies, Policy Department B, Structural and Cohesion Policies, Fisheries. http://www.europarl.europa.eu/RegData/etudes/etudes/join/2009/431569/IPOL-PECH_ET(2009)431569_EN.pdf
5. Bostock J, McAndrew B, Richards R, Jauncey K, Telfer T, Lorenzen K, Little D, Ross L, Handisyde N, Gatward I, Corner R (2010) Aquaculture: global status and trends. Philos Trans R Soc B 365(1554): 2897–2912. http://rstb.royalsoci etypubli shing.org/content/365/1554/2897.full
6. Bronzi P, Rosenthal H, Gessner J (2011) Global sturgeon aquaculture production: an overview. J Appl Ichtyol 27(2):169–175. doi:10.1111/j.1439-0426.2011.01757.x
7. Doukakis P, Pikitch EK, Rothschild A, DeSalle R, Amato G, Kolokotrionis S-O (2012) Testing the effectiveness of an international conservation agreement: Marketplace forensics and CITES caviar trade regulation. Plos One 7(7) e40907, 1–9. doi:10.1371/journal.pone.0040907
8. EATIP (2012) The future of European aquaculture—Our Vision: a strategic agenda for research & innovation. European Aquaculture Technology and Innovation Platform. http://www.eatip.eu/default.asp?SHORTCUT=92
9. EC (2008) Synopsis of fisheries & aquaculture research projects in the 6th Framework Programme. Directorate General for Maritime Affairs and Fisheries &

Directorate General for Research. European Commission, Brussels. http://cordis.europa.eu/food/docs/2008-0699_en.pdf
10. EC (2010) Ex-post evaluation of the Financial Instrument for Fisheries Guidance (FIFG) 2000–2006. Executive summary. European Commission, Brussels. http://ec.europa.eu/fisheries/documentation/studies/fifg_evaluation/fifg_evaluation_executive_summary_en.pdf
11. EC (2013) Strategic Guidelines for the sustainable development of EU aquaculture. Communication from the Commission to The European Parliament, The Council, The European Economic and Social Committee and the Committee of the Regions. COM (2013) 299 final. http://ec.europa.eu/fisheries/cfp/aquaculture/official_documents/com_2013_229_en.pdf
12. Edwards P (1993) Environmental issues in integrated agriculture-aquaculture and wastewater fed systems. In: Pullin RSV, Rosenthal H, Maclean JL (eds) Environment and aquaculture in developing countries. Worldfish, pp 139–170
13. Ernst and Young (2011) Interim Evaluation of the European Fisheries Fund (2007–2013) Final report. http://ec.europa.eu/fisheries/documentation/studies/eff_interim_evaluation_en.pdf
14. EUMOFA (2013) Case study: price structure in the supply chain for fresh cod in the United Kingdom. European Market Observatory for fisheries and aquaculture products. European Commission. http://www.eumofa.eu/documents/20178/0/Price+structure+Cod+in+the+United+Kingdom/707cbc82-e17c-4331-b9a1-463fc5e77bb1
15. EUMOFA (2014) The EU fish market, 2014th edn. European Market Observatory for fisheries and aquaculture products, European Commission
16. European Court of Auditors (2014) The effectiveness of European Fisheries Fund support for aquaculture. Special report. Publications Office of the European Union, Luxembourg. doi:10.2865/24917. http://www.eca.europa.eu/Lists/ ECADocuments/SR14_10/QJAB14010ENC.pdf
17. Eurostat (2011) Fishery statistics. European Commission. http://ec.europa.eu/eurostat/statistics-explained/index.php/Fishery_statistics
18. FAO (2014a) FishStat—Software for fishery statistical time series. United Nations Food and Agriculture Organisation, Rome. http://www.fao.org/fishery/statistics/software/fishstatj/en
19. FAO (2014b) The state of world fisheries and aquaculture—opportunities and challenges. Food and Agriculture Organization of the United Nations. UN FAO, Rome. http://www.fao.org/3/a-i3720e.pdf
20. FEAP (2014) European aquaculture production report 2001–2013. Federation of European Aquaculture Producers. http://www.feap.info/shortcut.asp?FILE=1274
21. Fisheries and Oceans Canada (2010) Socio-economic impact of aquaculture in Canada. Report prepared by GardnerPinfold Consulting Economists Ltd. Fisheries and Oceans Canada, Ottawa, DFO/2010-1645. http://www.dfo-mpo.gc.ca/aquaculture /ref/aqua-es2009-eng.pdf
22. Gudmundsson E, Asche F, Nielsen M (2006) Revenue distribution through the seafood value chain. FAO Fisheries Circular No. 1019, FIIU/C1019 (En). Food and Agriculture Organization of the United Nations, Rome. http://www.fao.org/docrep/009/a0564e/a0564e00.htm
23. Guillen J, Natale F, Fernández Polanco JM (2015) Estimating the economic performance of the EU aquaculture sector. Aquacult Int 23(6):1–14. doi:10.1007/s10499-015-9891-x
24. IMANI/SRSL (2014) An assessment of the benefits of aquaculture to Scotland. Marine Scotland, Government of Scotland. http://www.gov.scot/Resource/0045/00450799.pdf
25. Impact Fish (2005) Impact assessment of the FP4 and FP5 research programmes on fisheries, aquaculture and seafood processing research area and the fisheries industry. EC FP6 Project (FP6-2003-SSP3-513651) involving the European Aquaculture Society and Oceanic

Development. http://www.easonline.org/component/content/article/1/7-impact-of-eu-research
26. Jensen F, Nielsen M, Nielsen R (2014) Increased competition for aquaculture from fisheries: Does improved fisheries management limit aquaculture growth? Fish Res 159:25–33. doi:10.1016/j.fishres.2014.05.004
27. Klinger D, Naylor R (2012) Searching for solutions in aquaculture: charting a sustainable course. Annu Rev Environ Resour 37:247–276. doi:10.1146/annurev-environ-021111-161531
28. Lane A, Hough C, Bostock J (2014) The long-term economic and ecologic impact of larger sustainable aquaculture. Study for the European Parliament, Directorate General for Internal Policies, Policy Department B, Structural and Cohesion Policies—Fisheries. http://www.europarl.europa.eu/RegData/etudes/STUD/2014/529084/IPOL_STU(2014)529084_EN.pdf
29. Lazard J, Lemcomte Y, Stomal B, Weigel JY (1991) Pisciculture en Afrique Subsaharienne: Situations et Projects dans les Pays Francophones. Propositions d'Action, Ministere de Cooperation et de Developpement, Paris
30. Marine Scotland Science (2014) Scottish Fish Farm Production Survey 2013. The Scottish Government. http://www.gov.scot/Publications/2014/10/7776/0
31. Martenot C, Oden E, Travaillé E, Malas JP, Houssin M (2011) Detection of different variants of Ostreid Herpesvirus 1 in the Pacific oyster, Crassostrea gigas between 2008 and 2010. Virus Res 160(1-2): 25–31. doi:10.1016/j.virusres.2011.04.012
32. Martins CIM, Eding EH, Verdegem MCJ, Heinsbroek LTN, Schneider O, Blancheton JP, Roque d'Orbcastel E, Verreth JAJ (2010) New developments in recirculating aquaculture systems in Europe: a perspective on environmental sustainability. Aquac Eng 43(3):83–93. doi:10.1016/j.aquaeng.2010.09.002
33. Natale F, Hofherr J, Fiore G, Virtanen J (2013) Interactions between aquaculture and fisheries. Mar Policy 38:205–213. doi:10.1016/j.marpol.2012.05.037
34. Norman-Lopez A (2009) Competition between different farmed and wild species: the US tilapia market. Mar Resour Econ 24(3):237–251. doi:10.5950/0738-1360-24.3.237
35. Norwegian Directorate of Fisheries (2015) Aquaculture Statistics. http://www.fiskeridir.no/English/Aquaculture/Statistics
36. Olsen Y, Otterstad O, Duarte CM (2008) Status and future perspectives of marine aquaculture, Chap 10. In: Holmer M, Black K, Duarte CM, Marbà N, Karakassis I (eds) Aquaculture in the ecosystem. Springer, Netherlands, pp 293–319. doi:10.1007/978-1-4020-6810-2
37. Petton B, Bruto M, James A, Labreuche Y, Alunno-Bruschia M, Le Roux F (2015) Crassostrea gigas mortality in France: the usual suspect, a herpes virus, may not be the killer in this polymicrobial opportunistic disease. Frontiers Microbiol 6:686. doi:10.3389/fmicb.2015.00686
38. Regnier E and Bayramoglu B (2014) Competition between farmed and wild fish: the French sea bass and sea bream markets. In: Presented at 16th annual BIOECON conference on biodiversity, ecosystem services and sustainability. 21–23 September 2014, Kings College, Cambridge, UK. http://www.bioecon-network.org/pages/16th_2014/Bayramoglu.pdf
39. Renault T (2011) A review of mortality outbreaks in the Pacific oyster, *Crassostrea gigas*, reported since 2008 in various European Union Member States and the related implementation of Council Directive 2006/88/EC. Bulletin de l'OIE 2011 No 3: 51–52. http://www.oie.int/doc/ged/D11038.PDF
40. Scottish Government (2015) Input-Output Tables 1998–2012—Latest Year (2012). http://www.gov.scot/Topics/Statistics/Browse/Economy/Input-Output/Downloads/IO1998-2012Latest
41. STECF (2013) The economic performance of the EU aquaculture sector (STECF 13-29). JRC Scientific and Policy Reports. Scientific, Technical and Economic Committee for Fisheries (STECF). Publication Office of the European Union, Luxembourg. EUR 26366 EN, JRC 86671. doi:10.2788/41898

42. STECF (2013a) The 2013 Annual Economic Report on the EU Fishing Fleet (STECF-13-15). 2013. European Commission, Joint Research Centre, Scientific, Technical and Economic Committee for Fisheries (STECF). Publications Office of the European Union, Luxembourg, EUR 26158 EN, JRC 84745. doi:10.2788/23331
43. STECF (2014) The economic performance of the EU aquaculture sector (STECF 14-18). JRC Scientific and Policy Reports. Scientific, Technical and Economic Committee for Fisheries (STECF). Publication Office of the European Union, Luxembourg. EUR 27033 EN, JRC 93169. doi:10.2788/15501
44. Szücs I, Stündl L, Váradi L (2008) Carp farming in Central and Eastern Europe and a case study in multifunctional aquaculture, Chap 26. In: Leung PS, Lee CS, O'Bryen PJ (eds) Species and system selection for sustainable aquaculture. Wiley, New York, pp 389–413. doi:10.1002/9780470277867.ch26
45. Verdegem MCJ, Bosma RH, Verreth JAJ (2006) Reducing water use for animal production through aquaculture. Int J Water Resour Dev 22(1):101–113. doi:10.1080/07900620500405544
46. Vezzulli L, Pezzati E, Stauder M, Stagnaro L, Venier P, Pruzzo C (2015) Aquatic ecology of the oyster pathogens *Vibrio splendidus* and *Vibrio aestuarianus*. Environ Microbiol 17(4):1065–1080. doi:10.1111/1462-2920.12484/pdf

CHAPTER 8

Effect of Production Quotas on Economic and Environmental Values of Growth Rate and Feed Efficiency in Sea Cage Fish Farming

M. Besson[1,2*], I. J. M. de Boer[3], M. Vandeputte[2,4], J. A. M. van Arendonk[1,] E. Quillet[2], H. Komen[1], J. Aubin[5]

[1] *Animal Breeding and Genomics Centre, Wageningen University, Wageningen, the Netherlands,*

[2] *GeÂneÂtique animale et biologie inteÂgrative, INRA, AgroParisTech, UniversiteÂ Paris-Saclay, Jouy-en-Josas, France,*

[3] *Animal Production Systems group, Wageningen University, Wageningen, the Netherlands,*

[4] *IFREMER, Chemin de Maguelone, Palavas-les-Flots, France,* **5** *INRA, Agrocampus Ouest, UMR1069 Sol*

ABSTRACT

In sea cage fish farming, production quotas aim to constrain the impact of fish farming on the surrounding ecosystem. It is unknown how these quotas affect economic profitability and environmental impact of genetic improvement. We combined bioeconomic modelling with life cycle assessment (LCA) to calculate the economic (EV) and environmental (ENV) values of thermal growth coefficient (TGC) and feed conversion ratio (FCR) of sea bass reared in sea cages, given four types of quota commonly used in

Europe: annual production (Qprod), annual feed distributed (Qannual_feed), standing stock (Qstock), and daily feed distributed (Qdaily_feed). ENV were calculated for LCA impact categories climate change, eutrophication and acidification. ENV were expressed per ton of fish produced per year (ENV(fish)) and per farm per year (ENV(farm)). Results show that irrespective of quota used, EV of FCR as well as ENV(fish) and ENV(farm) were always positive, meaning that improving FCR increased profit and decreased environmental impacts. However, the EV and the ENV(fish) of TGC were positive only when quota was Qstock or Qdaily_feed. Moreover, the ENV(farm) of TGC was negative in Qstock and Qdaily_feed quotas, meaning that improving TGC increased the environmental impact of the farm. We conclude that Qstock quota and Qdaily_feed quota are economically favorable to a genetic improvement of TGC, a major trait for farmers. However, improving TGC increases the environmental impact of the farm. Improving FCR represents a good opportunity to balance out this increase but more information on its genetic background is needed to develop breeding programs improving FCR.

INTRODUCTION

The production of fish in sea cages releases, without being filtered, nutrients to the surrounding environment. The accumulation of organic matter stemming from the cages on the benthos may cause eutrophication, which affects the natural ecosystem [1, 2]. In all European countries, therefore, producing fish in sea cages has to comply with regulatory measures to limit the environmental impact. These measures require an environmental impact study of biotic and abiotic change3s due to the farming process [3, 4]. The aim of this environment impact study is to explore how much fish can be produced based on the carrying capacity of the natural ecosystem. The estimations can be supported by modeling tools. In Norway, for example, the Modelling—On growing fish farms—Monitoring (MOM) [5–7] is legally required by the Directorate of Fisheries for site selection [8]. The environmental impact study determines the delivery of the farming authorization accompanied by prescriptions that set a specific quota to the farm. The nature of the quota varies across European countries. In France, Greece or Spain, quotas constrain the annual production of fish or the annual feed distributed per farm [9, 10], whereas in Denmark, the quota is based on annual emission of nitrogen [4]. In Ireland, the production is constrained by the density of fish [4]. In Norway, the production of salmon is limited by the standing biomass per license and per site [11]. The main goal of these quotas is to limit the environmental impact of the farm to an acceptable level.

Nevertheless, fish farming is growing due to an increasing demand for fish products; the challenge, therefore, is to reach the demand while constraining or reducing environmental impacts. So far, the reduction of environmental impacts have been achieved through improving farming technologies and practices. For instance, Ytrestøyl et al. [12] showed that salmon is now a net contributor of marine ingredients due to a lower inclusion of fish meal and dish oil in the feed. Genetic improvement could also be a tool to decrease environmental impacts as it acts at the source of emission by generating cumulative changes in animal performances [13, 14]. The economic (EV) and environmental value (ENV) of genetic improvement of traits included in the breeding goal can be estimated using bio-economic models. These values represent the economic or environmental impacts of a change in one trait keeping the other traits constant [15]. Environmental values were first calculated in dairy systems using the principles of life cycle assessment (LCA) [14, 16, 17]. In fish farming, we combined a bioeconomic model with an LCA to estimate the economic and environmental values of thermal growth coefficient (TGC) and feed conversion ratio (FCR) of African catfish (*Clarias gariepinus*) produced in a recirculating aquaculture system, RAS [18]. The results showed that in dairy as well as in fish farming, genetic change of a trait can simultaneously reduce environmental impacts per unit of products and increase economic farm profit by improving production efficiency or production level. An improved production efficiency decreases the amount of resources needed per unit of product. A higher production level dilutes fixed environmental impacts over more production. However, as we showed in previous studies [18, 19], the economic and environmental values depended on the factors limiting production, which in the case of RAS are fish rearing density and nitrogen treatment capacity. In sea cages, the limiting factor is, most of the time, the production quota. The variety of quotas applied in sea cage farming suggests that genetic improvement might lead to a variety of economic and environmental responses, depending on which quota is being applied.

This study aims to investigate how different types of quota affect the economic and environmental impacts of genetic improvement in sea cage system. Using our bioeconomic / LCA model, we calculated the economic and environmental values of TGC and FCR in a sea cage system producing sea bass. First, we express ENV as a change of environmental impacts per ton of fish produced. This functional unit emphases the change in environmental efficiency of producing fish after genetic improvement. However, in the context of quotas such a functional unit cannot describe the dynamics of environmental impacts at farm level. Therefore, we also calculate ENV at farm level.

MATERIAL AND METHODS

Bioeconomic model

1. The bioeconomic model developed in R [20] was based on the model described in Besson et al. [19] and Besson et al. [21]. In the present study the model was adapted to estimate the production of sea bass (*Dicentrarchus labrax*) in a hypothetical sea cage farm (based on a real farms data from Gloria Maris) constrained by quota. The farm was composed of 34 circular cages of 600 m^3 for pre-growing and 34 circular cages of 1,800 m^3 for on-growing. Fish were stocked at 10 g and sold at a fixed harvest weight of 400 g. Stocking occurred all year round. The bioeconomic model was divided in four model parts.

2. The fish model estimates individual fish growth using the thermal growth coefficient (TGC) corrected for the concave relationship between growth rate and temperature [22]. The time to reach harvest weight, therefore, varied according to the daily temperature encountered by the fish and thus according to the stocking date. Feed conversion ratio (FCR) was modelled by combining a specific seabass model from Person-Le Ruyet et al. [23] with a model from [24]. The fish model also estimates the individual emission of nutrient based pollutants using mass-balance [25, 26]. Further details about the fish model are given in S1 Table.

3. The batch model estimates the average stocking density of a batch depending on individual fish performances (from fish model) and mortality. A batch is defined as the group of fish stocked at the same time in the same pre-growing cage.

4. The farm model estimates the number of batches produced to calculate annual fish production, emission of pollutants, and annual feed consumption. At farm level, a quota is applied which constrain farm production. In the reference scenario the production of the farm was set to 1,000 tons per year and four different quotas were tested:

   - -. Production quota (Qprod). The production of the farm was limited to 1,000 tons per year.
   - -. Feed quota (Qannual_feed). The total amount of feed distributed per year per farm was limited to 2,050 tons.

- -. Standing stock (Qstock). The instant biomass present on site at any day of the year was constrained to 435 tons.
- -. Daily feed distribution (Qdaily_feed). The amount of feed distributed per day per farm was limited to 4 tons.

The values of each quota was set to allow the farm to produce 1,000 tons in the reference scenario. In this study, we considered that O_2 availability was never limiting. In every quota scenario, the density of stocking was considered fixed along the year. The outputs of the bioeconomic model were used to generate inventory data for the LCA.

1. Finally, in the economic model, annual profit is calculated by combining results of the farm model with economic parameters.

Life cycle assessment

Goal and scope. LCA is a standardized method conceived to calculate the environmental impact of a production chain, from raw material extraction up to the product's end-of life [27]. In this study, we applied LCA according to the main specifications of ILCD standards.

The system was defined from cradle-to-farm-gate and included five distinct sub-systems: (1) production of purchased feed, including production of ingredients, processing, and transportation; (2) production of energy expended at farm level (electricity, gas and petrol); (3) production of farming facilities and equipment; (4) Chemical used, including the production and the use of anti-fouling for nets; (5) farming operation, including nutrient based pollutants emission from biological transformation of feed.

The functional units in which environmental impacts were expressed was (a) per ton of fish produced on a basis of one year of routine production (impact_fish) and (b) per farm on a basis of one year of routine production (impact_farm). impact_fish and impact_farm were used to calculate environmental values.

Life cycle inventory.

1. Production of purchased feed—Crop-derived ingredients originated from Brazil and France (e.g. soybean meal from Brazil and wheat from France), whereas fish-derived ingredients originated from the Peruvian fish milling industry (Biomar, pers com, 2014). The chemical composition of the diet and the origin of the ingredients are given in S2 Table. The exact composition is not

given to respect confidentiality. Economic allocation was used to calculate the environmental impacts of processes yielding multiple products in the feed production industry [28]. Ingredients were transported to the feed manufacture in France (Aquitaine) by transoceanic ship and by lorry (>32t), whereas the transport of feed from feed mill to the fish farm in southern France was by lorry (>32t). Transport distances and other data required to compute the environmental impact of feed ingredients were based on Boissy et al. [29] and Pelletier and Tyedmers [30], presented in detail in S2 and S3 Tables.

2. Production of energy expended on farm—The energy consumed by the farm was considered fixed. The energy consumption was set at 25,000 l of diesel per year, 55,000 l of petrol per year and 110,000 kWh of electricity per year. The electricity used by the farm was coming from the French energy mix proposed by EcoInvent v3 database. Contribution analysis is available in S4 Table.

3. Production of farming facilities and equipment used—We considered the construction of a building of 650 m^2 with a life span of 30 years. The equipment includes cages (64 in total), vehicles (2 boats and 6 barges, 2 trucks, 1 car), two ice making machines and other small equipment (i.e. plastic buckets). The use of building and equipment was considering fixed per year at farm level. The background processes stem from EcoInvent v3. Contribution analysis is available in S5 Table.

4. Chemical used—This sub-system includes the emission from the production and the use of anti-fouling for nets. We considered that the nets were treated every nine months with water-based anti-fouling with copper dioxide at 24%.

5. Farm operation—The farm operation sub-system includes the emission of pollutants from biological transformation of the feed distributed to the fish. The amount of nitrogen (N), phosphorus (P) and chemical oxygen demand (COD) of the dissolved organic matter excreted by the fish directly into the sea were calculated through the bioeconomic model. Contribution analysis is available in S6 Table.

Life cycle impact assessment. Each flow observed in the system was assigned to different impact categories relatively to its potential environmental effects. We chose to investigate eutrophication, acidification and climate change [27], because fish farming contributes most to these environmental impact categories [31, 32]. The characterization factors from CML2 Baseline 2000 version 2.04 were used for eutrophication and climate change. The impact categories were calculated using Simapro 7.0 software.

Economic and environmental values

Economic (EV) and environmental values (ENV) represent the change in profit and environmental impacts due to genetic improvement in a trait, while keeping other traits in the breeding goal constant. We calculated EV and ENV for two important traits representing production performance of a farm: TGC and the FCR. EV and ENV were calculated for each quota scenario. The genetic improvement implemented was one genetic standard deviation (σg) from the mean (μ). μTGC = 2.25, σg-TGC = 0.23 [33] and μFCR = 2.0, σg-FCR = 0.38. The genetic standard deviation of FCR was calculated according to Sutherland [34] using data from Kause et al. [35]. EVs were calculated as the difference between the annual profits of the farm after genetic change minus before genetic change. We calculated EVs at farm scale because the aim of selecting breeding is to increase farmer's income. Then, the EVs were rescaled per ton of fish for better visualization. This rescaling is done before genetic gain to account for the economic gain of an increase of production after genetic change [36]. The EV was expressed in euro per ton of fish produced:

$$EV = \frac{\text{Profit}_{\mu+\sigma_g} - \text{Profit}_{\mu}}{\text{Production}_{\mu}} \quad (1)$$

Environmental values were first calculated per ton of fish produced; ENV(fish). ENV(fish) were calculated as environmental impacts (category level) per ton of fish produced before genetic improvement minus environmental impacts per ton of fish produced after genetic improvement:

$$\text{ENV}(\text{fish}) = \text{Impact\_fish}_{\mu+\sigma_A} - \text{Impact\_fish}_{\mu} \quad (2)$$

ENV(fish) express, therefore, the capacity of a trait to improve environmental efficiency of fish production. However, the aim of the quota is mainly to limit the impact of the farming process on the local environment. Hence, we also calculated environmental values at farm level; ENV(farm). ENV(farm) estimate the capacity of a trait to affect environmental impacts of a specific farming site. They were calculated similarly as economic values i.e. difference between the impacts at farm level after genetic change minus the impacts at farm level before genetic change divided by annual production before genetic change. It means that the ENV(farm) were rescaled to impacts per farm per ton of fish to be able to compare with ENV(fish).

$$\text{ENV(farm)} = \frac{\text{Impact\_farm}_{\mu+\sigma_A} - \text{Impact\_farm}_{\mu}}{\text{Production}_{\mu}} \tag{3}$$

Each trait has an ENV(fish) and an ENV(farm) for each of the three impact categories investigated. ENV(fish) and ENV(farm) were also expressed in percentage of change.

$$\text{ENV(fish\%)} = \frac{\text{ENV(fish)} \times 100}{\text{Impact\_fish}_{\mu}} \tag{4}$$

$$\text{ENV(farm\%)} = \frac{\text{ENV(farm)} \times 100}{\text{Impact\_farm}_{\mu}} \tag{5}$$

RESULTS

Genetic improvement can affect the production level (i.e. tonnes of fish produced per year) and/or the production efficiency of the farm (i.e. quantity of input used per unit of fish produced) (Table 1). These changes affect the economic and environmental values of thermal growth coefficient (TGC) and feed conversion ratio (FCR).

Table 1. Summary of the consequences of genetic improvement in TGC (thermal growth coefficient) and FCR (feed conversion ratio) on technical performances of a sea cage farm constrained by different quota.

Quota type	Improving TGC	Improving FCR
Qprod	none	Production efficiency
Qannual_feed	none	Production + Production efficiency
Qstock	Production	Production efficiency
Qdaily_feed	Production	Production + production efficiency

Qprod is the quota on annual production, Qannual_feed is on annual feed distributed, Qstock is on the daily biomass present on site and Qdaily_feed is on daily feed distributed.

Economic values, EV (Tables 1 and 2)

Table 2. Effect of different values of Thermal Growth Coefficient (TGC) and Feed Conversion Ratio (FCR) on fish production parameters in different quota scenarios.

Quota	TGC	FCR	Days to reach harvest weight (d)	Number of batch produced (#)	Production per batch (t)	Production at farm (t)	Feed consumption per farm (t)	Incomes (€ × 1000)	Feed cost (€ × 1000)	Juveniles cost (€ × 1000)	Fixed cost (€ × 1000)	Profit (€)	EV TGC (€/kg)	EV FCR (€/kg)
Qprod	2.25	2.02	573	30.87	32.4	1000	2047	5598	2662	692	2245	-0.03	0	0.5
	2.48	2.02	528	35.64	28.1	1000	2047	5598	2662	692	2245	-521.88		
	2.25	1.64	573	30.87	32.4	1000	1660	5598	2159	692	2245	503048.8		
Qannual_feed	2.25	2.02	573	31.22	32.0	1000	2047	5597	2661	692	2245	0.56	0	1.14
	2.48	2.02	528	36.43	27.4	1000	2047	5597	2661	692	2245	-521.19		
	2.25	1.64	573	28.97	42.5	1232	2047	6900	2661	853	2245	1141326		
Qstock	2.25	2.02	573	32.02	31.2	1000	2046	5596	2659	692	2245	0.46	0.12	0.5
	2.48	2.02	528	34.49	30.6	1055	2160	5904	2809	730	2245	120354.6		
	2.25	1.64	573	32.02	31.2	1000	1657	5596	2155	692	2245	504751.8		
Qdaily_feed	2.25	2.02	573	31.12	32.1	1000	2046	5597	2660	692	2245	0.59	0.08	0.95
	2.48	2.02	528	35.06	29.5	1035	2118	5792	2753	716	2245	77713.2		
	2.25	1.64	573	29.34	39.6	1162	1929	6506	2508	804	2245	949212.7		

Qprod is the quota on annual production, Qannual_feed is on annual feed distributed, Qstock is on the daily biomass present on site and Qdaily_feed is on daily feed distributed.

Economic values, EV (Tables 1 and 2)

EV of TGC.

Improving TGC decreased the time to reach harvest size and increased daily feed intake. In consequence, production should increase as more batches can be produced (increasing batch rotation). When the quota was on annual production (Qprod), higher batch rotation was balanced by lower stocking density to comply with the quota. This situation was similar with a quota on annual feed distributed (Qannual_feed). Thus, the economic value of TGC was null in Qprod and Qannual_feed. Conversely, when the quota was on daily standing stock (Qstock) or daily feed distributed (Qdaily_feed), increasing TGC led to more batches, but without a proportional decrease in stocking density. The resulting annual production was higher. Therefore, the economic value of TGC was positive: 0.12 €/kg of fish for Qstock and 0.08 €/kg of fish for Qdaily_feed.

EV of FCR.

Improving FCR decreased the amount of feed required per unit of fish produced. When the quota was on feed distributed (Qannual_feed or Qdaily_feed) better FCR increased, therefore, production efficiency but production could also be increased until the feed quota was reached. Consequently, the economic value of FCR was 1.14 €/kg for Qannual_feed and 0.95 €/kg for Qdaily_feed. When Qprod or Qstock were the quotas, improving FCR improved only production efficiency. Thus, less feed was consumed at farm level and the economic value of FCR was 0.50 €/kg.

Environmental value at fish level, ENV(fish)

More details about the results of the life cycle assessment are given in S7, S8 and S9 Tables.

ENV(fish) of TGC (Tables 3 and 4).

Table 3. Effect of different values of Thermal Growth Coefficient (TGC) and Feed Conversion Ratio (FCR) on annual emission of pollutants and environmental impacts in different quota scenarios.

Quota	TGC	FCR	Production per farm (t)	Nitrogen emission (t)	COD emission (t)	Phosphorus emission (t)	Climate change (kg CO_2-eq / ton of fish)	Eutrophication (kg PO_4-eq / ton of fish)	Acidification (kg SO_2-eq / ton of fish)
Qprod	2.25	2.02	1000	114.92	2614.15	16.44	3636.53	168.62	21.77
	2.48	2.02	1000	114.98	2616.06	16.45	3636.53	168.73	21.77
	2.25	1.64	1000	88.29	1959.10	12.18	2995.38	127.90	18.33
Qannual_feed	2.25	2.02	1000	114.87	2613.11	16.43	3636.02	168.59	21.76
	2.48	2.02	1000	114.93	2615.01	16.45	3636.02	168.69	21.76
	2.25	1.64	1232	108.84	2414.94	15.02	2949.75	127.66	17.66
Qstock	2.25	2.02	1000	114.79	2611.13	16.42	3635.28	168.51	21.76
	2.48	2.02	1055	121.20	2757.49	17.34	3622.52	168.57	21.57
	2.25	1.64	1000	88.11	1954.71	12.16	2991.65	127.68	18.31
Qdaily_feed	2.25	2.02	1000	114.86	2612.80	16.43	3635.76	168.58	21.76
	2.48	2.02	1035	118.94	2706.36	17.02	3627.54	168.67	21.64
	2.25	1.64	1162	102.59	2276.24	14.16	2960.63	127.67	17.83

Qprod is the quota on annual production, Qannual_feed is on annual feed distributed, Qstock is on the daily biomass present on site and Qdaily_feed is on daily feed distributed.

Table 4. Environmental Value (ENV) of TGC (thermal growth coefficient) at fish and farm level in different quota.

Quota	TGC	FCR	ENV at farm level			ENV at fish level		
			Climate Change (kg CO_2-eq)	Eutrophication (kg PO_4-eq)	Acidification (kg SO_2-eq)	Climate Change (kg CO_2-eq)	Eutrophication (kg PO_4.eq)	Acidification (kg SO_2-eq)
Qprod	2.25	2.02	0	-0.11	0	0	-0.11	0
	2.48	2.03	(0%)	(-0.06%)	(0%)	(0%)		(0%)
Qannual_feed	2.25	2.02	0	-0.11	0	0	-0.11	0
	2.48	2.03	(0%)	(-0.06%)	(0%)	(0%)	(-0.06%)	(0%)
Qstock	2.25	2.02	-186.84	-9.35	-1	12.76	-0.06	0.19
	2.48	2.03	(-5.14%)	(-5.55%)	(-4.61%)	(0.35%)	(-0.04%)	(0.86%)
Qdaily_feed	2.25	2.02	-118.74	-5.97	-0.63	8.22	-0.09	0.12
	2.48	2.03	(-3.25%)	(3.54%)	(-2.91%)	(0.23%)	(-0.05%)	(0.55%)

Qprod is the quota on annual production, Qannual_feed is on annual feed distributed, Qstock is on the daily biomass present on site and Qdaily_feed is on daily feed distributed. Between brackets is the percentage a change in environmental impacts. A negative sign means that the environmental impact considered increased after genetic change.

ENV(fish) of TGC were null for all impact categories when the quota was Qprod or Qannual_feed because in this situation TGC did not increase production or production efficiency. However, when the quota was Qstock or Qdaily_feed, improving TGC increased production. When production increased, the fixed environmental impacts are diluted over more fish produced. The fixed environmental impacts are the production of energy, the use of chemical and the production of equipment and infrastructure. These impacts represent 16.4% of acidification and 6.7% of climate change. Consequently, the environmental impact per ton of fish produced decreased and the ENV(fish) of TGC were positive for acidification and climate change categories. However, fixed environmental impacts represent less than 1% of eutrophication. The 99% remaining are caused by feed production and fish excretion which increased with higher production in Qstock and Qdaily_feed. Therefore, ENV(fish) of TGC for eutrophication is close to zero in Qstock and Qdaily_feed.

ENV(fish) of FCR (Tables 3 and 5).

Table 5. Environmental Value (ENV) of FCR (feed conversion ratio) at fish and farm level in different quota.

Quota	TGC	FCR	ENV at farm level			ENV at fish level		
			Climate Change (kg CO_2-eq)	Eutrophication (kg PO_4-eq)	Acidification (kg SO_2-eq)	Climate Change (kg CO_2-eq)	Eutrophication (kg PO_4-eq)	Acidification (kg SO_2-eq)
Qprod	2.25	2.02	641.15	40.72	3.44	641.15	40.72	3.44
	2.48	2.03	(17.63%)	(24.15%)	(15.80%)	(17.63%)	(24.15%)	(15.80%)
Qannual_feed	2.25	2.02	0	11.23	0	686.27	40.93	4.11
	2.48	2.03	(0%)	(6.66%)	(0%)	(18.87%)	(24.28%)	(18.87%)
Qstock	2.25	2.02	643.63	40.83	3.45	643.63	40.83	3.45
	2.48	2.03	(17.71%)	(24.23%)	(15.86%)	(17.71%)	(24.23%)	(15.86%)
Qdaily_feed	2.25	2.02	194.12	20.16	1.04	675.13	40.91	3.94
	2.48	2.03	(5.34%)	(11.96%)	(4.79%)	(18.57%)	(24.27%)	(18.09%)

Qprod is the quota on annual production, Qannual_feed is on annual feed distributed, Qstock is on the daily biomass present on site and Qdaily_feed is on daily feed distributed. Between brackets is the percentage a change in environmental impacts. A positive value means that the environmental impact considered decreased after genetic change.

The ENV(fish) of FCR in all quotas is positive, i.e. improving FCR decreases environmental impacts, because improving FCR decreased the amount of inputs need to produce one tonne of fish. The decrease was higher for eutrophication because eutrophication also includes the reduction in pollutants emission from the fish.

Environmental values at farm level, ENV(farm)

ENV(fish) of TGC (Tables 3 and 4).

ENV(farm) of TGC was null or close to null for all impact categories when quota was Qprod or Qannual_feed because TGC did not increase production or production efficiency. When the quota was Qstock or Qdaily_feed, production increased as well as feed consumption and emission of pollutants, which are the main contributors to environmental impacts.

Consequently, the ENV(farm) of TGC was negative for all impact categories, meaning that increasing TGC increased environmental impacts at farm level. The increase of environmental impacts was higher in Qstock quota than in Qdaily_feed.

ENV(fish) of FCR (Tables 3 and 5).

When the quota was Qprod or Qstock, improving FCR improved production efficiency. Therefore, less feed was consumed at farm level and the ENV(farm) of FCR were positive for all impacts categories, meaning that environmental impacts decreased. When the quota was Qannual_feed, improving FCR increased production efficiency and production, which kept annual consumption of feed constant. Consequently the ENV(farm) of FCR for climate change and acidification was null. However, with better FCR, less nutrient based pollutants were emitted. It resulted in a decrease in eutrophication by 6.66% because the nutrients-based pollutants emitted to water were considered to remain in water and thus they were contributing to eutrophication only. Thus, for Qannual_feed, the ENV(farm) of FCR for eutrophication was positive. When the quota was Qdaily_feed, improving FCR increased at the same time production efficiency and production but less feed was consumed at farm level. It resulted in a positive ENV(farm) for FCR. The ENV(fam) of FCR was also higher for eutrophication than other impact categories because improving FCR reduces the amount of pollutants emitted.

DISCUSSION

In sea cage farming, quotas are implemented to limit the environmental impacts of farm operations, such as the deposition of organic matter under the cages and the emission of dissolved nutrients in water, leading to eutrophication. In the present study, we investigated the effect of these quotas on economic and environmental responses to genetic improvement of sea bass. This is the first time the effect of quotas on genetic improvement of fish is tested. Economic values are used to maximize the economic response in the breeding goal. These values weigh the traits of a breeding goal according to their impacts on farm profit. They are specific to production systems according to the quota applied [37] or the environmental conditions [38]. When EVs differ from one production system to another, a single breeding program may not be enough to maximize economic response in all production systems [38]. It is important to remember that EVs and ENVs represent the economic and environmental change due to an improvement in one trait while keeping the other traits of the breeding goal constant. Therefore, EVs and ENVs are informative about the economic and environmental importance of a particular trait, but they do not fully predict the response to selection of a breeding program. This could be particularly the case if, in our case, FCR and TGC were genetically correlated. However, this is poorly documented in fish at the present time, and while some studies show an improvement of FCR when TGC is genetically improved [39] others show none [40, 41].

In sea cage farming system, EVs of TGC and FCR varied across quota scenarios. When the quota was standing stock (Qstock) or daily feed distributed (Qdaily_feed), both TGC and FCR had a positive EV, but when the quota was annual production (Qprod) and annual feed (Qannual_feed), only FCR had a positive EV. Thus, in Qprod and Qannual_feed, economic gain would be achieved only if FCR could be improved, either by direct selection or by a genetically correlated response to improvement of TGC, meaning that increasing TGC would decrease FCR. However, there are no practical ways to directly select for FCR in fish as it is difficult to measure this trait in individual fish. Moreover, the existence and the magnitude of the genetic correlation between FCR and TGC is still debated [35, 39–41]. Conversely, Qstock and Qdaily_feed had positive values for EV of TGC. Consequently these are the quotas that will generate farm profit from selective breeding on TGC, which is easily achieved in fish breeding programs [42]. Additionally, Qstock and Qdaily_feed have a large positive EV for FCR, which could promote the inclusion of this trait in future breeding programs if efficient selection methods for this trait are developed. The different EVs observed across quotas imply, however, that different breeding programs would be needed to optimize economic response in each quota system.

The concerns about environmental impacts of aquaculture are increasing and, in the future, the objective of breeding programs might also shift towards decreasing environmental impacts instead of maximizing economic profit. To do so, environmental values could be used in breeding programs to orient them towards the reduction of environmental impacts. In this study, the environmental values were calculated per ton of fish produced (ENV(fish)) and per farm (ENV(farm)).

We calculated ENV(fish) because the aim of selective breeding could be to produce a kg of product with the lowest environmental impact. This type of production-related functional unit has been used for evaluating the environmental impacts of genetic improvement in dairy farming [15, 18] as well as in fish farming in RAS [19]. In this latter study, we found a positive ENV(fish) of TGC for climate change and eutrophication because faster growing fish could increase production, which diluted fixed environmental impacts (i.e. energy use). In RAS, fixed environmental impacts represent a high proportion of the total impacts, i.e. 42.4% of acidification and 26.9% of climate change in the above-mentioned study. This is because RAS is a highly technological production system requiring a lot of energy and equipment [43]. In a sea cage system, we also observed the same trend when Qstock and Qdaily_feed were the quotas. Nevertheless, ENV(fish) of TGC for acidification and climate change were small and even null for eutrophication because in sea cage system, there are less fixed environmental impacts to dilute with higher production. Fixed environmental impacts represented only 6.7% of climate change, 16.4% of acidification and 0.8% of eutrophication. Hence, the ENV(fish) are very sensitive to the type of system and to the proportion of fixed environmental impacts. Regarding FCR, the ENV(fish) were always positive for all impact categories meaning that environmental impacts decreases per ton of fish produced. According to these results, genetic improvement of TGC or FCR affects environmental impacts per ton of fish produced similarly in all quotas. Therefore, a single breeding program using ENV(fish) of TGC and FCR would minimize environmental impacts per ton of fish produced in all quota scenarios. In Europe, the production of sea bass was 156,449 t in 2014 and the production was expected to be about 178,000 t in 2017, which represent a growth of 13.77% [44]. If nothing is done, the subsequent increase in eutrophication would be about 13.5% as 98.2% of eutrophication is caused by variable environmental impacts (feed production and nutrient excretion). Thus, if we consider the eutrophication to be 168.57 kg of $PO_{4\text{-eq}}$ per t of fish produced, the eutrophication at European scale would increase from 26,373 to 29,933 t of $PO_{4\text{-eq}}$. Keeping eutrophication constant with the same increase in production would require a 11.9% improvement of FCR over 3 years (1 generation), which could be achieved by a 20% selection pressure (i = 1.4) on FCR if its heritability is 0.17 as suggested in Kause et al. [35].

In addition, the increase in production could also increase environmental impacts of farming site and the role of quota system is to prevent this increase. Thus, we chose to calculate ENV(farm) to estimate the capacity of a trait to affect environmental impacts of a specific farming site in compliance with the aim of quota system. At farm level, mitigation does not exist anymore and an increase in production increases feed distribution and hence nitrogen emission. Therefore, the ENV(farm) of TGC are negative in Qstock and Qdaily_feed. It means that, in Qstock and Qdaily_feed, the environmental response to selection on TGC can be interpreted differently whether we look at farm level or at fish level. This has been shown in dairy where genetic improvement in milk yield would increase emission of CO_2-eq. at herd level but reduce emission per kilogram of milk produced [17]. Regarding FCR, genetic improvement would always decrease environmental impacts at farm level in every quota scenario. However, the ENV(farm) of FCR in Qannual_feed and Qdaily_feed are lower than in Qprod and Qstock because improving FCR in Qannual_feed and Qdaily_feed increases production level together with production efficiency. The variation of ENV(farm) of TGC and FCR suggests that a single breeding program including ENV(farm) would not minimize environmental response in all quotas. It also suggests that breeding objectives using ENV(farm) and ENV(fish) would not lead to the same response. Therefore, the choice between ENV(farm) and ENV(fish) depends on the objectives of the breeding program.

In our model, the only limiting factor was one of the four production quota, which implied a constant stocking density through the time. This stocking density was calculated to reach the quota limitation without overtaking it. In the reference scenario (without genetic improvement) we considered that the oxygen availability was not limiting the number of fish stocked. However, in the scenario with genetic improvement, increasing TGC increases individual oxygen consumption due to higher feed intake and increasing FCR decreases individual oxygen consumption due to lower feed intake. Therefore, if you considered oxygen limitation, changing TGC and FCR could impact the number of fish stocked in cages to avoid hypoxia. In Besson et al. [21], we showed that the economic value of TGC, when oxygen was the limiting factor, was null when the average temperature was 18°C and increased with higher average temperatures. These results suggest that the ENV(fish) of TGC would also increase with higher temperature when oxygen is the limiting factor. However, ENV(farm) would decrease with higher temperature. Regarding FCR, both the EV and ENV(fish) would be positive because better FCR decreases feed intake and oxygen consumption which would increase production and increase production efficiency. However, ENV(farm) would also be positive be only due to better production efficiency.

The economic and environmental sustainability of fish farming is already a main target of fish breeders. Since the 70's and the development of the first breeding program for salmon, new traits have been included in breeding goals such as disease resistance or filet yield beside growth rate [42]. Nevertheless, the lack of knowledge on the potential economic and environmental impacts of these traits could generate economic shortfall for farmers and increase their environmental impacts. The advantages of using EV or ENV to weigh the traits in a breeding program is that they maximize the economic gain or the environmental gain of a breeding program. However, it is not possible to combine EV and ENV in a common breeding objective as they are not expressed with the same unit. EV is in euros while ENV is, for instance, in CO2-eq. Consequently, the economic breeding objective may have adverse effects on the environment. In that case, environmental values could be calculated differently. van Middelaar et al. [14] suggested to estimate the ENV of a trait using linear programming to minimize environmental response while keeping profit constant after genetic change.

CONCLUSION

This is the first study investigating the influence of different quotas on the economic and environmental value of production traits in a selective breeding program. The results show that the economic and environmental responses change across quota scenarios, which suggest that each quota might need a specific breeding program in order to maximize profit or to minimize environmental impacts. It also suggests that policy makers could choose the quota depending on the objectives to achieve. For instance, standing stock quota and daily feed quota are economically favorable to a genetic improvement of growth rate, a major trait for farmers. However, in these quotas, improving growth increases the environmental impact of the farm. Improving FCR represents a good opportunity to balance out this increase in environmental impacts but we need more information on its genetic background to develop breeding programs including FCR. FCR is the most important trait to increase economic profit and to decrease environmental impacts in all scenarios.

ACKNOWLEDGMENTS

M. Besson benefited from a joint grant from the European Commission and IMARES, within the framework of the Erasmus-Mundus joint doctorate "EGS-ABG". The authors thank Gloria Maris and Biomar for their collaboration.

REFERENCES

1. Folke C, Kautsky N, Troell M (1994) The Costs of Eutrophication from Salmon Farming: Implications for Policy. J. Environ. Manage. 40: 173–82.
2. Mente E, Pierce GJ, Santos MB, Neofitou C (2006) Effect of feed and feeding in the culture of salmonids on the marine aquatic environment: a synthesis for European aquaculture. Aquacult. Int. 14: 499–522.
3. Read P, Fernandes T (2003) Management of environmental impacts of marine aquaculture in Europe. Aquaculture 226: 139–63.
4. Holmer M, Hansen PK, Karakassis I, Borg JA, Schembri PJ. Monitoring of environmental impacts of marine aquaculture. In: Holmer M, Black K, Duarte CM, Marbà N, Karakassis I, editors. Aquaculture in the Ecosystem. Springer; 2008. pp. 47–85.
5. Ervik A, Hansen PK, Aure J, Stigebrandt A, Johannessen P, Jahnsen T (1997) Regulating the local environmental impact of intensive marine fish farming I. The concept of the MOM system (Modelling-Ongrowing fish farms-Monitoring). Aquaculture 158: 85–94.

6. Hansen PK, Ervik A, Schaanning M, Johannessen P, Aure J, Stigebrandt A (2001) Regulating the local environmental impact of intensive, marine fish farming: II. The monitoring programme of the MOM system (Modelling–Ongrowing fish farms–Monitoring). Aquaculture 194: 75–92.
7. Stigebrandt A, Aure J, Ervik A, Hansen PK (2004) Regulating the local environmental impact of intensive marine fish farming: III. A model for estimation of the holding capacity in the Modelling–Ongrowing fish farm–Monitoring system. Aquaculture 234: 239–61.
8. Lundebye A-K. Aquaculture site selection and carrying capacity for inland and coastal aquaculture in Northern Europe. In: Ross LG, Telfer TC, Falconer L, Soto D, Aguilar-Manjarrez J, editors. Site selection and carrying capacities for inland and coastal aquaculture. FAO; 2013. pp. 171–181.
9. FAO (2014) Vue generale du secteur aquacole national. http://www.fao.org/fishery/countrysector/naso_france/fr.
10. FAO (2014) National Aquaculture Legislation Overview. http://www.fao.org /fishery/legalframework/nalo_greece/en#tcNB008E.
11. Asche F, Bjorndal T. The economics of salmon aquaculture. John Wiley & Sons; 2011.
12. Ytrestøyl T, Aas TS, Åsgård T (2015) Utilisation of feed resources in production of Atlantic salmon (*Salmo salar*) in Norway. Aquaculture 448: 365–74.
13. Wall E, Simm G, Moran D (2010) Developing breeding schemes to assist mitigation of greenhouse gas emissions. Animal 4: 366–76. pmid:22443941
14. Van Middelaar CE, Berentsen PBM, Dijkstra J, van Arendonk JAM, de Boer IJM (2014) Methods to determine the relative value of genetic traits in dairy cows to reduce greenhouse gas emissions along the chain. J. Dairy Sci. 97: 5191–205. pmid:24881792
15. Hazel LN (1943) The genetic basis for constructing selection indexes. Genetics 28: 476–90. pmid:17247099
16. Bell MJ, Wall E, Russell G, Morgan C, Simm G (2010) Effect of breeding for milk yield, diet and management on enteric methane emissions from dairy cows. Anim. Prod. Sci. 50: 817–26.
17. Bell MJ, Wall E, Russell G, Simm G, Stott AW (2011) The effect of improving cow productivity, fertility, and longevity on the global warming potential of dairy systems. J. Dairy Sci. 94: 3662–78. pmid:21700056
18. Besson M, Aubin J, Komen H, Poelman M, Quillet E, Vandeputte M, et al. (2016) Environmental impacts of genetic improvement of growth rate and feed conversion ratio in fish farming under rearing density and nitrogen output limitations. J. Clean. Prod. 116: 100–9.
19. Besson M, Komen H, Aubin J, de Boer IJM, Poelman M, Quillet E, et al. (2014) Economic values of growth and feed efficiency for fish farming in recirculating aquaculture system with density and nitrogen output limitations: a case study with African catfish (*Clarias gariepinus*). J. Anim. Sci. 92: 5394–405. pmid:25414104
20. R Development Core Team (2008) R: A language and environment for statistical computing. R Foundation for Statistical Computing, Vienna, Austria.
21. Besson M, Vandeputte M, van Arendonk JAM, Aubin J, de Boer IJM, Quillet E, et al. (2016) Influence of water temperature on the economic value of growth rate in fish farming: The case of sea bass (*Dicentrarchus labrax*) cage farming in the Mediterranean. Aquaculture 462: 47–55.
22. Mallet JP, Charles S, Persat H, Auger P (1999) Growth modelling in accordance with daily water temperature in European grayling (*Thymallus thymallus* L.). Can. J. Fish. Aquat. Sci. 56: 994–1000.
23. Person-Le Ruyet J, Mahé K, Le Bayon N, Le Delliou H (2004) Effects of temperature on growth and metabolism in a Mediterranean population of European sea bass, *Dicentrarchus labrax*. Aquaculture 237: 269–80.
24. Lanari D, D Agaro E, Ballestrazzi R (2002) Growth parameters in European sea bass (*Dicentrarchus labrax* L.): effects of live weight and water temperature. Ital. J. Anim. Sci. 1: 181–6.

25. Cho CY, Kaushik SJ (1990) Nutritional energetics in fish: energy and protein utilization in rainbow trout (Salmo gairdneri). World Rev. Nutr. Diet. 61: 132–72. pmid:2408253
26. Cowey CB, Cho CY (1991) Nutritional strategies and aquaculturewaste. Proc. 1st Int. Symp. Nutr. Strategies Manage. Aquaculture Waste. Univ. Guelph, Guelph, Ontario, Canada.
27. Guinée J. Handbook on life cycle assessment operational guide to the ISO standards. Springer; 2002.
28. De Vries M, de Boer IJM (2010) Comparing environmental impacts for livestock products: A review of life cycle assessments. Livest. Sci. 128: 1–11.
29. Boissy J, Aubin J, Drissi A, van der Werf HMG, Bell GJ, Kaushik SJ, et al. (2011) Environmental impacts of plant-based salmonid diets at feed and farm scales. Aquaculture 321: 61–70.
30. Pelletier N, Tyedmers P (2007) Feeding farmed salmon: Is organic better? Aquaculture 272: 399–416.
31. Pelletier NL, Ayer NW, Tyedmers PH, Kruse SA, Flysjo A, Robillard G, et al. (2007) Impact categories for life cycle assessment research of seafood production systems: review and prospectus. Int. J. Life Cycle Assess. 12: 414–21.
32. Aubin J (2013) Life Cycle Analysis as applied to environmental choices regarding farmed or wildcaught fish. CAB Reviews: Perspectives in Agriculture, Veterinary Science, Nutrition and Natural Resources 8: 1–10
33. Vandeputte M, Garouste R, Dupont-Nivet M, Haffray P, Vergnet A, Chavanne H, et al. (2014) Multi-site evaluation of the rearing performances of 5 wild populations of European sea bass (*Dicentrarchus labrax*). Aquaculture 424–425: 239–48.
34. Sutherland T (1965) The correlation between feed efficiency and rate of gain, a ratio and its denominator. Biometrics 21: 739–49. pmid:5858102
35. Kause A, Tobin D, Houlihan D, Martin S, Mäntysaari E, Ritola O, et al. (2006) Feed efficiency of rainbow trout can be improved through selection: different genetic potential on alternative diets. J. Anim. Sci. 84: 807–17. pmid:16543557
36. Groen A (1989) Economic values in cattle breeding. II. Influences of production circumstances in situations with output limitations. Livest. Prod. Sci. 22: 17–30.
37. Gibson JP (1989) Economic weights and index selection of milk production traits when multiple production quotas apply. Anim. Sci. 49: 171–81.
38. Rose G, Mulder HA, Thompson AN, van der Werf JHJ, van Arendonk JAM (2015) Breeding objectives for sheep should be customised depending on variation in pasture growth across years. Animal 9: 1268–1277. pmid:25857317
39. Thodesen J, Gjerde B, Grisdale-Helland B, Storebakken T (2001) Genetic variation in feed intake, growth and feed utilization in Atlantic salmon (Salmo salar). Aquaculture 194: 273–81.
40. Sanchez M-P, Chevassus B, Labbé L, Quillet E, Mambrini M (2001) Selection for growth of brown trout (Salmo trutta) affects feed intake but not feed efficiency. Aquat. Living Resour. 14: 41–8.
41. Mambrini M, Médale F, Sanchez MP, Recalde B, Chevassus B, Labbé L, et al. (2004) Selection for growth in brown trout increases feed intake capacity without affecting maintenance and growth requirements. J. Anim. Sci. 82: 2865–75. pmid:15484936
42. Janssen K, Chavanne H, Berentsen P, Komen H (2016) Impact of selective breeding on European aquaculture. Aquaculture. http://dx.doi.org/10.1016 /j.aquaculture. 2016.03.012Kause
43. Aubin J, Papatryphon E, van der Werf HMG, Chatzifotis S (2009) Assessment of the environmental impact of carnivorous finfish production systems using life cycle assessment. J. Clean. Prod. 17: 354–61.
44. FAO (2014) Cultured Aquatic Species Information Programme: *Dicentrarchus labrax*. http://www.fao.org/fishery/culturedspecies/Dicentrarchus_labrax/en

CHAPTER 9

Economic Risk and Efficiency Assessment of Fisheries in Abu-Dhabi, United Arab Emirates (UAE): A Stochastic Approach

Eihab Fathelrahman [1,*], Aydin Basarir [1,†], Mohamed Gheblawi [1,†], Sherin Sherif [2,†] and James Ascough II [3,†]

[1] *Department of Agribusiness and Consumer Sciences, College of Food and Agriculture, United Arab Emirates University, P.O. Box 15551, Al Ain, United Arab Emirates;*

[2] *Economics and Agribusiness Department, College of Agriculture, Alexandria University, El-Shatby, Alexandria, Egypt;*

[3] *United Sates Department of Agriculture (USDA), Agricultural Research Services (ARS), Northern Plain Area (NPA), Agricultural Systems Research Unit (ASRU), 2150 Centre Avenue, Building D, Suite 200, Fort Collins, CO 80526, USA;*

ABSTRACT

The fishing industry in Abu-Dhabi, United Arab Emirates (UAE), plays an important role in diversifying food sources in order to enhance national food security. The fishing industry is facing an increasing risk that may impact the sustainability (*i.e.*, quantity and quality) of the fish caught and consumed in the UAE. Therefore, the main objective of this study is to analyze common Abu-Dhabi fishing management alternatives using various stochastic dominance techniques (*i.e.*, first/second degree stochastic dominance, stochastic dominance with respect to a function and stochastic efficiency with respect to a function) to assess the risk facing UAE fishermen. The techniques represent a risk assessment continuum, which can provide a ranking of management alternatives to improve decision making outcomes and help maintain long-term UAE fishing sustainability. Data for the stochastic dominance analyses were obtained from a cross-sectional survey conducted through face-to-face interviews of Abu Dhabi, UAE, fishermen. Analysis of fishing methods, trap sizes and trap numbers using stochastic efficiency with respect to a function (SERF) showed that fishermen efficient practices were not the same for risk-neutral fishermen compared to risk averse fishermen. Overall, the stochastic dominance results illustrated the importance of considering both attitude towards risk and economic inefficiencies in managing UAE fishery practices and designing successful fishery policies, as well as improving decision-making at the fishermen level.

Keywords: fisheries; technical efficiency; allocative efficiency; stochastic dominance; stochastic efficiency with respect to a function (SERF); socio-economics

1. INTRODUCTION

Fishing is one of the oldest traditional industries in the United Arab Emirates (UAE) and is typically inherited from former generations. Pearl extraction and fishing were dominant sources of income before oil and natural gas discoveries in the country. The majority of the fish catch originates from the Emirate of Abu Dhabi, which includes more than 65% of the UAE sea area in its borders. The Food and Agriculture Organization (FAO) [1] reported that UAE fishery policy objectives can be summarized as follows: (1) maintain traditional fishing and produce fresh fish for local markets to serve national food security policy goals; (2) control and support research for the sustainable use of fishery stocks; (3) prohibit destructive fishing practices to protect the marine environment; (4) enhance local fish populations, build artificial reefs and re-stock important local species; and (5) subsidize local fish producers. These policy objectives were translated to Fishery Law No. 23, which was enacted in 1999 to regulate fishery operations in the UAE [1,2].

In order to control fishing efforts, the Emirate of Abu Dhabi introduced new regulations in 2003 [2]. The Emirate has the right to restrict fishing in certain fishing areas, seasons of the year and certain fishing methods, such as traps, nets and threads. Traps in the UAE might consist of a frame of thick steel wire in different shapes, such as triangular cages, cones or igloo-shaped traps. Traps are locally called "Gargor". Nets are made of acrylic fiber, forming different sizes and shapes of grids. Threads include different types of lines and hooks. The Emirate of Abu Dhabi introduced a restriction policy on the number of fish traps per vessel and issued new commercial fishing licenses in 2003. Furthermore, the Emirate required all fish boats to carry a government-issued tag, and at least one UAE national person has to be physically present on vessels during fishing operations.

In order to estimate UAE fishing self-sufficiency ratios, time series for fish production, export, import and consumption have been analyzed. The self-sufficiency ratio (SSR) is a percentage of the locally-produced quantity of product over the total supply available for the consumption of such a product. The total quantity of supply available for consumption is calculated by subtracting product exports from the total of locally-produced product and adding product imports from abroad. The fish catch in UAE increased steadily from 70,000 metric tons in 1982 to 115,000 metric tons in 1999. Production then started declining until it reached 77,700 metric tons in 2010. Meanwhile, fish consumption in the UAE steadily increased from 76,400 metric tons in 1982 to 192,640 metric tons in 2010. The differences between consumption and production are filled with imports from other countries to balance the food production in addition to imports (supply) and consumption (demand) in the UAE. Fish imports in the UAE increased from 8780 metric tons in 1982 to 132,720 metric tons in 2010. The UAE fish self-sufficiency ratio (SSR) historically increased from 92% in 1982 to 129% in 1991, when the country was exporting fish to neighboring countries. However, the SSR has declined in recent years (40% in 2010) to the lowest level since 1982 (Figure 1), due to several external and internal factors [3]. Factors external to the fishing industry include changes in federal and Abu-Dhabi Emirate regulations in relation to fishing capacity and seasons, increased water pollution in over 40 ports and sites where the fishing industry has expanded in the last three decades (especially after the first Gulf War in 1991 and the invasion of Iraq in 2003), and the reduction or disappearance of favored commercial fish species, such as Hamour, Kan'ad-Chabbat, Sultan Ibrahim and Safi Arabi. Factors internal to the fishing industry include resource management to meet federal and UAE regulations, nationalization and the optimal mix of fishing equipment, boat horsepower and manpower to maximize profit [4]. Figure 2 shows the total fish catch originating from the Emirate of Abu-Dhabi during the period 2001–2010.

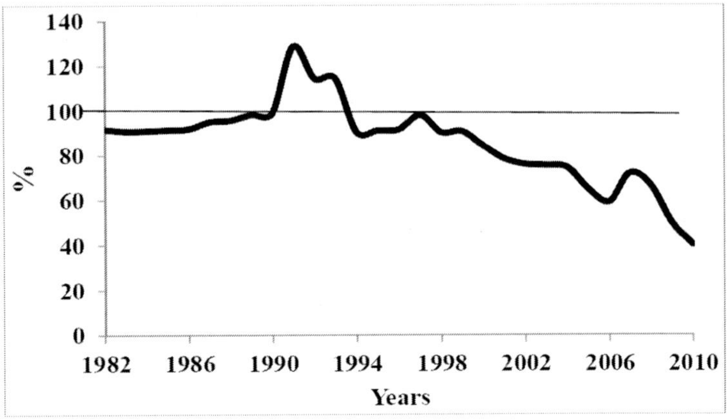

Source: Arab Organization for Agricultural Development. Statistical Yearbook (1982–2012) [3].

Figure 1. UAE fish self-sufficiency ratio (1982–2010).

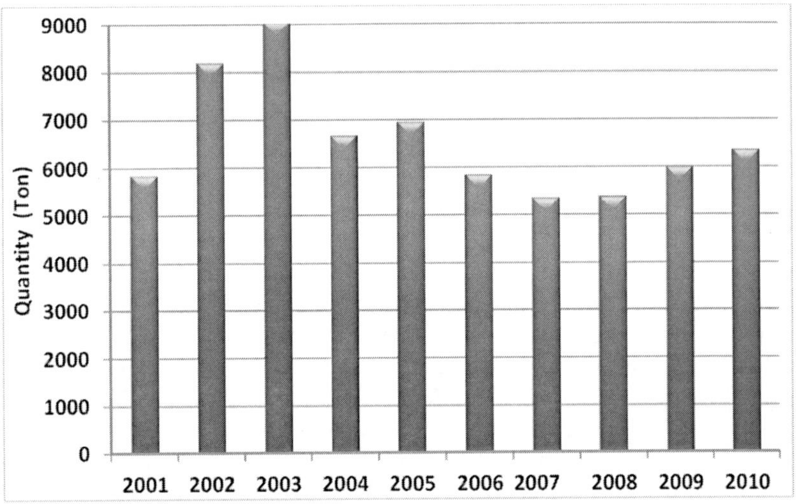

Source: Abu-Dhabi Environment Agency [5].

Figure 2. Total fish catch originating from Abu-Dhabi Emirate, UAE (2001–2010).

Since 2004, the total fish catch has stayed below the level of 7000 metric tons annually in the Emirate of Abu-Dhabi after the annual increase up to 2003 (Figure 2). The changes in the UAE fishing industry over the past three decades have increased uncertainty regarding sustainability (*i.e.*, quantity and quality) of fish caught and consumed. The primary UAE fishing objectives are to maintain adequate sources of protein options for UAE consumers and to satisfy increasing consumer demand for healthier seafood products. Increasing fishermen resource use efficiency through the use of less risky fishing management practices may help contribute towards attaining UAE food security objectives.

Efficiency differences in fisheries are often attributed to differences in technology and the skill of the skipper and crew, as shown in Kirkley *et al.* [6] and Sharma and Leung [7]. However, Pascoe and Coglan [8] concluded that skipper skill is the major factor affecting differences in efficiency, with other factors explaining relatively little of the efficiency differences. Sherif and Gheblawi [9] examined the revenue determinants and assessed the input allocative efficiency of Abu-Dhabi fishermen. The main determinants of fishermen revenues were found to be labor, fishing effort, boat capacity, boat boarding expenses and expenditures on oil and lubricants. Moreover, price efficiency was found not to be achieved for any of the aforementioned five prominent inputs. More specifically, effort, boat capacity and labor were under-utilized, while boarding of the fishing crew and expenses on oil and lubricants were both found to be significantly over-utilized. Such inefficiency contributes towards increased risk and the decreased sustainability of the overall fishing industry in UAE. Sherif and Gheblawi [9] concluded that the increased risk faced by Abu-Dhabi fishermen causes declining revenue and leads fishermen to seek jobs in other sectors of the economy.

Aisyah *et al.* [10] used data envelopment and Tobit analysis to determine the technical efficiency level and factors influencing technical efficiency among Malaysian fishermen. The mean technical efficiency for the sample was estimated to be 55% for the peak season and 40% for the non-peak season. Variables related to management, such as planning, staffing and controlling, and variables related to demographics, such as family size and formal education, were found to exert positive effects on the technical efficiency of fisheries in Kuala Terengganu, Malaysia. Meanwhile, Ahmed *et al.* [11] used data envelopment analysis and also used Tobit models and found that integrated rice-fish farming was not popular among farmers in Bangladesh. The reasons for not adopting the technically efficient system were the lack of technical knowledge of farmers, high production costs and risks associated with floods and droughts. Espino *et al.* [12] utilized data envelopment analysis to study the capacity utilization of the fishing fleet in the Strait of Gibraltar. The results highlight the role increased technical efficiency has played in the persistence of overcapacity well after the implementation of recovery plans. The above group of studies addressed the technical efficiency of various fishery industries and analyzed fishermen socio-economic characteristics, which impact their technical efficiencies.

Singh *et al.* [13] developed a dynamic fishery model, which simultaneously incorporates random stock growth and costly capital adjustment, such that marginal net harvest benefits diminish with the catch (which provides an incentive to smooth harvest over time). The results demonstrate the importance of jointly determining harvest and capital investment. However, the value of the immediate harvest value rises as the fleet size increases. The results of Herrero and Pascoe [14] support the idea that the objective of fishermen is to maximize the value of the catch per trip rather than the quantity, while fishing strategies are affected by the presence of risk and uncertainty. Targeting high-value species by technically-efficient fishermen was found to be a more beneficial tactic than aiming for a high quantity catch of low-value fish species. In contrast, when the uncertainty of catching high-value fish is high, this tactic becomes costly to the inefficient fisherman. Danielsson [15] utilized a general bio-economic model to analyze the efficiency of different methods of fisheries management when there are some risks involved. The two management practices were the volume of catch in a catch-quota system and the volume of effort in an effort-quota system. It was found that when the price elasticity of demand is high and the variability in the growth of the stock is great compared to the variability in the catch per unit of effort (CPUE), management with effort quota was superior to management with catch quotas. However, if the price-elasticity of demand is low and the variability in the CPUE is large compared to the variability in the growth of the stock, management with catch quotas is superior to management with effort quotas.

Singh *et al.* [13], Herrero and Pascoe [14] and Danielsson [15] considered capital costs, the value of the catch and various fishing management alternatives, respectively. This group of studies considered the fishermen profit maximization problem under risk and uncertainty and discussed various fishery management alternatives. Bastardie *et al.* [16] developed a spatially explicit individual-based model for fishing vessel movements. The factors used in the model included vessel-based fuel consumption, energy efficiency defined as the quantity of fish caught per liter of fuel used and profitability. Fishermen adaptation to these factors was evaluated against three alternative effort allocation scenarios: (a) preferring nearby fishing grounds rather than distant grounds with potentially larger catches and higher values; (b) shifting to other fisheries targeting resources located closer to the harbor; and (c) allocating effort towards optimizing the expected area-specific profit per trip. The trade-off between fuel savings and energy efficiency improvements when effort is displaced closer to the harbor compared to reductions in total landing amounts and profit were the outcomes of Scenarios a and b. Scenario c indicated that the historical effort allocation was actually sub-optimal, because increased profit from decreased fuel consumption and larger landings could have been obtained by applying a different spatial effort allocation. Data envelopment analysis was utilized by Maravelias and Tsitsika [17] to estimate and assess capacity utilization and efficiency in the Eastern Mediterranean. Longer vessels were found more efficient than shorter ones, and there was evidence of overcapacity. It was found that fishing vessels could potentially improve earnings by approximately 20% by increasing variable inputs. Andersen [18] used data on Danish seiners and data envelopment analysis to estimate efficiency and by accounting for fish stock development. The findings showed that estimations based on the catch per unit effort stock measures differ from the estimations based on independent stock measures and are independent of the choice of time horizon and the choice of input/output measures. Kirkley *et al.* [6] provided a comparative analysis of the deterministic data envelopment analysis and the stochastic production frontier capacity estimates for a variety of possible capacity concepts using a panel data set. The capacity output measures from data envelopment analysis were higher than the corresponding stochastic production frontier measures. However, the two approaches provided similar guidance about overall and even relative boat-specific capacity levels under certain circumstances.

Tingley *et al.* [19] compared the results from both data envelopment analysis and stochastic production frontier techniques for segments of English Channel fisheries. The factors most affecting technical efficiency were found to include vessel and skipper characteristics, and the results from both methods showed general consistency. The age of the vessel in the English Channel was found by Pascoe and Coglan [8] to be the most important measurable factor affecting efficiency. Immeasurable factors accounted for about 65% of the variation in efficiency and as much as 9% of the total variation in catches between boats. The factors contributing the most to those differences were found to be skipper and crew skills; however, "luck" accounted for around 11% of the variation in catches between boats. Del Hoyo *et al.* [20] estimated four econometric models measuring the technical efficiency of vessels operating in the Gulf of Cadiz, Spain. Since the approaches provided different estimates, the authors argued for different policy implications depending on the estimates adopted. They concluded that the useful life of vessels was inversely proportional to the technical efficiency, and the total allowable catch regime based only on vessels larger than 12 m was unsuitable for that particular fishery. Greenville *et al.* [21] utilized a stochastic frontier to examine the technical efficiency in input-controlled fisheries exemplified by the New South Wales (NSW) ocean prawn trawl fishery. They showed that input controls had a negative effect on the technical efficiency of fishermen, while such controls did not appear to have had a lasting effect on productivity.

The above studies show the importance of decision-making in relationship to different management alternatives at the fishermen decision-maker

(DM) level under various risk and uncertainty conditions. Successful fishing practices were found to be those that considered the impact of boat capacity, fishing effort, size of capital investment, boat length, labor training and management, as well as the number and sizes of traps. Previous studies also highlighted the importance of diversification, economies of scale and the catch of high-value fish management options on fishing profitability and both technical and economic efficiency.

The overall goal of this research is to address concerns related to risk and uncertainty in the fishing industry in the Emirate of Abu-Dhabi, UAE. The specific objective of this paper is to quantify and rank UAE fishermen preferences concerning various fishing management alternatives at different levels of attitudes towards risk. To achieve the specific research objective, several statistical and stochastic analysis techniques were employed, including descriptive statistic and stochastic dominance techniques (first/second degree stochastic dominance, stochastic dominance with respect to a function and stochastic efficiency with respect to a function) to rank the fishing management alternatives (methods of fishing, sizes of traps and number of traps) based on a well-accepted economic indicator (profitability represented by gross margin). The stochastic dominance techniques were applied across a spectrum of risk attitudes and preferences ranging from risk neutral to risk averse. The significance of this research stems from the DM need to identify less risky management alternatives, the adoption of which will hopefully retain more UAE fishermen in the business, attract a new generation of fishermen and augment the sustainability of this important food sector to ensure a sufficient and long-term supply of fish in the UAE.

This paper is divided into four sections. The first section summarized the overall UAE fisheries' socio-economic indicators, a literature review of previous studies and the research objective, which focuses on quantifying and ranking UAE fishermen preferences concerning various fishing management alternatives at different levels of attitudes towards risk. The second section of this paper describes the methods used, including stochastic first and second degree stochastic dominances, stochastic dominance with respect to a function and stochastic efficiency with respect to a function, to assess the risk facing UAE fishermen. The third section includes the results and discussion, starting with the analysis of the Abu-Dhabi fishermen survey results and continues with the ranking of the fishermen management alternatives using various ranking methods. The fourth section of this paper includes a summary and conclusion, which summarizes the importance of considering both decision-maker's attitude towards risk and economic inefficiencies in managing UAE fisheries that can be used for designing successful fisheries policies.

2. METHODS

2.1. Standard Stochastic Dominance Techniques

In economic systems, there is generally risk associated with production quantities, input prices, output prices, interest rates, quantities demanded by consumers, market share, government regulations and weather conditions, among others. For a business, a manager does not have control over many of these variables, *i.e.*, risk is the part of a business decision the manager cannot control. Risk is often quantified using probability theory, *i.e.*, a stochastic process involving random variables that is often used to represent the evolution of a system over time or space. Stochastic dominance is a methodology based on expected utility maximization theory. A detailed discussion of the usefulness of stochastic dominance decision criteria can be found in Robison and Barry [22]. Boggess and Ritchie [23] and Williams *et al.* [24] also present the rationale and application of various techniques. The theoretical attractiveness of stochastic dominance analysis lies in its non-parametric orientation, *i.e.*, it does not require a full parametric specification of the preference of the decision-maker and the statistical distribution of the choice alternative [25,26,27].

As previously discussed, several well-known decision criteria for stochastic dominance exist, including first degree stochastic dominance (FSD), second degree stochastic dominance (SSD) and stochastic dominance with respect to a function (SDRF). Hardaker *et al.* [28] stated that FSD assumes that the decision-maker (DM) has positive marginal utility for the performance measure (*i.e.*, more is preferred to less). Thus, given two alternatives, A and B, each with a probability distribution of outcomes defined by a Cumulative Distribution Function (CDF), A dominates B in the FSD sense if the CDF of A is always below and to the right of the CDF of B (*i.e.*, FA(X) ≤ FB(X) for all X), where F(X) denotes the distribution describing the random variable (X). SSD holds for those decision makers who are risk neutral or risk averse for all values of X; thus, the applicable range of the absolute risk aversion coefficient, r_a (a measure of how much a person would pay to avoid risk), for the SSD criterion is from 0 to $+\infty$. SSD assumes a utility function of a positive, but decreasing slope, *i.e.*, $U^1(X) > 0$ and $U^2(X) < 0$ with SSD, and A is preferred to B if $\int_{-\infty}^{x} F_A(X)dx \le \int_{-\infty}^{x} F_B(X)dx$ for all values of X. Graphically, this means SSD requires that the curve of the cumulative area under the CDF for a dominant alternative lies everywhere below and to the right of the corresponding curve for the dominated alternative. Although more powerful than FSD, SSD often leaves a large number of choices as being risk-efficient. To improve the discriminating power of SSD, Meyer [29] proposed SDRF, which is a more general notion of stochastic dominance. This rule helps to identify risk-efficient options for the class of decision-makers whose risk aversion coefficients are bounded by lower and upper values. The smaller the range of risk aversion coefficients, the more powerful is the criterion. The SDRF criterion orders the choices by defining intervals using the r_a absolute risk aversion coefficients. These risk-preference intervals are bounded by a lower risk aversion coefficient, r_{aL}, and an upper risk aversion coefficient, r_{aU}, which characterize the general degree of risk aversion for a manger. A risk-efficient set of strategies will include the choices preferred by each manager having risk preferences consistent with the restrictions imposed by the lower to upper interval. Furthermore, Hardaker *et al.* [30] argued that eliciting decision-makers (or inferring) the bounds on their risk aversion coefficients may be simpler than eliciting a complete utility function. Further discussion of SDRF is provided by Cochran [31], King and Robison [32] and Robison and Barry [24].

2.2. Stochastic Efficiency with Respect to a Function (SERF)

A more discriminatory procedure than SDRF that considers the full range of decision-maker preferences is stochastic efficiency with respect to a function (SERF). SERF is more efficient than other stochastic dominance techniques, because it can be applied for any utility function form based on the full range (*i.e.*, from negative to positive) of an absolute or relative risk aversion coefficient (r_a). A positive r_a indicates that the DM is risk averse, while a negative r_a indicates that the DM is a risk taker (a risk-neutral DM is more concerned about the expected return on their investment, not on the risks they may be assuming). Furthermore, SERF is a preferable method, due to the fact that it is one of few risk analysis techniques that can be used to easily visualize the stochastic frontier across the entire r range, where preferences for a particular alternative may be illustrated.

SERF identifies and orders utility efficient alternatives in terms of certainty equivalents (CEs) for a specified risk preference. Hardaker *et al.* [28] state that the SERF procedure can potentially find a smaller set of preferred strategies (*i.e.*, it has stronger discriminating power) compared to stochastic dominance approaches in addition to being more transparent and easier to implement. The CE of a risky alternative (in this study, the trap, net and thread fishing methods) is the amount of money at which the decision-maker is indifferent between the certain dollar value and the risky alternative. That is, the CE is the sure amount of money with the same utility as the expected utility of a risky alternative as suggested by Keeney and Raiffa [33] and can be calculated by taking the inverse of the utility function, U:

$$CE\ (w, r(w)) = U^{-1}\ (w, r(w)) \tag{1}$$

where *w* is the initial wealth and *r(w)* represents the risk aversion coefficient with respect to wealth. The utility function, U, can be represented as:

$$U\ (w,r(w)) = \int u\ (w,r(w))\ dF(w) \tag{2}$$

where *U (w,r(w))* is the utility as a function of the random variable, *w*, and a set of risk aversion coefficients, r. Strategies with higher CEs are preferred to those with lower CEs, and the interpretation of the CEs is straightforward, because, unlike utility values, they may be expressed in monetary terms [34]. To calculate the CEs using SERF, various types of utility functions can be used (e.g., power, negative exponential, quadratic, log-log). In this study, similar to that of Pendell *et al.* [34], we assume a negative exponential form for the utility function:

$$U(w) = -\exp(-\ r_a\ (w)w) \tag{3}$$

where $r_a(w)$ is the absolute r_a with respect to wealth. Given a random sample of size n from alternative w (*i.e.*, a management alternative, such as fishing method) with i possible outcomes (*i.e.*, the fisheries' gross margin), the estimated CE can be defined as:

$$CE(w, r_a(w)) = -\frac{1}{r_a(w)} \ln\left\{\frac{1}{n}\sum_i^n \exp(-r_a(w)w_i)\right\} \tag{4}$$

A negative exponential utility function conforms to the hypothesis that managers prefer less risk to more given the same expected return and assumes managers have constant absolute risk aversion, as suggested by Lien *et al.* [34]. Under this assumption, managers view a risky strategy for a specific level of risk aversion the same without regard for their level of wealth, such as in Pendell *et al.* [35], and Babcock *et al.* [36] state that this functional form is often used to analyze decisions-maker's alternatives under risk. The decision rule for SERF is to rank the risky alternatives (within the decision-makers' specified risk aversion coefficient) from the most preferred (*i.e.*, the highest CEs at specified levels of risk aversion) to the least preferred (*i.e.*, the lowest CEs at specified levels of risk aversion). There is a general agreement on the valid range for the r_a that a DM should use with SERF [37].

In this study, UAE fishing management alternatives (methods, size of traps and number of traps) were considered for the descriptive statistic and stochastic dominance analyses. Fishermen catch determines revenues, because the fishermen are a price taker in the market. Subtracting the cost of operations from revenues determines the gross margin or revenue over variable cost. The fishing management alternatives are analyzed under various levels of fishermen attitudes towards risk. SERF is a practical method for analyzing the fishing management alternative problem, because SERF considers both the variability of gross margins and DM attitude towards risk. As stated above, SERF calculates a CE, which is the sure amount of money with the same utility as the expected utility of a risky alternative [37].

SERF is a procedure for ranking risky alternatives based on the certainty equivalent (CE) for risk alternatives and the absolute risk aversion coefficients (ARAC). Hardaker *et al.* [28] described three steps that are considered the computation steps for the SERF discrete CE points. In the first step, selected points on each CDF from the set of w values are determined. Then, the second step is to convert the selected w values to their utility function (*i.e.*, the negative exponential function in this study) value, as well as the values representing the risk aversion coefficient are selected. The third step involves multiplying each finite utility by its corresponding probability to calculate the weighted average of the utility outcome. As a result, only alternatives that have higher CE values for some specified range of $r(w)$ are utility efficient. The Simetar© 2008 risk analysis software implements the steps above, and they are graphically illustrated in the software's user manual [37]. The next section in this paper further describes the stochastic dominance and stochastic efficiency analyses adopted for this paper.

2.3. Risk Simulation Analyses

The Simetar © 2008 risk analysis software [37] was used to perform the stochastic dominance (FSD, SSD and SDRF) and SERF analyses. For the SDRF stochastic dominance analysis, two points represented the spectrum of decision-maker attitude towards risk. A lower absolute risk aversion coefficient (r_{aL}) equal to 0.0 (representing a neutral decision-maker) and an upper absolute risk aversion coefficient (r_{aU}) equal to 0.004 (representing a risk-averse decision-maker) were used to perform the analysis. For the SERF analysis, gross margin CE curves by fishing management alternatives (fishing methods, size of traps and number of traps) were produced by calculating 25 CE values for each curve over the entire range (0.0 to 0.004) of absolute risk aversion (r_a). Stochastic efficiency with respect to a function (SERF) is considered a sufficient ranking method. Hardaker [30] showed that SERF will potentially identify a smaller efficient set than SDRF, because it picks only the utility efficient alternatives, compared to other alternatives simultaneously, at each point of the DM attitude towards risk.

2.4. Study Data Source

The data for this study were extracted from face-to-face interviews with Emirate of Abu-Dhabi, UAE, fishermen. The United Arab Emirates Ministry of Water and Environment (MOEW) [38] showed that the Emirate of Abu-Dhabi area was targeted because of the on-going intensive fishing activities taking place across the Emirate's shores. The number of boats increased from 856 in 2001 to 1100 in 2009 or about 30% [38]. One hundred thirty one fishermen (representing about 9% of the total population of fishermen in the area) participated in the survey, which was divided into four sections. The first section focused on measuring the socioeconomic characteristics of the fisheries. The second section investigated the asset values (*i.e.*, fixed costs) of the fisheries, such as boat market values, as well as equipment values. The third section collected data on variable costs, including the quantities of inputs used and input prices (e.g., labor, material and supplies costs). Finally, the fourth section investigated the output (fish catch) and revenue (fish catch multiplied by output prices) of the fisheries. Data from the survey were entered into a spreadsheet, coded and prepared for analysis. For complete details of the survey, see Sherif and Gheblawi [9].

3. RESULTS AND DISCUSSION

3.1. Fisheries Survey Descriptive Statistics

Based on the 131 fishermen interviewed, the survey sample results indicated that the average fishing boat age was about 22 years with a standard deviation (SD) of nine years. The average fishing boat market price was 142,000 Dirhams (AED) with an SD of 93,000 Dirhams. The United Arab Emirate Dirham (AED) is the official currency of the UAE (one USD = 3.675 AED). A boat license costs UAE fishermen about 220 AED on average, and survey results showed that there was on average four workers per boat earning a salary of 712 AED per worker monthly. The average boat capacity was around 208 horsepower with an SD of 67.5 horsepower; the average boat length was 32 feet (9.8 m) with an SD deviation of 14.4 feet (4.4 m). The number of fishing trips per month ranged from two to 22 trips. On average, each fisherman went on six fishing trips per month, with each trip taking four days on average. Investigation of the socio-economic characteristics of Abu-Dhabi fishermen showed that 97% of the fishermen were boat owners with only 3% renting boats for fishing. The survey results showed that 53% of the fishermen were UAE nationals, while 47% were citizens of other countries (e.g., India, Iran and Oman). When survey respondents were asked if they purchased boat insurance, only nine fishermen (or 7%) responded yes. Furthermore, only 37 survey respondents (or 28%) indicated that their sons were interested in pursuing fishing as a profession and career. These findings confirm the fact that the next generation does not consider fishing in Abu-Dhabi as a secure source of income (and would not want to enter the business and sustain it for the future). Five different species of fish were identified as common in the Abu-Dhabi fishermen daily catch, including Hamour, Kan'ad-Chabbat, Sultan Ibrahim, and Safi Arabi. However, Abu-Dhabi fishermen are not specialized in fishing of a certain species in favor of other species. Their daily catch is mostly a mix of different species based on stock availability and seasonal fish migration. Only 18 survey respondents (or 14%) were found to target and find shrimp in their catch.

The survey results also focused on the estimation of the gross margin per boat on an annual basis. The gross margin was calculated by subtracting the variable costs, which are costs that vary with the quantity of the fish catch and include the costs of labor, fuel, maintenance and boarding (such as food, nets, water and ice). Labor and fuel costs combined represent 89% of the total variable costs for an average Abu-Dhabi fisherman. Labor and fuel represent 62% and 27% of the variable costs, respectively. Laborers are commonly foreigners; it costs boat owners about 4370 AED on average to hire a worker from abroad. The economic results of the Emirate of Abu-Dhabi, UAE, fishermen survey are presented in Table 1. The average annual total revenue is about 298,000 AED, subtracting average annual variable costs of 159,000 AED from the total revenue results in an average gross margin of 139,000 AED. The SD of the total revenue, annual variable costs and gross margin is 186,000, 52,000 and 174,000 AED, respectively. The higher SD for the gross margin reflects the large economic variability in the survey sample. The coefficient of variation (CV, calculated as the SD divided by the mean) is a more accurate measure of variability compared to the SD. The annual gross margin CV of the sample is 1.25, which indicates that the SD is higher than the mean for the Abu-Dhabi, UAE, fisheries. The total revenue, variable costs and gross margin standard deviations in Table 1 are high, which indicate the large variability of both input costs and revenue. This finding indicates the need to use an alternative quantitative approach (e.g., stochastic dominance analysis) to further investigate the variability and the overall impact of the various fishing management practices on the gross margin.

Table 1. Descriptive statistics of selected economic variables based on the Emirate of Abu-Dhabi, UAE, fisheries survey.

Variable	Annual total revenue (TR) (AED)	Annual variable costs (VC) (AED)	Annual gross margin (GM) = TR − VC (AED)
Mean	298,173	159,112	139,061
Median	276.000	157,680	125,320
Standard deviation (SD)	185,899	52,415	174,042
Coefficient of variation (SD/mean)	0.62	0.33	1.25
Number of observations	131	131	131

Table 2 summarizes descriptive statistics for the gross margin (total revenue (TR) – variable costs (VC)) for the different fishing methods in the Emirate of Abu-Dhabi, UAE. Based on the mean, the traps method had the highest average gross margin (about 168,000 AED annually) compared to the thread and net methods. However, the higher average for the traps method was also accompanied by the highest standard deviation (about 176,000 AED) for all of the fishing alternatives. The CV results, commonly used to indicate risk, show that the most risky alternative would be the thread method of fishing. In theory, a manager could examine only the mean standard deviation and CV risk-tradeoff results and simply decide which fishing method alternative is best; however, this may be difficult in practice in that these criteria often exhibit high variability and can result in contradictive and inconclusive rankings. The application of a more sophisticated risk-based methodology, such as stochastic dominance or SERF, can help fishermen and decision-makers see these tradeoffs more clearly with very few additional assumptions.

Table 2. Descriptive statistics of annual gross margins (AED) for various methods of fishing based on the Emirate of Abu-Dhabi, UAE, fisheries survey.

Statistical measure	Traps	Thread	Nets
Mean	167,551	65,078	67,274
Median	159,360	66,430	45,280
Standard deviation (SD)	175,891	169,834	142,885
Coefficient of variation (SD/mean)	1.05	2.61	2.12
Number of observations	94	10	27

3.2. First Degree Stochastic Dominance (FSD)

First degree stochastic dominance (FSD) can be determined by visually considering the Cumulative Distribution Functions (CDFs) between various management alternatives. As previously stated, Alternative A completely dominates Alternative B if the CDF of Alternative A is always below and to the right of the CDF of B (*i.e.*, the CDFs do not intersect). Figure 3, Figure 4 and Figure 5 depict Abu-Dhabi, UAE, fisheries' annual gross margin CDFs for the fishing method, trap size and number of traps, respectively, based on the fisheries sample. All three graphs show inconclusive results that do not show a dominant alternative, because all CDFs cross each other either at the upper or the lower tails of the CDF. In Figure 3, the CDF for the trap fishing method is further to the right compared to the CDFs for the thread and net fishing methods. However, the trap method CDF crosses the net method CDF in the upper tail (when the decision-maker is more risk averse). Hardaker [30] showed that if CDF A crosses CDF B, where A and B are the probability distributions of the outcome, then neither dominates the other in the first degree sense. In other words, A dominates B only if $F_A(x) \leq F_B(x)$ with at least one strong inequality. Figure 3, Figure 4 and Figure 5 below indicate that no clear dominance of the fishing management alternatives exists. The absence of a dominant alternative for the fishing method, trap size or number of traps necessitates conducting a second degree stochastic dominance analysis.

Figure 4 shows FSD results for the gross margin based on trap size, and Figure 5 shows FSD results for the gross margin based on the number of traps used by fishermen in Abu-Dhabi. Again, both figures show non-conclusive results, because the CDFs intersect at the upper tails of the graph. These results indicate that the DM cannot rely on FSD to rank the risky alternatives.

To further investigate the statistical differences between the three sets of management alternative distributions (fishing methods, traps sizes and number of traps), Kolmogorov-Smirnov tests were performed with results reported in Table 3. The test results showed no significance differences between cumulative probability distributions for fishing method (traps, threads and sets) pairs. Similar to fishing methods, no significance differences were found between cumulative probability distributions for the number of trap pairs. However, the Kolmogorov-Smirnov test showed the presence of significant differences between the cumulative probability distributions for trap size (large, medium and small). These results are different than those obtained from the visual inspection of the CDFs (FSD) in ranking the risky management alternatives. The analysis proceeds using second degree stochastic analysis in the search for more conclusive stochastic analysis results.

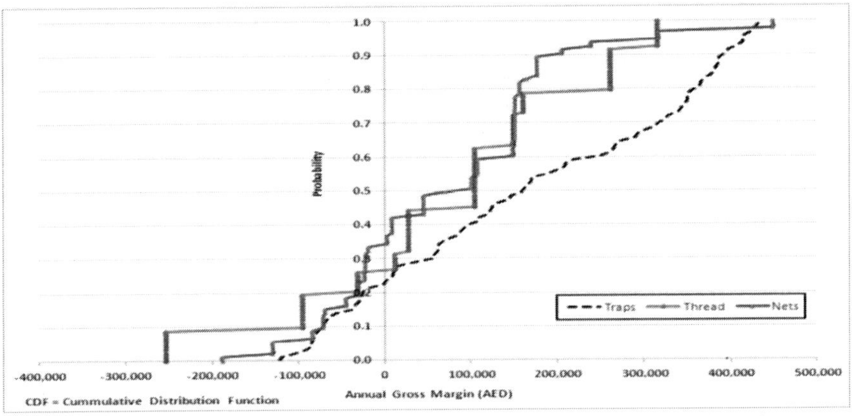

Figure 3. Sample annual gross margin (AED) cumulative distribution functions (CDFs) for different methods of fishing.

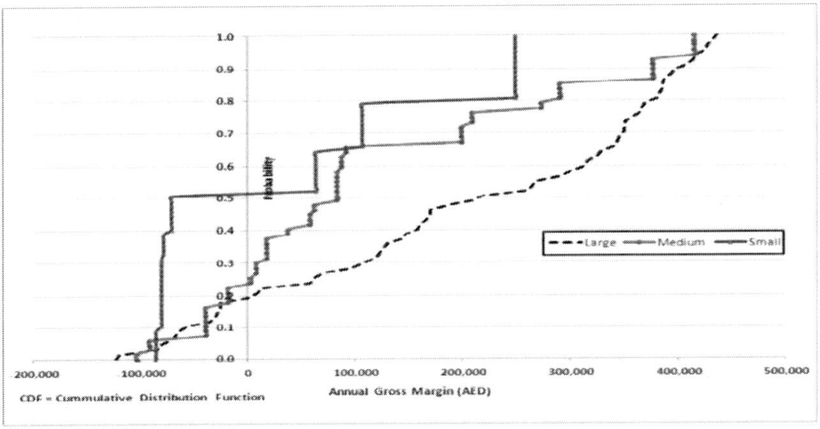

Figure 4. Sample annual gross margin (AED) cumulative distribution functions (CDFs) for trap sizes.

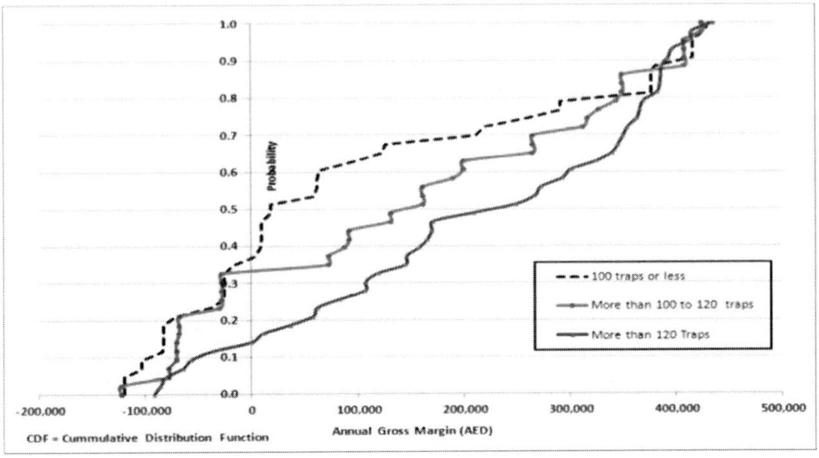

Figure 5. Sample annual gross margin (AED) cumulative distribution functions (CDFs) for the number of traps.

Table 3. Kolmogorov–Smirnov (K-S) statistical significance tests of cumulative distribution functions (CDFs) for pairs of fishing management alternatives (methods, size of traps and number of traps).

Pairs of fishing management alternatives	K-S test values	D values	Conclusion
Fishing methods			D value is greater than the absolute value of the greatest difference between each pair: cannot conclude the presence of significant difference between the three distributions.
Traps/Threads	−0.034		
Traps/Nets	0.041	0.140	
Threads/Nets	0.074		
Trap size			D value is less than the absolute value of the greatest difference between each pair: can conclude the presence of significant difference between the three distributions.
Large/Medium	−0.256		
Large/Small	0.322	0.166	
Medium/Small	0.389		
Number of traps			D value is greater than the absolute value of the greatest difference between each pair: cannot conclude the presence of significant difference between the three distributions.
Less than 100/100–120	0.173		
Less than 100/More than 120	0.19	0.205	
100–120/More than 120	−0.144		

3.3. Second Degree Stochastic Dominance (SSD)

Table 4 below illustrates the second degree stochastic dominance (SSD) results for methods of fishing, types of traps and numbers of traps, respectively, based on the gross margin. SSD results indicate that fishery operation profitability risk is higher when fishermen are using threads and nets. The risk is also higher when fishing operations use medium- and small-sized traps and less than 120 traps. The SSD results indicate that: (1) the trap fishing method dominates both the thread and the net methods; (2) large-sized traps are dominant over medium- and small-sized traps; and (3) using more than 120 traps dominates the lesser number of trap alternatives.

Table 4. Second degree stochastic dominance (SSD): gross margins based on indicators for the methods of fishing, the size of traps and the number of traps.

Fishing methods	Traps	Threads	Nets	SSD dominance ranking
Traps	--	Dominant	Dominant	1
Threads	Not Dominant	--	Not Dominant	3
Nets	Not Dominant	Dominant	--	2
Trap size	**Large**	**Medium**	**Small**	**SSD dominance ranking**
Large	--	Dominant	Dominant	1
Medium	Not Dominant	--	Dominant	2
Small	Not Dominant	Not Dominant	--	3
Number of traps	**100 traps or less**	**More than 100 to 120 traps**	**More than 120 traps**	**SSD dominance ranking**
100 traps or less	--	Not Dominant	Not Dominant	3
More than 100 to 120 traps	Dominant	--	Not Dominant	2
More than 120 Traps	Dominant	Dominant	--	1

3.4. Stochastic Dominance with Respect to a Function (SDRF)

Stochastic dominance analyses include specific assumptions related to DM attitude towards risk. Table 5 summarizes the stochastic dominance with respect to a function (SDRF) results for fishing methods, trap size and number of traps. Among the various fishing methods, threads is preferred over other fishing methods (traps and nets) for both the risk-neutral and the extreme risk-averse DM. The risk-neutral DM, based on the SDRF risk analysis method, would select a large trap size over other sizes (medium, small). In contrast to the risk-neutral DM, the risk-averse DM would prefer small trap sizes over the large size, potentially to avoid large losses, as large trap sizes are usually accompanied by higher spending on inputs, such as labor and fuel, and require a larger catching effort. Considering the SDRF for risk analysis, the DM would prefer more than 120 traps over less than 120 traps for both risk-neutral and risk-averse conditions.

Table 5. Stochastic dominance with respect to a function (SDRF) rankings for the fishing management alternatives based on the Emirate of Abu-Dhabi, UAE, fisheries survey.

Management alternatives	Efficient set ranking with an absolute risk aversion coefficient = 0.0	Efficient set ranking with an absolute risk aversion coefficient = 0.004
Fishing methods		
Thread	Most Preferred	Most Preferred
Traps	2nd Most Preferred	2nd Most Preferred
Nets	3rd Most Preferred	3rd Most Preferred
Trap size		
Large	Most Preferred	3rd Most Preferred
Medium	2nd Most Preferred	2nd Most Preferred
Small	3rd Most Preferred	Most Preferred
Number of traps		
More than 120 traps	Most Preferred	Most Preferred
More than 100 to 120 traps	2nd Most Preferred	2nd Most Preferred
100 traps or less	3rd Most Preferred	3rd Most Preferred

3.5. Stochastic Efficiency with Respect to a Function (SERF)

Figure 6, Figure 7 and Figure 8 illustrate the stochastic efficiency with respect to function (SERF) results for the three ranking efficiency indicators (*i.e.*, fishing methods, size of traps and number of traps). The advantage of SERF over the other risk analysis methods is that SERF enables a DM to rank the various alternatives at different levels of attitude towards risk. Figure 8 shows that a DM would prefer traps over threads and nets when the gross margin CE is used at all positive levels of attitude towards risk. In other words, across all ranges of attitudes towards risk, including risk neutral and risk averse, the DM would choose traps over the other fishing methods. Figure 6 also shows that the net alternative is dominant over the thread alternative. Figure 7 shows that the large trap alternative dominates the medium- and small-sized trap alternatives when the DM is risk neutral. This implies that the variability/stochastic impact is higher among fishermen who are using large traps compared to those who use medium- or small-sized traps. This stochastic impact is due to the fact that large traps are used by large boats, which are the ones that usually take longer and more frequent trips and, so, are exposed to relatively higher levels of uncertainty and variability.

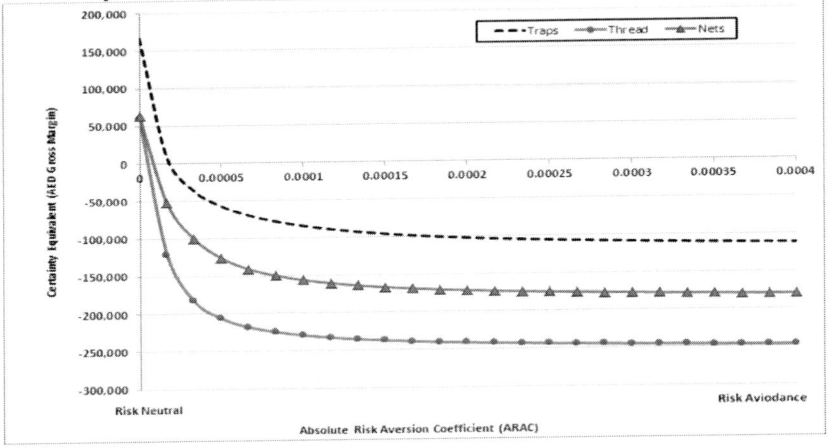

Figure 6. Stochastic efficiency with respect to a function (SERF) for fishing methods (negative exponential utility function).

Figure 8 shows the SERF results based on the gross margin for the number of traps. Consistent with the SSD results, the SERF analysis indicates that using the more than 120 traps alternative is dominant over the smaller number of traps alternative across all levels of attitude towards risk. Using the stochastic efficiency approach and analysis reveals that using traps is more efficient than other fishing alternatives (threads and nets) and using large traps is more efficient than using medium or small traps. Furthermore, the analysis reveals that 120 traps or more is more efficient than using less than 120 traps. One could also conclude that the thread and net fishing methods are not efficient relative to using traps (*i.e.*, they are more risky alternatives). Using medium and small traps are risky alternatives, as well, in addition to using less than 120 traps based on the gross margin SERF analysis. However, the SERF results showed that large-sized traps would be the choice of a risk-neutral DM. Alternatively, small-sized traps would be the choice of a risk-averse DM. This result indicates the importance of risk taking for a business to be successful, *i.e.*, no profitable business exists without taking a certain level of risk.

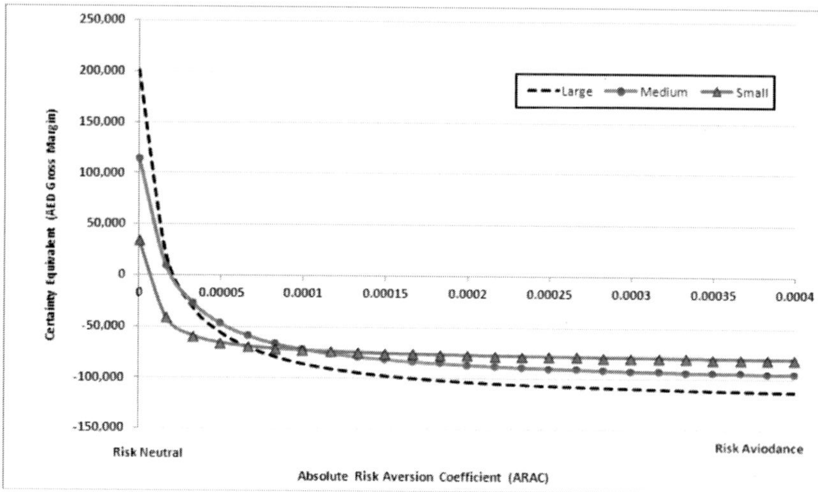

Figure 7. Stochastic efficiency with respect to a function (SERF) for the size of traps (negative exponential utility function).

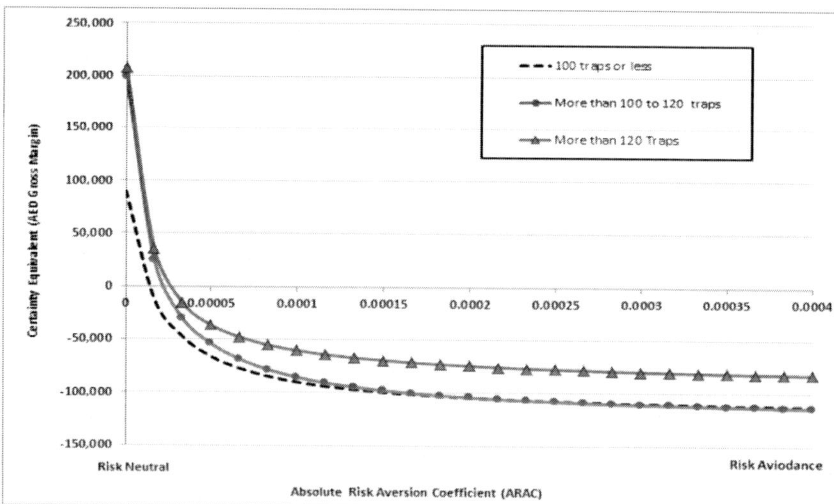

Figure 8. Stochastic efficiency with respect to a function (SERF) for the number of traps (negative exponential utility function).

The Emirate of Abu-Dhabi, UAE, fisheries survey of risk and uncertainty analysis showed both agreement and disagreement based on the analysis of the fishing management alternatives. All stochastic dominance risk analysis methods showed that the trap fishing method alternative was preferred over the net and thread alternatives at all levels of attitude towards risk. Furthermore, all stochastic dominance risk analysis methods showed that the large number of traps (100 traps or more) alternative was preferred over the smaller and medium number of traps alternatives. However, the SERF analysis results showed that a DM who is risk neutral would prefer large-sized traps over medium- or small-sized traps. This is not the case for a risk-averse DM, who would prefer the small-sized trap alternative over the medium- and large-sized alternatives.

4. SUMMARY AND CONCLUSIONS

The Emirate of Abu-Dhabi, UAE, fishermen operate under extreme risk and uncertainty conditions. The UAE fishery sector has experienced multiple challenges in the last three decades (1980–2010). To analyze the sources of risk that face both fishermen and fisheries in the Emirate of Abu-Dhabi, UAE, two major risk analysis methods were considered: standard stochastic dominance techniques (first/second degree stochastic dominance and stochastic dominance with respect to a function) and stochastic efficiency with respect to a function (SERF). Stochastic dominance techniques and SERF allow the DM to rank risky fishing management alternatives (*i.e.*, fishing methods, size of traps and number of traps). The risk analysis (based on survey information obtained from 131 Emirate of Abu-Dhabi, UAE, fishermen) showed that the coefficients of variation (CV) for the annual gross margin and the monthly number of fishing trips were estimated at 1.25 and 0.61, respectively. First degree stochastic dominance (FSD) and second-degree stochastic dominance (SSD) were utilized to display gross margin (AED) CDFs based on the fishing method, trap size and number of traps. Furthermore, stochastic efficiency with respect to a function (SERF) under a negative exponential utility function between fishing methods, trap sizes and trap numbers was implemented and showed that fishermen's efficient management choices were not the same for risk-neutral decision-makers compared to risk-averse decision-makers. SERF results showed that large-sized traps would be choice of a risk-neutral DM; however, small-sized traps would be the choice of a risk-averse DM. The results illustrate the importance of considering both the attitude towards risk and economic inefficiencies in UAE fishery practices management and in designing successful fishery policies, as well as considering decision-making at the fisherman level. The risk and uncertainty analyses carried out in this paper included analysis of DM behavior using a continuum of stochastic methods across a spectrum of risk preference from risk neutral (*i.e.*, willing to consider some risk) to risk averse. The results show that fishermen who own small-sized traps face more risk compared to fishermen who own large- and medium-sized traps. The results also showed that DM-selected management alternatives based on the size of traps and the number of traps vary from those selected by a risk-neutral DM compared to those selected by a risk-averse DM. The stochastic dominance analysis showed that traps as a management alternative were preferred over both nets and threads for all techniques. Furthermore, results indicate that policy-makers should target small- and medium-sized fishery operations (those that typically own small-sizes traps and/or a small number of traps) to provide technical and business efficiency support and to also increase their ability to manage risk. Finally, the findings are among the first in the literature utilizing stochastic dominance techniques to identify and rank risky fishing management alternatives and also indicate that the adoption of less risky management alternatives by fishermen may augment the sustainability of this important food sector and ensure a sufficient long-term supply of fish in the UAE.

REFERENCES

1. FAO. Information on Fisheries management in the United Arab Emirates. Available online: http://www.fao.org/fi/oldsite/FCP/en/ARE/body. html (accessed on 1 June 2013).
2. Ministry of Environment and Water (MOEW). Exploitation, protection and development of the living aquatic resources in the waters of the state of the United Arab Emirates. Available online: http://www.moew.gov.ae/Portal/en/laws-and- legislations /federal-laws.aspx#page=1 (accessed on 15 June 2013).
3. Arab Organization for Agricultural Development (AOAD). Arab Agricultural Statistical Database. Available online: http://www.aoad.org/ publications.htm (accessed on 25 June 2013).
4. Ministry of Environment and Water (MOEW). Stop the new registration of speedy fishing boats. Available online: http://www.moew.gov.ae/ Portal/en/laws-and-legislations/executive-by-laws.aspx (accessed on 15 June 2013).
5. Abu-Dhabi Environment Agency. Fishers Statisical Bulletin. Available online: https://www.ead.ae/en/portal/fisheries.statistical.bulletin.aspx (accesse d on 15 July 2013).
6. Kirkley, J.; Paul, C.J.M.; Squires, D. Deterministic and stochastic capacity estimation for fishery capacity reduction. *Mar. Resour. Econ.* **2004**, *19*, 271–294.
7. Sharma, K.R.; Leung, P. Technical efficiency of the longline fishery in Hawaii: An application of a stochastic production frontier. *Mar. Resour. Econ.* **1999**, *13*, 259–274.
8. Pascoe, S.; Coglan, L. The contribution of unmeasurable inputs to fisheries production: An analysis of technical efficiency of fishing vessels in the English Channel. *Am. J. Agr. Econ.* **2002**, *84*, 585–597.
9. Sherif, S.A.; Gheblawi, M.S. Revenue determinants for Abu-Dhabi fishermen and assessment of input allocative efficiency. *Agr. Mar. Sci.* **2009**, *14*, 17–25.
10. Aisyah, N.; Arumugam, N.; Hussein, M.A.; Latiff, I. Factors affecting the technical efficiency level of inshore fisheries in Kuala Terengganu; Malaysia. *Int. J. Agr. Manage. Dev.* **2012**, *2*, 49–56.

11. Ahmed, N.; Zander, K.K.; Garnett, S.T. Socioeconomic aspects of rice-fish farming in Bangladesh: Opportunities, challenges and production efficiency. *Aust. J. Agr. Resour. Econ.* **2011**, *55*, 199–219.
12. Espino, D.C.; Del Hoyo, J.J.G.; Sharp, B.M.H. Capacity and Capacity Utilization of the "Voracera" Fleet in the Strait of Gibraltar. *Mar. Resour. Econ.* **2005**, *20*, 367–384.
13. Singh, R.; Weninger, Q.; Doyle, M. Fisheries management with stock growth uncertainty and costly capital adjustment. *J. Environ. Econ. Manage.* **2006**, *52*, 582–599.
14. Herrero, I.; Pascoe, S. Value versus volume in the catch of the Spanish South-Atlantic Trawl Fishery. *J. Agr. Econ.***2003**, *54*, 325–341.
15. Danielsson, A. Efficiency of catch and effort quotas in the presence of risk. *J. Environ. Econ. Manage.* **2002**, *43*, 20–33.
16. Bastardie, F.; Nielsen, J.R.; Andersen, B.S.; Eigaard, O.R. Effects of fishing effort allocation scenarios on energy efficiency and profitability: An individual-based model applied to Danish fisheries. *Fish. Res.* **2010**, *106*, 501–516.
17. Maravelias, C.D.; Tsitsika, E.V. Economic efficiency analysis and fleet capacity assessment in Mediterranean fisheries. *Fish. Res.* **2008**, *93*, 5–91.
18. Andersen, J.L. The inclusion of stocks in multi-species fisheries: The case of Danish Seiners. *Mar. Resour. Econ.* **2005**, *20*, 163–184.
19. Tingley, D.; Pascoe, S.; Coglan, L. Factors affecting technical efficiency in fisheries: Stochastic production frontier versus data envelopment analysis approaches. *Fish. Res.* **2005**, *73*, 363–376.
20. Del Hoyo, J.J.G.; Espino, D.C.; Toribio, R.J. Determination of technical efficiency of fisheries by stochastic frontier models: A case on the Gulf of Cadiz (Spain). *ICES J. Mar. Sci.* **2004**, *61*, 416–421.
21. Greenville, J.; Hartmann, J.; Macaulay, T.G. Technical efficiency in input-controlled fisheries: The NSW Ocean Prawn Trawl Fishery. *Mar. Resour. Econ.* **2006**, *21*, 159–179.
22. Robison, L.J.; Barry, P. *The Competitive Firm's Response to Risk*; Macmillan Publishing Co.: New York, NY, USA, 1987.
23. Boggess, W.G.; Ritchie, J.T. Economicandriskanalysisofirrigation decisions in humid regions. *J. Prod. Agric.* **1988**, *1*, 116–122.
24. Williams, J.R. A stochastic dominance analysis of tillage and crop insurance practices in a semiarid region. *Am. J. Agr. Econ.* **1988**, *70*, 112–120.
25. Williams, J.R.; Llewelyn, R.V.; Mikesell, C.L. An economic risk analysis of conservation tillage systems for wheat, grain sorghum, and soybeans in the Great Plains. *J. Soil Wat. Cons.* **1989**, *34*, 234–239.
26. Greene, W.H. *Econometric Analysis*, 5th ed.; Prentice-Hall International Inc.: New York, NY, USA, 2003.
27. Weersink, A.; Walker, M.; Swanton, C.; Shaw, J.E. Economic comparison of alternative tillage systems under risk. *Can. J. Agr. Econ.* **1992**, *40*, 199–217.
28. Hardaker, J.B.; Richardson, J.W.; Lien, G.; Schumann, K.D. Stochastic efficiency analysis with risk aversion bounds: A simplified approach. *Aust. J. Agr. Resour. Econ.* **2004**, *48*, 253–270.
29. Meyer, J.; Richardson, W.; Schumann, K.D. Stochastic efficiency analysis with risk aversion bounds: A correction. *Aust. J. Agr. Resour. Econ.* **2009**, *53*, 521–525.
30. Hardaker, J.B.; Huirne, R.B.M.; Anderson, J.R.; Lien, G. *Coping with Risk in Agriculture*, 2nd ed.; CABI Publishing: Cambridge, MA, USA, 2004.
31. Cochran, M.J. Stochastic dominance: The state of the art in agricultural economics. In Proceedings of Southern Region Project S-180 Seminar: An Economic Analysis of Risk Management Strategies for Agricultural Production Firms, Tampa, FL, USA, 23–26 March 1986.
32. King, P.K.; Robison, L.J. An interval approach to measuring decision maker preferences. *Am. J. Agric. Econ.* **1981**, *63*, 510–520.

33. Keeney, R.L.; Raiffa, H. *Decisions with Multiple Objectives: Preferences and Value Tradeoffs*; Wiley Publishers: New York, NY, USA, 1976.
34. Lien, G.; Stordal, S.; Hardaker, J.B.; Asheim, L.J. Risk aversion and optimal forest replanting: A stochastic efficiency study. *Eur. J. Oper. Res.* **2007**, *181*, 1584–1592.
35. Pendell, D.L.; Williams, J.R.; Boyles, S.B.; Rice, C.W.; Nelson, R.G. Soil carbon sequestration strategies with alternative tillage and nitrogen sources under risk. *Appl. Econ. Perspect. Pol.* **2007**, *29*, 247–256.
36. Babcock, B.A.; Choi, E.K.; Feinerman, E. Risk and probability premiums for CARA utility functions. *J. Agr. Res. Econ.* **1993**, *18*, 17–24.
37. Richardson, J.W.; Schumann, K.D.; Feldman, P.A. *Simulation & Econometrics to Analyze Risk: Simetar© Inc. User Manual*; Texas A & M University: College Station, TX, USA, 2008.
38. United Arab Emirates Ministry of Water and Environment (MOEW). Fisheries Statistics. Available online: http://www.moew.gov.ae/en/knowledge-and-statistics/ statistics.aspx#page=1 (accessed on 12 May 2014).

CHAPTER 10

Dynamics of Regional Fisheries Organization's and Foreign Trade: Emphasis on the Environment and Restructuring in Mexico

Hidalgo Gallardo Amada, Hidalgo Gallardo Ruth Leticia, Ávila Hidalgo Sofía Elizabeth

Department of Economic and Management Science, University of Hidalgo State, Pachuca, México

ABSTRACT

This article has the purpose to give a perspective related with actions about climate change, fishing, fisheries regional organizations and the States regarding trading sector, considering that fishing is a vulnerable but important sector for food security. However, it is important to analyze the applications of external politics in internal national politics about restructuring fishing, safeguard and respect the resource natural. According to the theme, this study has been suitable to analyze it from the theories of the interdependence complex (Keohane and Nye, 1977 [1]), and vulnerability and sensitivity theories (Katzenstein, 1976 [2], Dhal, 1957 [3], Tucker, 1977 [4] and Holsti, 2006 [5]) in the application of external policies, and make the most vulnerable States as well as the sensitivity in their sectors. In order to verify the implementation of all policies also to prove the climate change governance, it has been determined to analyze the case of Mexico (2006-2013) who is facing this phenomenon. The results show that while Mexico has implemented programs and projects, the achievements will be given in a long term, due to the lack of infrastructure, technology, innovation and technical capacity. The strategies of Mexico government go through links with the Regional Fisheries Organization in order to achieve the best strategies for maintaining fishery productivity and trade sector.

Keywords: Environmental Regulation, Foreign Trade, Regional Fisheries Management Organizations (RFMOs), Fishing, Mexico

1. INTRODUCTION

Today the International Community (IC) is addressing challenges such as: economic interdependence, natural resources and environment. These challenges concern to humanity who looks how to ensure the ecosystem, food and life.

Environmental pollution in the seas has confronted the States and International Organizations (IGOs) to discuss and process the strategies which are necessary for its containment. This international scene, the most catastrophic of recent times, is making rapid changes related to markets, exploitation and forms of governance in fisheries.

So, on one hand there is the vulnerability in fishing activity, and on the other the sensitivity in states to abide by the international policies on restructuring and resource, which generally affects to the less developed countries that at the same time present overfishing. Here that States adhere to the determinations of International Organizations (IGOs) with the purpose to provide mechanism and support to safeguard this economic activity, without affecting production or foreign trade sector.

The International Community (IC) proposal on it goes regarding the Regional Fisheries Management Organizations (RFMOs) who are responsible for providing information to States about the dynamics of regional and global fisheries. The aim is that with this information States may take decisions on their strategies of exploitation, domestic consumption and trading expectations. In Mexico, the fishing activity is important because of its geographic location but is also highly susceptible of climate events; before these two aspects the support of the RFMOs for Mexico is essential because it can generate mechanisms of protection, at same time of be assertive in taking decisions.

In this context, International Community (IC) tends to search strategies to face the shortage of fisheries, displayed to next generations but it is clear that currently the demand of fish is intensified, prices tend to increase, production diminishes and fishing activity is each time more globalized.

2. METHODOLOGY

This investigation is orientated to the application of the descriptive analytical method and substantiated on the Complex Interdependence Theory of [1], both recognize that economic acceleration have contribute to the global warming phenomenon who has taken to unexpected consequences in all aspects. This situation exposes States to look for methods to address this global overview. Furthermore the opinions of [2] [3] [4] [5] are taking into account regard to the conditions of sensibility and vulnerability that States address derived from the foreign policies applicable on their domestic policies. This case of study verifies Mexico's sensibility owing to the fishing restructuration through international system; also it is interpreted fishing vulnerability, transformation, threat and habitat change which are derived from the environmental phenomenon.

On this basis, is resorted to the recompilation and analysis of data and information from documents and literature surveys from the Regional's Fisheries Management Organizations (RFMOs) as well as the guidelines of specific organizations for environmental treatment and their consequences on fishing trade. This analysis specifies the implemented actions by Mexico on their domestic policies between 2006-2013 with the purpose of counteracting the international rate with national application. Documents such as the Development National Plan, the Official Journal of the Federation and National Regulation are considered.

Furthermore, documents, strategies, reports and data of Mexican government agencies are reviewed for the frame of reference to present a complete analysis; transcendental for the phenomenon of study.

2.1. Regional Fisheries Management Organizations (RFMOs) as the Way for Industry Restructuring

In accordance with Sarukhán, et al., (2009) [6] the international approach; Mexico is a mega diverse country, since it has between 60 and 70% of the planet diversity, taking into account privileged geographic location (11,592 km of shore, 3 million km^2 of Exclusive Economic Zone (EEZ), 2.9 million of continental water has) which contains many ecosystems and species with the most varied forms of life but SEMARNAT (2007) [7] has mentioned that the future with the changing climate affects in increase of the level of the sea with impacts to the ecosystem coastal and marine; according with the OECD regarding its situation, the Mexican fishing activity is important for national and international economy. The Mexican government has a long term view to encourage national development, competitiveness and fishing strategic planning, articulated on the (PND) Development National Plan (OECD, 2013) [8].

The international policies on fishing restructuring focus on the RFMOs as an essential part on the guideline ruling. It is relevant to connect the RFMOs since it arises from international agreements where States decide to bring them the management of natural resource from a determined area. Therefore, these are mechanisms created to provide statistic and analytical data information with the aim of help both, governments and Intergovernmental Organizations, to make pertinent decisions about the responsible use of resources towards conservation, management and development of fishing resources[1].

Similarly there are other organisms dedicated on coordination and cooperation as Regional Organizations of Fishing Management (OROP) who then, have implemented on their faculties and procedures the application of international legal instruments. Such instruments ought to exercise efficiency in the resources' conservation and order ensuring sustainability[2].

In this approach, it is denoted the recapitalization of the industry which is amplified and transformed with the aim to answer the problematic of the activity. One more time it is confirmed that the variables to balance such as: overexploitation, capitalization, technology, fleet renewal, subsidies, competence between manufacturing and artisanal fishing are multidimensional and hard to harmonize (Fernández, 2006) [9].

Given this scenario, Mexico joins into the international arrangements as Law of the Seas and Responsible Fishing, regulations which are the guide for the IC to make the appropriate considerations on their governments to safeguard the environment and the natural resource. Mexico is part of this commitment at the time on signing multilateral conventions such as, UNCLOS in 1982, FAO in 1995 and COFI. Additionally, Mexico belongs and contribute with global and interoceanic RFMOs as the International Whaling Commission (IWC), and the Latin-American Organization for Fishing Development (OLDEPESCA)[3].

In the Atlantic, Mexico participates in the International Commission for the Conservation of Atlantic Tunas (ICCAT) and the Western Central Atlantic Fishery Commission (WCAFC). In the pacific, the country is member of the Inter-American Tropical Tuna Commission (IATTC) and regard to the fishing on inland waters; it belongs to the Aquaculture and Fishing Commission of Latin America and the Caribbean (COPESCAAL). Therefore, for Mexico the collaboration with RFMOs offers a panoramic to consider in the future.

It is regarded that fishing represents today's major threatening for open sea, but another potentially dangerous activities-such as: biosprection, mining, energetic production and CO_2retention-could constitute future threats for marine biodiversity that would make necessary the instauration of a complex set of management agreements.

Therefore, the impact of Climate Change will confront regard to the industry trends and the situations presented on fishing activity, thus, International Com- munity has reaffirm the decisive importance of RFMOs and the need to strength and modernize them[4].

2.2. The Environmental Influence on Mexican Fishing

The fishing industry is facing dramatic consequences derived of various circumstances such as; climate change, pollution and acidification of seas, infrastructure and net fishing. These aspects have contributed to fishing sensibility and on their uncertain future regard on climate variability. The same fishing industry contributes on greenhouse effect derived from their activities on capture, transport, elaboration and storage. Considering that fishing ships consume and throw fuel into the sea, and also that in many cases, fishing periods are long.

Consequently, actually it has increase health deterioration of aquatic ecosystems on seaside areas, this area is objective of an excessive exploitation or an excessive waste feeds. As seaside ecosystems are essentials for reproduction and growth of numerous marine species, the impact of degradation over the volume of marine resources including those of commercial importance is direct and negative[5].

In the case of Mexico, the government on the National Development Plan (PND) 2006-2012, confirms that fishing managements on Climate Change Special Program (PECC)includes variations on climate factors, for that reason, the adaptive capacity of the fisheries of the country is strengthened. Within the timeframe 2009-2012, it has been elaborated that more than 60% of fishing programs considered on the objectives, being in process the totality of the committed programs[6].

The National Charter for Fisheries (CNP) determines that from 75 fishing handling units, 45 (60%) have reached their maximum capacity, being complete exploited 20 (26.6%) record overexploitation of 86.6% from fishing marine units. This means that they are no more susceptible to an increase of production. On the contrary, many of them require management and protection actions to stop deterioration. National Commission for

Aquaculture and Fisheries CONAPE- SCA attests to the above estimating that only 14.2% of fisheries have potential development [6].

Mexico faces a considerable environmental pressure; the estimations indicate that the 68% of population and the 71% of GDP are very exposed to climate change risks. This situation affects agriculture production, water availability, deforestation, biodiversity and health. Even if does not exist attributable alterations on climate change, it is generally estimated that it is necessary to reduce the fishing capacity and effort on most of the fisheries (Daw, et al., 2009) [10]. Nevertheless, the reduction of greenhouse effects is still an important challenge since, if they continue increasing, natural catastrophes will heat coastlines and consequently the productivity of marine natural resource [8].

Mexico bases their fishing strength on their vast coastlines but, due to its geographic location, is facing natural phenomena as "El Niño" which generates changes in the position of pelagic fisheries, higher losses due to bad weather and risks for fishermen because of population migration and the obtaining of the resource. In any case, procedures of fisheries governance have to be malleable to changes and oriented to equal and sustainable fishing and on an eco-systemic approach with can increase the capacity of fishing adaptation [10].

2.3. Production and Marketing of Mexican Fishing before RFMOs Restructuring, Environmental and Guideline Measurements

The Mexican government adjusts their domestic and management policies in order to restructure the fishing activity through rules and bodies which collaborate to an optimum management of the resource. Mexico's position on fishing activity is framed on aspects of fishing Mexican legislation, institutionalization and the industry intervention of RFMOS, considering that the importance of management and institutional incentives has increased.

Internally, the Mexican government has based their management policies and regulations under the General Fishing Law (GFL 1992)[7], Regulation of the Fisheries Law (1999)[8], and the Fishing National Charter (CNP 2000)[9]. The Charter works as an instrument on cartographic and written version that has an informative purpose about the fishing resource susceptible of exploitation[10].

In this way, it has been launched phases of control and regulation, promotion and development, social participation, political control and opening and deregulation. Legislation will still be the instrument which continues with the development plans of the country (Cruz-Ayala, and Igartúa-Calderón, 2006) [11].

Therefore, the legislation mentioned above, are three normative dimensions which lie on the management of the resource. These measures are executed to exercise control over the captures, using traditional methods as licenses and charges, the maximum quota of capture, fishing seasons, respect of closure or that one's of protection and special management, and alternative mechanisms that applies on incidental fishing and on individual fishing control[11].

The strategy on the administration and the procedure of the management precise: the granting of fishing and aquaculture concessions and licenses, the verification of arts and the allowed equipment to carry out the activity. The resources availability is ruled according to evaluations and technique, economic and social studies, but mainly in the granting of quotas of capture to people with Mexican nationality or to legal persons constituted under Mexican laws.

However to put on practice the legislation, it is necessary the institutional view, the Secretariat of Environment and Natural Resources (SEMARNAT) holds the fishing management through policies, strategies and actions for sustainable use with benefits for the whole industry (Marín, (2007) [12]. Currently, it is up to Secretariat of Agriculture, Livestock, Rural Development, Fisheries and Food (SAGARPA) to follow up management, managing the use of seas on a profitable and sustainable way, for both the internal and external-markets, str- engthening quality and responding to customer's requirements[12].

On the international aspect, the Mexican government through the ministerial meeting of Cancun in 1992 proposed a code of responsible practice for fishing, and at the heart of the summit was decided to establish a normative document of international fishing activities; signed in 1995[13].

Furthermore, Mexico attends the RFMOs ruling toward the Mexican fishing activity, making emphasis on the industry's disadvantages. These mismatches are found on fishing administration and on the geographical extent to manage. The capture is considerable situated in the northwest of the country, spear headed by sardine fishing, while the Gulf of Mexico presents a lower level of capture. Another aspect to highlight is the unbalance between the industrial and artisanal fleet.

Despite the regulatory and institutional reordering raised by Mexican government, international institutions suggest to the government the need to strengthen conservation and sustainable use of biodiversity. At the same time they recognized the effectiveness of the programs, rescue boats plans and licenses to ensure an improvement on the sustainable fishing management. However, the RFMOsen courage Mexico to reform the fishing sector to manage the ecosystem with more effectiveness and to obtain biological resources, which in turn will favor respectful environmental practices[14].

According to international remarks from RFMOs, the Mexican government promotes programs related to each other, raised a holistic way, that is, integrated due to the need of coordinate diagnostic actions and technique operation of protected and economically important species(Espinoza-Tenorio y Espejel, 2012) [13].

The established programs are subjected to rules of operation with fishing and aquaculture components focused on food security, emphasizing; investment for equipment and infrastructure, the natural resources sustainable program and transversal projects[15].

The programs derive from modernization projects of smaller vessels, incorporating modern and ecologic equipment that contribute to environmental care reducing fuel consumption. Within this program is the coverage for investment as a reconversion project from 2008-2011. Another variety of projects focuses on the decrease of fishing effort, mainly the removal of shrimp vessels to contribute on the use of sustainable resource.

The inspection and surveillance focus on fight against illegal fishing, among other media through satellite monitoring system. The project of systemic payment is applied in concessions and licenses on the frame of the productive reconversion of the Gulf of Mexico and the Caribbean. One last branch of the projects is focused on energetic resource such as: the use of marine diesel and the riverside gasoline, transport and trade of fishing resources and attention to natural disasters.

This modernization requires financial resources since they mean changes, improvement, provision, development and opening to address the climate change challenge. It is indispensable to count with economic resource in long term because as the modernization process is deepened, the financial requirement for infrastructure needs, equipment, machinery, operation supplies, human resource, services and technology will increase (Cebreros y Guarneros, 2006) [14].

Modernization will have to attach to social aspects, the environmental conservation and on the available financial resources. Considering that this strategy is developed in long term, is required to arrange investment both private and public. With all these, Mexico pretends to manage resource for international investment and financial services which require productive modernization and commercial opening.

Despite the disadvantages mentioned above, the Mexican fishing activity presented a rebound of 1.4% annual between 2006-2011, especially because of the sardine species, which were the principal captured products, with 1.4 m/t produced. SAGARPA, via CONAPESCA, applied a resource of approximately 3 million dollars for the programs described above and other resources in the presentation of projects that helped with fishing reordering. These programs are annually reviewed, but the sector still requires more reforms as; fiscal, infrastructure and innovation to achieve a major productivity (OMC, 2013) [15].

The Table 1 indicates the productivity and composition of the Mexican fishing activity.

In another perspective and taking into account that Mexico is member of the Inter-American Tropical Tuna Commission (IATTC) its regional participation is relevant through its tuna fleets, emphasizing the yellow fin tuna. This tuna is industrialized and is addressed to both the domestic and export markets. Furthermore, tuna is fished for two purposes; for fattening in sea corrals and for sale to international markets, which generates 50 million US dollars annually. Exports go the United States and Japan, which generates direct and indirect jobs (MSNBC.com 2016a) [16]. Therefore, the IATTC includes 6,600 tons of bluefin tuna catch quotas (2017 and 2018) and the capture goes according to the stability of the industry. The measure is drastic, but seeks to maintain a control on mortality in addition to bluefin tuna continues to have problems of exploitation (MSNBC.com 2016b) [17]. According to the IATTC (2016) proposals for the protection of natural resources, specifically for tuna in the Pacific Ocean are on edge of capture either by individual ship and/or days of veda.

Table 1. Main appropriation lines of the Mexican fishing industry by 2010.

Appropriation Line/Year	2003	2010	Observations
Vessels	239,539	217,176	
Employment	271,714	293,803	
Capture	1,279,060	1,357,900	
Financial resources provided by the Mexican government.	1,285,042	931,965	The first resource was in 2004 and the last in 2007.

Source: Compiled from data of the OECD document Review of fisheries: country statistics 2012.

All the above mentioned have influenced the international trade since the products demand has increased strongly; 40 per cent of the total value and 33 per cent of the total volume of the produced fish entering international trade. The half of this volume is exported by developing countries (Delgado et al., 2003) [18], representing more than any other foodstuff export income

(Thorpe et al., 2006) [19]. In this respect, fishing position tends to fluctuations and consequently becomes difficult in relation to commitments to international agreements with respect to the marketing of the product as well as partnerships between national and foreign fleets, considering that its activity can only be exercised in its EEZ so that trade would be affected.

According with these procedures, it can realize that climate change, international systems and governance are creating situations of vulnerability in the fisheries sector. These circumstances causes uncertainty in the fish markets facing an increasing demand due to increasing globalization, these commercial disorders will continue to be altering fisheries restructuring.

3. DISCUSSION

The trading and restructuring fisheries in Mexico lies in three important aspects; the opinions and support by statistics and proposals generated by the RFMOS, the climatological phenomenon, and industrial waste. According to the former approach, the panoramic Mexican fishing notes that it serves and engages in international regulation, adapting them to its internal policies in regulatory, institutional aspects and application. Through them it fits and creates the implementation of programs and projects in favor of fishing. However, the fishing sector in Mexico, faces climate events that are intensified each time more derivative the greenhouse effect as well as the treatment of waste and modernization for the ravages of climate change. On the other hand, modernization provides through computer applications, processes administrative, customs, port infrastructure, fleet and transport in general, which can act on the exploitation of the resource, surveillance, transport and marketing. He is additionally subject to the provisions of capture, veda and banks which determines the RFOS. In summary the situation in Mexico affects the capitalization of programs and projects in the long term, requiring technologies, financial resources, treatment of waste water, meteorology and, finally, resources to be able to achieve the goal.

4. CONCLUSIONS

First at all, international agencies are cooperating in the fisheries restructuring; however, not all countries have the same infrastructure to follow up such a program. On the other hand, the environment is a decisive factor in the lack of resources, is something that cannot be controlled.

Therefore, for developing countries, it is very difficult to confront these two processes since they require technology to respond requirements of international fisheries restructuring policies also to the environmental challenge.

Mexico faces these consequences, but the government is working on it; however there is another relevant sector that is being damaged because of the above, the fishing marketing sector. It is very hard to make political actions when necessary to cover market internal demand and give follow-up to its marketing with others countries.

Mexico is trying to create programs and projects that support all this kind of fisheries restructuring and marketing process but, they are in a long term.

REFERENCES

1. Keohane, R. and Nye, J. (1977) Power and Interdependence: World Politics in Transition. Little Brown, Boston.
2. Katzenstein, P. (1976) International Relations and Domestic Structures: Foreign Economic Policies of Advanced Industrial States. International Organization, 30, 1-45. https://doi.org/10.1017/S0020818300003726
3. Dahl, R. (1957) The Concept of Power. Systems Research and Behavioral Science, 2, 201-215. https://doi.org/10.1002/bs.3830020303
4. Tucker, R. (1977) The Inequality of Nations. Basic Books, New York, 97.
5. Holsti, O.R. (2006) Theories of International Relations. Making American Foreign Policy. Taylor and Francis Group, New York, 313-344.
6. Sarukhán, J., et al. (2009) Síntesis: Conocimiento actual, evaluación y perspectivas de sustentabilidad, Comisión Nacional para el Conocimiento y Uso de la Biodiversidad. Gobierno Federal México, Capital Natural de México, México, 21-63.
7. SEMARNAT (2007) "México ante el cambio climático" Instituto Nacional de Ecología. SEMARNAT, Secretaría de Medio Ambiente y Recursos Naturales. http://www2.inecc.gob.mx/publicaciones/gacetas/154/cclimatico.html
8. OECD Review of Fisheries: Policies and Summary Statistics 2013 (2013) OECD Publishing, Paris, Cap. 26.
9. Fernández, J. (2006) Características y problemas de la pesca en México. Pesca Acuacultura e Investigación en México, Comisión de Pesca, Centro de Estudios para el Desarrollo Rural Sustentable y la Soberanía Alimentaria CEDRSSA, Cámara de Diputados, 157-169.
10. Daw, T., Adger, W.N., Brown, K. and Badjeck, M.C. (2009) El cambio climático y la pesca de captura: Repercusiones potenciales, adaptación y mitigación. In: Cochrane, K., De Young, C., Soto, D. and Bahati, T., Eds., Consecuencias del cambio climático para la pesca y la acuicultura: visión de conjunto del estado actual de los conocimientos científicos, FAO Documento Técnico de Pesca y Acuicultura No. 530, Roma, 119-168.
11. Cruz-Ayala, M. and Igartúa-Calderón, L.E. (2006) La transformación de la legislación pesquera en México: Un acercamiento en el contexto político-económico (1925-1992). In: Guzmán Amaya, P. and Fuentes Castellanos, D.F., Coord., Pesca, acuacultura e investigación en México, Cámara de diputados, Comisión de pesca, CEDRESSA, México, 233-247.

12. Marín, G. (2007) Pesca artesanal, comunidad y administración de recursos pesqueros. Experiencias en la costa de Michoacán, Gazeta de Antropología, Número 23, México, 23-20.
13. Espinoza-Tenorio, A. and Espejel, I. (2012) Investigación sobre el manejo holístico de la pesca en México: Prioridades gubernamentales para el siglo XXI. Ciencia Pesquera, 20, 91-96.
14. Cebreros, A. and Guarneros, J. (2006) Mecanismos financieros para la modernización de la pesca en México. In: Guzmán Amaya, P. and Fuentes Castellanos, D.F., Coord., Pesca, acuacultura e investigación en México, Cámara de diputados, Comisión de pesca, CEDRESSA, México, 135-143.
15. OMC (2013) Examen de Políticas Comerciales. México, 14-138.
16. Ensenada.Net (2016) Reeligen a Guillermo Compean al frente de la CIAT. órgano rector de la pesca atunera en el Pacífico, Noticia del 30 de Junio 2016 publicada por Elizabeth Vargas. http://www.ensenada.net/noticias/nota.php?id=45563
17. Ensenada.Net (2016) Mantendrá CIAT cuotas de alea azul 2017 y 2018. 3,300 toneladas anuales para ranchos atuneros. Noticia del 03 de noviembre 2016 publicada por Karla Lorena Lamas. http://www.ensenada.net/noticias/nota.php?id=47025
18. Delgado, C.L., Wada, N., Rosegrant, M.W., Meijer, S. and Ahmed, M. (2003) Fish to 2020: Supply and Demand in Changing Global Markets. World Fish Center, Penang.
19. Thorpe, L.E., Gwynn, R.C., et al. (2006) Community Health and Nutrition Examination Surveys for Public Health Surveillance: The New York City Experience. Preventing Chronic Disease, 3, 1-8.

FOOT NOTES

[1]The Royal Institute of International Affair. Buenas prácticas recomendadas para las OROP 2007.

[2]Comisión Permanente del Pacífico Sur.-Ordenamiento Jurídico Internacional de la Pesca 2009. http://www.pnuma.org/gobernanza/cd/Conferencias/3%20Jueves%207/Ordenamiento%20Jur%EDdico%20de%20la%20Pesca%20en%20Alta%20Mar..pdf

[3]FAO. ORP. Áreas de competencia pesquera: México.

[4]FAO (2012). "La ordenación pesquera—Las áreas marítimas protegidas y la pesca" AMP y Redes de AMP en Alta Mar. Anexo 1 pag. 175-220 www.fao.org/docrep/015/i2090s/i2090s.pdf.

[5]FAO. Comité de Pesca 1997.

[6]Comisión Intersecretarial de Cambio Climático. Informe de Avances del PECC 2009-2012. pp 50.

[7]General fishing law.-In Mexico it was created, authorized and implemented by the Chamber of Deputies, announced in the Official Gazette of the Fisheries Act in 1992; this law has had updates in 1999, 2001, 2004 and 2007 according to national and international changes. This law ensure the conservation, preservation and use the resources fishery to establish them for suitable promotion and administration. Everything is applicable according to the NDP see.

http://www.diputados.gob.mx/LeyesBiblio/abro/lpes/LPes_abro.pdf

[8]Regulation of the fishing law. It was elaborated by the SEMARNAP to manage and regulate fisheries, this regulation works together with General Law of fisheries and PNC, this regulation is applied to optimize the natural resource to the management of commercial and civil society

in their fishing activities see http://www.sagarpa.gob.mx/saladeprensa /Documents/RLGPAS3DEJULIO.pdf.

[9]The National Charter for fishing was founded in the year 2000, according to the changes given by the same process in the sector; the Charter has undergone updates in the years 2004, 2006, 2010, 2012 and 2014. The main changes were made on scientific, regulatory and operational bases, making to the activity safe in terms technical, economic and social. To ensure these terms, it was necessary the guide by letters national of fishing, plans of management fishing and aquaculture, standards official Mexican and technical opinions of these instruments, provide information detailed on the sector through technical and scientific works in order to take of decisions. This work covers the natural and coastal species both can be new as updated looking at areas of the Pacific and the Gulf of Mexico. All the evaluation and results about research scientific and technology is available in the book red called "sustainability and fishing responsible in Mexico" which contains data on assessments biological fishing, diagnostics of production, quality of the water, developments technological and projects to develop, these data are thrown of it application in different areas coastal and for different species. The updates and the reforms to the fishing sector goes chained to the Plan of National Development, of course regarding the international requirements. All this information can be consulted in INAPESCA:

http://www.inapesca.gob.mx/portal/component/ content /article/17-publicaciones/74-carta-nacional-pesquera.

[10]DOF. SEMARNAP. Autorización CNP 2000.

[11]AIDA. Herramientas para la pesca sostenible y el manejo costero. pp 9, 23.

[12]Gestión Pesquera en México. Coordinado por la Secretaría de Industria y Comercio de 1972, la Secretaría de Pesca de 1982, Secretaría de Medio Ambiente y Recursos Naturales y Pesca (SEMA- RNAP) en 1991 y la SAGARPA a partir del 2001. Asimismo, se fortalecen los asuntos pesqueros me- diante las actividades de la Comisión Nacional de Acuicultura y Pesca (CONAPESCA), el Centro Informático para la Pesca (INFOPESCA) y el Instituto Nacional de Pesca (INAPESCA).

[13]FAO. Código de Conducta de la pesca Responsable 1995.

[14]OECD. Evaluaciones de la OCDE sobre el desempeño ambiental: México 2013. pp 17.

[15]SAGARPA. Programas sujetos a reglas de operación. Acuerdo anunciado en el DOF-2010.

CHAPTER 11

Benefits of Rebuilding Global Marine Fisheries Outweigh Costs

Ussif Rashid Sumaila[1*], William Cheung[2], Andrew Dyck[1], Kamal Gueye[3], Ling Huang[4], Vicky Lam[1], Daniel Pauly[2], Thara Srinivasan[5], Wilf Swartz[1], Reginald Watson[2], Dirk Zeller[2]

[1]*Fisheries Economics Research Unit, Fisheries Centre, University of British Columbia, Vancouver, British Columbia, Canada,*

[2]*Sea Around Us Project, Fisheries Centre, University of British Columbia, Vancouver, British Columbia, Canada,*

[3]*The United Nations Environment Programme, Geneva, Switzerland,*

[4]*Department of Economics, University of Connecticut, Storrs, Connecticut, United States of America,*

[5]*Pacific Ecoinformatics and Computational Ecology Lab, Berkeley, California, United States of America*

ABSTRACT

Global marine fisheries are currently underperforming, largely due to overfishing. An analysis of global databases finds that resource rent net of subsidies from rebuilt world fisheries could increase from the current negative US$13 billion to positive US$54 billion per year, resulting in a net gain of US$600 to US$1,400 billion in present value over fifty years after rebuilding. To realize this gain, governments need to implement a rebuilding program at a cost of about US$203 (US$130–US$292) billion in present value. We estimate that it would take just 12 years after rebuilding begins for the benefits to surpass the cost. Even without accounting for the potential boost to recreational fisheries, and ignoring ancillary and non-market values that would likely increase, the potential benefits of rebuilding global fisheries far outweigh the costs.

INTRODUCTION

Fish are among the planet's most important renewable natural resources. Beyond playing a crucial role in marine ecosystems, fish support human well-being through employment in fishing, processing, and retail services [1]–[3], as well as food security for the poor, particularly in developing countries [4]. Overexploitation [3], [5], [6] and rising ocean temperatures threaten global fisheries [7]–[9]. As demonstrated by the collapse of northern cod off Newfoundland, the depletion of fish stocks can have devastating effects on human well-being [10], [11]. As human populations continue to grow, the future benefits that fishery resources can provide will depend largely on how well they are rebuilt and managed. However, policy makers often perceive that rebuilding fisheries is too expensive in the short-term and therefore avoid taking the necessary actions to sustainably manage fish stocks. Therefore, a crucial question for policy makers is what is the potential net economic benefit of rebuilding global fisheries? Here, we address this question on a global scale.

Fisheries economists use resource rent (i.e., what remains after fishing costs and subsidies are deducted from revenue) as an indicator of fisheries performance [12], although others argue that this is inadequate because it does not capture all the benefits derived from marine fisheries [13]. Here, we adhere to using resource rent as our primary indicator of economic performance, but we also report payments to labor (i.e., wages) and earnings to fishing companies as additional indicators of fisheries benefits. With these additional indicators, we recognize that fishing capacity is not often converted to other uses easily (i.e., it is non-malleable) and that the opportunity cost of fishing labor (i.e., the alternative wages that fishers can earn if they did not fish) in many fishing communities is low due to a dearth of alternative employment. Even with these additional indicators, other important contributions of fish populations to the economy, such as the value created through the production chain [1] and non-market values [14] are not captured.

Over the past decade, we have gathered data on the economics of global fisheries from a range of sources, including scientific, economic, governmental, inter- and non-governmental publications, to create several global databases on catch [15]; ex-vessel fish prices [16]; subsidies [17]; and fishing costs [18] (Tables S1, S2, S3, S4, S5 and S6). From these databases, we compile landed value of catch, cost of fishing, payments to labor, earnings of fishing companies, and fisheries subsidies for 144 maritime countries of the world. We then compute both current and potential maximum resource rent, wages, and earnings to fishing enterprises.

RESULTS

1. Gains from Rebuilding

Global marine fisheries landings are projected to average 89 million t per year (range 83–99 million t) (Table 1) when rebuilt [19], with a corresponding mean landed value of US$101 billion per year (range US$93–116 billion). The wide ranges help to address uncertainties about the magnitude of global overfishing currently debated in the literature (as discussed in Materials and Methods). The cost of fishing in this rebuilt scenario is estimated at US$37 (US$29–44) billion compared to US$73 (US$50–96) billion per year currently. Returns to capital invested (i.e., normal profit) and payments to labor would amount to US$3 (US$2–4) billion and US$16 (US$12–19) billion per annum, respectively, while resource rent from rebuilt global fisheries would be US$54 (US$39–77) billion per year (summary of current resource rent is displayed by country in Figure 1, with details in Table 1). (The *Sunken Billions* report of the World Bank [20], which estimated economic rent without addressing the cost of reform, arrived at a potential resource rent of US$50 billion per year, using a different approach.) Gains in resource rent from the current situation to a rebuilt global fishery would be US$66 (US$51–89) billion a year, while wages and returns to capital will decrease to US$16 and US$3 billion, respectively (Table 1). Figure 2 summarizes the net gains in resource rent by maritime country.

2. COST OF REFORM

The real cost to society of rebuilding fisheries, once the elimination of an estimated US$19 billion per year of harmful and ambiguous subsidies is taken into account [17], is negative, implying that society as a whole will make money by engaging in rebuilding (Figure 3). However, fishing enterprises and fishers will lose profits and wages during rebuilding. Hence, to implement a rebuilding reform, governments may need to temporarily invest extra resources to mitigate these impacts.

Table 1. Key economic figures of global fisheries.

Key indicators, annual data (unit)	Current	Rebuilt fisheries		
		Lower bound	Mean	Upper bound
Catch (t)	80.2	82.7	88.7	99.4
Catch value (US$ billions)	87.7	92.6	100.5	116.3
Variable fishing cost (US$ billions)	73.0	43.9	36.6	29.3
Normal profit (US$ billions)	6.1	3.7	3.0	2.4
Wages (US$ billions)	31.0	18.6	15.5	12.4
Subsidies (US$ billions)	27.2	10.0	10.0	10.0
Rent net of subsidies* (US$ billions)	−12.5	39.0	54.0	77.0
Rent increase over current values (US$ billions)	–	51.2	66.4	89.4
NPV of resource rent increases (US$ billions)	-	665.2	972.0	1,428.1
Transition costs** (US$ billions)	–	129.9	202.9	292.2
NPV net of transition costs (US$ billions)	-	535.3	769.1	1,135.9

NPV: Net Present Value.
*The (resource) rent is the return to 'owners' of fish stocks, which is the surplus from gross revenue after total cost of fishing is deducted and subsidies taken into account.
**Transition costs include the costs to society of reducing current fishing effort to levels consistent with maximum sustainable yield and the payments governments may decide to employ to adjust capital and labour to uses outside the fisheries sector. Such payments may include vessel buyback programs and alternative employment training initiatives for fishers.

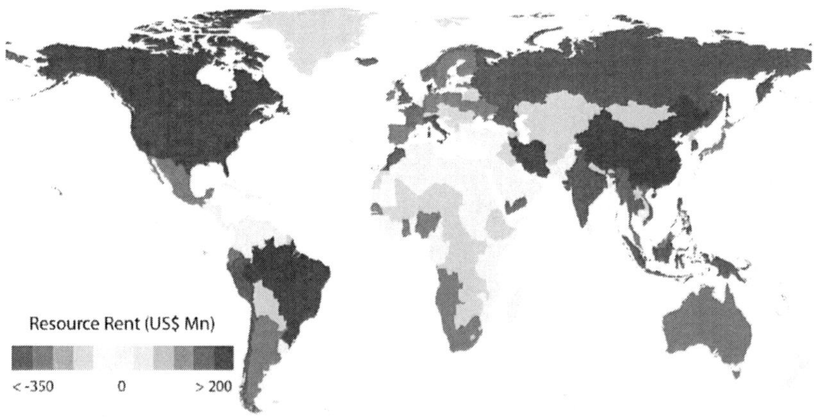

Figure 1. Summary of resource rent (adjusted for subsidies) from current fisheries. We see that several countries are in red once the full cost of fishing, including harmful subsidies are taken into account.

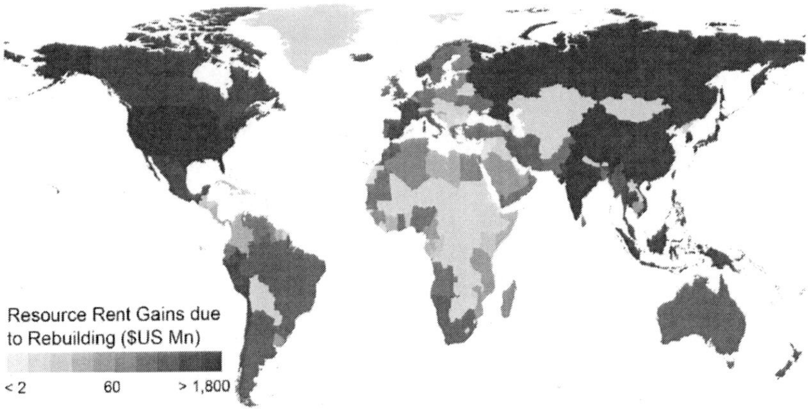

Figure 2. Summary of resource rent (adjusted for subsidies) from rebuilt fisheries (rent in all maritime countries increase after rebuilding).

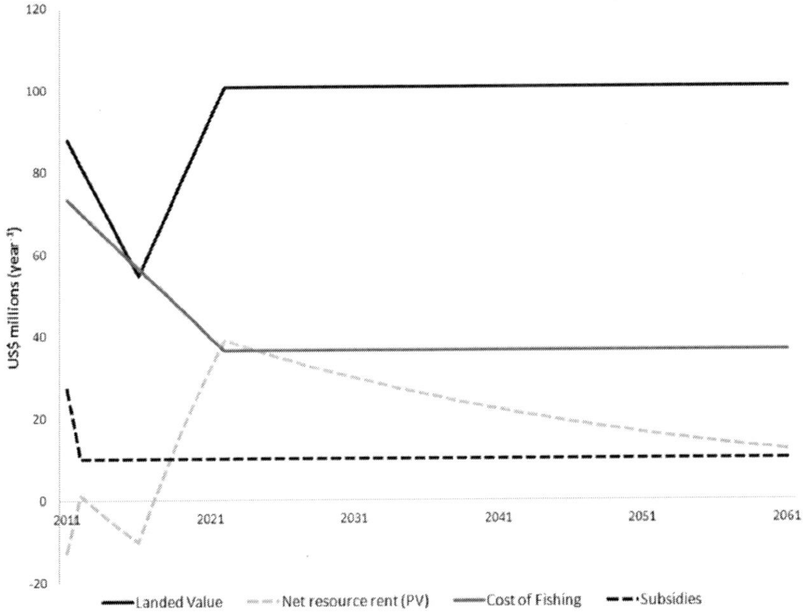

Figure 3. Transition time path of key rebuilding global fisheries variables.

The world's current fishing capacity is estimated to be up to 2.5 times more than what is needed to land the Maximum Sustainable Yield (MSY) [21]. This suggests that to rebuild global fisheries, we need to trim excess capacity from the current 4.3 million fishing boats [3]. Assuming that current capacity is between 1.5 and 2.5 times the level needed to maximize sustainable catch, fishing effort needs to be reduced by between 40 and 60 per cent, or up to 2.6 million boats. Fisheries currently employ more than 35 million people globally [3]. If we simplify by assuming linearity between boats and people, this implies that between 15 and 22 million fishers would need to be moved to other livelihood activities in order to rebuild global fisheries. This is a challenge, but one that is surmountable. For instance, even though in some fisheries most fishers may see fishing as a way of life and therefore may not want to exit fishing [22], it has been reported that up to 75% of fishers in Hong Kong would be willing to leave the industry if suitable alternatives or compensation were available [23]. Similar sentiments are likely to also occur in many other countries. In any case, it is better to undertake this transition as part of a rebuilding policy rather than having it forced upon us through loss of resources [10], [11].

Using the unit cost of reducing fishing effort calculated in Materials and Methods, the total amount that governments need to invest to rebuild world fisheries ranges between US$130 and US$292 billion in present value, with a mean of US$203 billion. This total transition cost would be spread over the time required to rebuild fisheries within each country.

3. NET GAIN FROM REBUILDING

Global fisheries are not living up to their revenue potential; the total cost of fishing is too high and governments provide harmful subsidies to the sector, which results in a negative resource rent (i.e., economic loss to society) of about US$13 billion per year (Table 1). Rebuilding would result in a gain in resource rent of US$66 billion per year, which when discounted over the next 50 years using a 3 per cent real discount rate, generates a present value of between US$660 and US$1,430 billion (Table 1), i.e., between 3 and 7 times the mean cost of fisheries rebuilding reform. Furthermore, it would likely take just 12 years after rebuilding efforts begin for the gains to exceed the costs of adjustment (Figure 3). A higher discount rate will reduce the present value of gain from rebuilding and increase the time needed to balance the gain with the costs of adjustment, and *vice versa* (see Materials and Methods for the justification of a 3% discount rate). Our results suggest that, even without accounting for the potential boost to recreational fisheries, processing, retail and non-market values that would likely increase, there is a substantial net economic benefit to be derived from rebuilding global fisheries, with net gains large enough to compensate for uncertainties in our assumptions and estimates. Rebuilding fisheries makes good business sense. The challenge is how to move global fisheries from their current dismal economic state to a more prosperous one.

DISCUSSION

Even though the overall results we present are consistent with other estimates about the extent of subsidies, excess fishing pressure and the potential for increased biological yield, the country-by-country analysis (Tables S1, S2, S3, S4, S5 and S6) may reveal results that differ from expectations. This is not surprising, as our analysis produces estimates with ranges, and therefore computing midpoint estimates may over- or underestimate numbers for some countries. This is more likely to happen for small developing countries where observed data are limited, and we therefore had to rely on statistical methods to produce estimates for these countries. The key to improving our estimates is for the collection of economic data for fisheries to be given priority by maritime countries.

There are also situations where large maritime countries (e.g., Peru, Chile, and Indonesia) may show counterintuitive estimates. Similarly unexpected results for countries such as Australia and Iceland, known to have good fisheries management regimes were found. In these cases too, the result of the statistical estimation required in the absence of observed or collected, publicly available country-specific data may be a reason. However, the reported results may indeed be correct yet unanticipated, as explained below.

A recent fishing industry report [24] provides financial data over 3 years (up to 2009), including pre-tax profit for the top 1000 commercial fishing companies worldwide. The numbers in this report support some of the counterintuitive outcomes of our study. These 1000 companies operate in 43 countries on all continents. The total annual sales value for all companies is about US$21 billion or 25% of estimated landed value worldwide. Of these 1000 companies, 339 reported negative annual pre-tax profits. Thirty-one of the 43 countries have at least one company reporting negative pre-tax profits, and of the 12 countries that report only positive pre-tax profits, nine countries had only one company in the dataset, suggesting that the optimistic results for these countries may be a result of limited data. Sixteen of the 43 countries for which data were reported had negative average pre-tax profits at the aggregate national level, at which the average ratio of pre-tax profit to sales volume is only marginally greater than zero (Figure 4). These data present an interesting and more micro-level view of the industry that is complementary to our estimates, showing that within the same country, some firms may be quite profitable, while others are much less so, resulting in negative aggregate profit at the national level.

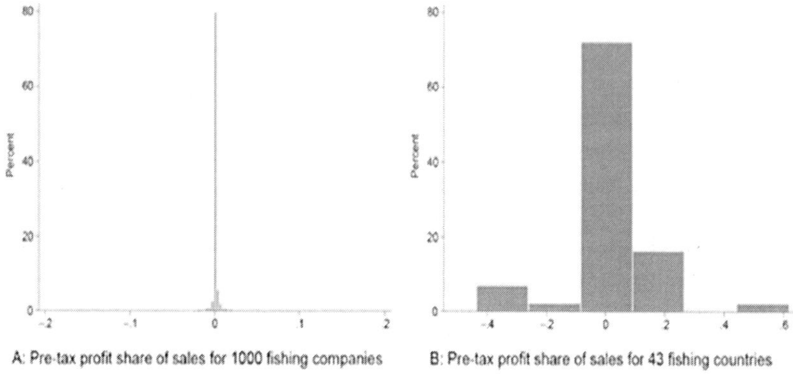

Figure 4. Histograms of pre-tax profit share of total sales for a sample of 1000 global fishing companies. Figure taken from [24].

MATERIALS AND METHODS

To estimate the potential gains from rebuilding global fisheries, we use estimates of catch loss [19], defined as the difference between current landings and Maximum Sustainable Yield (MSY) for those species that are considered to be over-exploited. It should be noted that MSY does not maximize economic yield (MEY) except when the stock size of fish does not affect the cost of fishing, and discount rate is zero. Still, we apply MSY in this analysis for practical and policy reasons, as it is a stipulated target or management reference point for many national legislations and international conventions. Other assumptions made in our analysis are: (i) the real ex-vessel fish price is constant through time (they have remained relatively stable since 1970) [16], [25]; (ii) during rebuilding, the costs of fishing change in proportion to changes in effort; (iii) the costs of fisheries management increase by 25% to US$10 billion a year, to support effective management under a rebuilt scenario; and (iv) the reported US$19 billion of annual harmful and ambiguous subsidies [17] are eliminated, since providing capacity-enhancing subsidies is fundamentally at odds with rebuilding fisheries. We also assume a rebuilding period of 10 years (e.g., Magnuson–Stevens Fishery Conservation and Management Act of the USA). Further support for this assumption is given by Costello et al. [26], who found that under an optimal rebuilding strategy, stock recovery requires between 4 and 26 years (with a mean of 11 years), depending on the fish species.

1. Estimating Global Fleet Size

The FAO estimates that there are currently 35 million people engaged in capture fisheries on either a part- or full-time basis [3]. The same report indicates that 90% of these fishers participate in the small-scale sector, while the remaining 10% can be classified as large-scale. The FAO also reports that the world's fishing fleet is comprised of 4.3 million vessels, 59% of which are motorized and 14% of motorized vessels (8% of all vessels) are greater than 12 meters in length [3]. In this study, we take a broad definition of large-scale vessels that includes all motorized vessels over 12 m in length, which is of sufficient size to represent considerable fishing pressure and potential impact on the environment. Under this criterion, we estimate the number of large-scale fishing vessels worldwide to be 355,000, with the remaining 3.94 million vessels are classified as small-scale.

2. Estimating Effort Reductions Required to Rebuild Global Fisheries

We model the global fishery using the Schaefer surplus-production model commonly applied to single-stock fisheries [27]. Since many fish stocks around the globe are either fully- or over-exploited, the global fishery is currently using more effort than needed to produce maximum sustainable yields (E_{MSY} in Figure 5), which we use as a global proxy for sustainable fisheries. We are cognizant of the diversity of fisheries management uses of biomass levels with MSY (either slightly above or below) as management reference or target points. In order to achieve maximum sustainable yields, effort will need to be reduced from current levels (e.g., E_0 in Figure 5) to a lower level that is consistent with maximum sustainable yield (E_{msy}). At E_{msy}, the total cost of fishing is reduced from TC_0 to TC_{msy}. For our calculations, we make the simplifying assumption that there is no substitution between labor and capital, so the shares of components of fishing costs (i.e., fuel, wages, etc.) remain constant.

Recognizing that large- and small-scale fisheries have different fishing power, and in order to minimize the effect of effort reductions on fishers (labor), who are predominantly in the small-scale sector, we weigh effort reductions more heavily on large-scale operations. We express total fishing effort in the global fishery as:

$$LSF * P_l + SSF * P_s = \delta_0$$

$$P_l = \gamma P_s \quad (1)$$

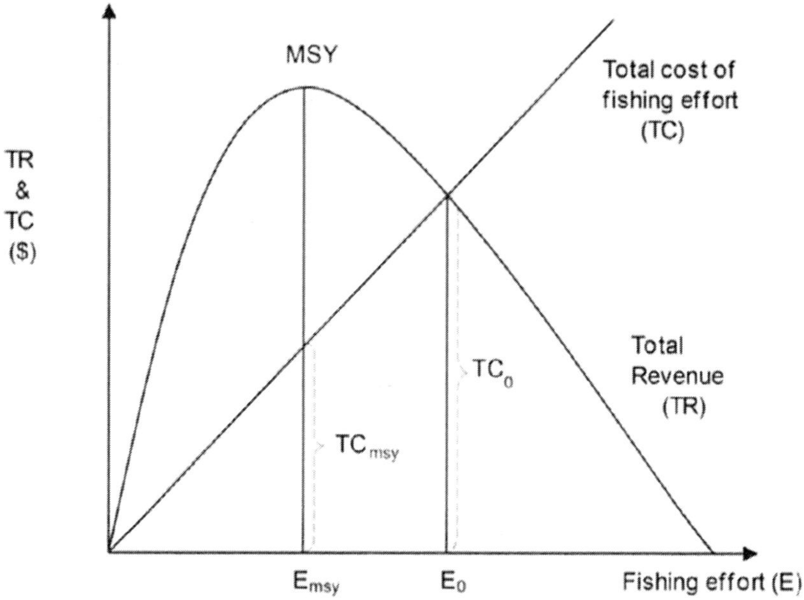

Figure 5. The Schaeffer surplus-production model, based on Gordon [27].

where *LSF* and *SSF* are the number of large- and small-scale fishers, respectively. The parameters P_l and P_s represent the fishing power of large- and small-scale fishers, while γ represents the power of large-scale fishers relative to small-scale fishers. Total current fishing effort is δ_0. By re-expressing *LSF, SSF* and δ_0 as terms that are relative to the total current fishing effort (i.e., dividing both LHS and RHS of eq. 1 by δ_0), we have:

$$LSF' * P_l + SSF' * P_s = 1 \tag{2}$$

Pauly [28] reports an estimate of γ, which places the fish catching power of large-scale fishers at 18 times that of their small-scale counterparts. This leaves us with a system of two equations with two unknowns that can be solved for P_l and P_s, which are used to estimate the proportions of large- and small-scale fishers required to reduce overall fishing effort:

$$LSF' * P_l * x + SSF' * P_s * y = \delta$$

$$w_l * LSF' * P_l * x = w_s * SSF' * P_s * \tag{3}$$

The parameters *LSF', SSF'*, P_l and P_s are defined as in the system of equations (1 & 2) above, while δ represents the ratio of current effort required to rebuild fisheries, while w_l and w_s represent the weight of effort cuts levied on large- and small-scale fishers, respectively. The parameters x and y, which represent the proportion of large- and small-scale fishing activity to be cut, are estimated from equations (3) and used to estimate the total reductions in large- and small-scale fishers as:

$$LSF_{cut} = LSF' * x$$

$$SSF_{cut} = SSF' * y \tag{4}$$

Lastly, we use our earlier estimates of the current number of large- and small-scale fishers and fishing vessels to estimate the number of large- and small-scale fishers and vessels that must be removed from the global fishery corresponding to our estimates of required reductions. We explore a range of weights (w_l and w_s) that represent equivalent total effort reductions. As can be seen in Figure 6, the trade-off between the cost of fishing effort reduction per fisher is non-linear, while the number of total fishers reduced is linear in the weighting placed on large-scale fishing effort. We suggest that by placing 80% of the weight of fishing effort reduction on large-scale fishing operations, it is consistent with cutting 60% of large-scale and 30% of small-scale fishing activity.

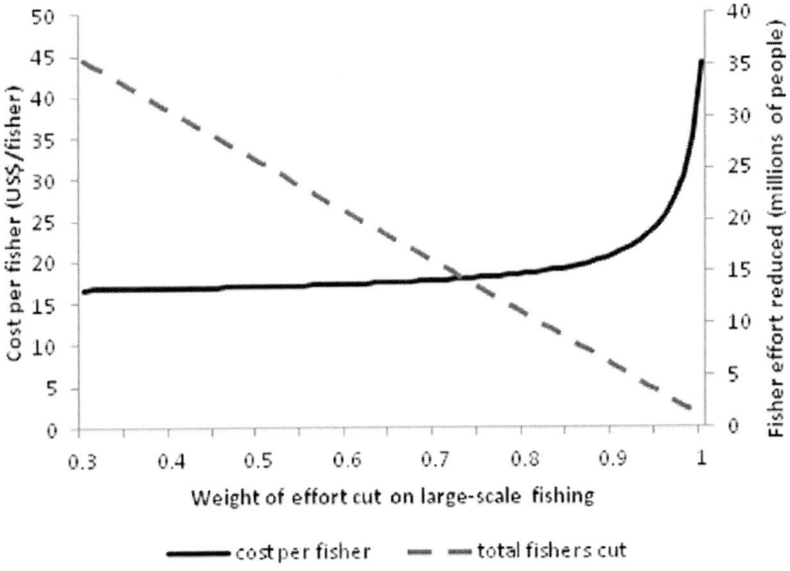

Figure 6. Trade-offs between reductions in cost of fishing effort and total fishing effort (in terms of number of fishers) reduced as the weight of effort cuts on large-scale fishing varies.

3. Estimating the Potential Value of Rebuilt Fisheries

For our present purposes, we assume that the estimated catch losses to overfishing reported by Srinivasan et al. [19] (Figure 7) may be fully regained after a period of rebuilding fisheries worldwide. To calculate potential catch losses, Srinivasan et al. [19] used catch time series from the *Sea Around Us* project for 1,066 taxa of fish and invertebrates in 301 EEZs, along with an empirical relationship they derived from catch data and stock assessments for 26 Northeast U.S. species from the U.S. National Oceanic and Atmospheric Administration (NOAA). The log-linear relationship that they found between a species' mean maximum catch C_{max} from catch data and its maximum sustainable yield (MSY) from stock assessment was robust ($R^2=0.84$, $p<0.001$), and has since been tested for 50 fully assessed stocks in the Northeast Atlantic, where variation in MSY accounted for 98% of the variability in C_{max} [29]. Therefore, given the dearth of detailed stock assessments for the majority of species in the world's fisheries, Srinivasan et al. [19] applied the relationship they derived (with a 50% prediction interval) to estimate MSY levels for all stocks they identified as overfished. By comparing with reported catch levels, they arrived at estimates of lost catch by mass, reporting that without overfishing, potential landings worldwide in the year 2000 may have been 9.1 million t higher than current landings (50% prediction interval: 3.6 to 19 million t higher).

To calculate the value of these potential landings under rebuilt global fisheries, Srinivasan et al. [19] used a database of ex-vessel fish prices by Sumaila et al. [16]. For each taxon-EEZ pair designated as overfished, a price-per-tonne p for the maximum sustainable yield (MSY) was set by taking a weighted average of the actual prices corresponding to catches of the taxon within ±30, ±50, or ±100% of the estimated MSY level, in order of preference depending on data availability. This approach was used to account for the impacts of overfishing, and thus scarcity, on price levels.

There is debate among fisheries scientists as to the reliability of overfishing estimates based on catch trends rather than stock assessments, with some arguing that catch-based approaches are prone to overestimate depletion [30]. Srinivasan et al. [19] were careful to avoid the biases described by Branch et al. [30], with the result that the former's estimate of the percentage of overfished stocks worldwide (16–31%) was similar to, but more conservative than, that reported by Branch et al. (28–33%), and similar also to a recent assessment by the FAO [31]. Indeed, Froese et al. [29] demonstrated that both stock- and catch-based assessments of overfishing in the Northeast Atlantic show the same trends, although the

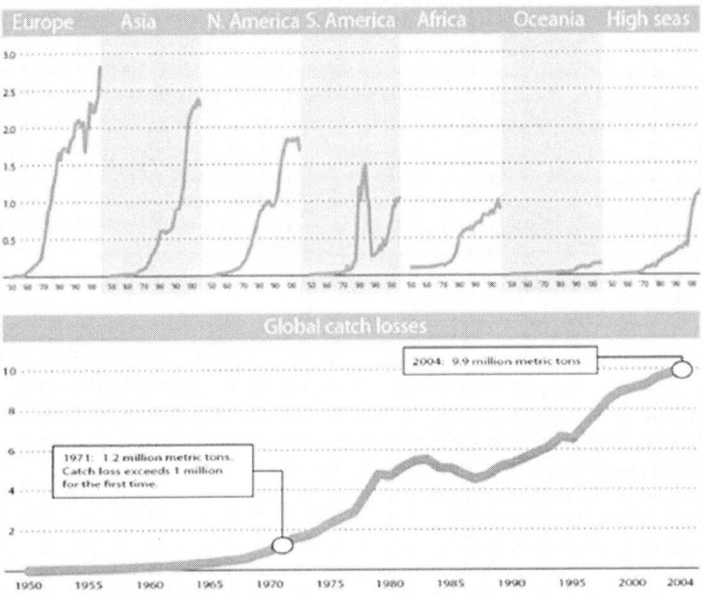

Figure 7. Lost catch potential due to overfishing for the six FAO regions of the world (top) and worldwide (bottom). Figure drawn using results reported in Srinivasan et al. [19].

catch-based methods were generally late to recognize declines in biomass. Thus, a catch-based method would underestimate lost catch, i.e., the direct opposite direction of the bias over which Branch et al. [30] have expressed concern. Moreover, Worm et al. [6] compared areas where there were both detailed stock assessment information and more general data including catch time series, and found that catches follow biomass trends, if belatedly.

Based on Costello et al. [26], who estimated the recovery time for 18 simulated fish species to be 11 years on average, with a range of 4–26 years depending on the species, we assume a rebuilding period of 10 years (t=0–9) in this study. During this period, we assume that the only gains to occur are those from a reduction in the current net resource rent loss from negative US$13 billion per year to zero. Following modelling work reported in UNEP's Green Economy Report [32], we also assume that global fisheries landings decline linearly from ~80 to 50 million tonnes per year from t=0–5 as fishing effort declines, but then rise linearly to the rebuilt level (~90 million tonnes) by t=9. Once global fisheries have been rebuilt, this potential gain in resource rent would recur annually into perpetuity; here we consider only the flow for the subsequent 40 years after rebuilding (t=10–49).

Table 2. Wages, normal profit and resource rent for current fisheries by FAO region.

Region	Wages	Earnings/Normal Profit*	Resource rent
	(US$ billions)		
Africa	1.51	0.60	−2.63
Asia	14.93	3.13	−4.73
Europe	6.77	0.77	−4.32
North America	4.72	1.12	1.13
Oceania	1.38	0.16	−0.58
South America	1.65	0.30	−1.80
World Total	30.96	6.08	−12.93

*Profit is defined here as the return to capital or normal profit, i.e., payments to owners of capital.

We estimate R, resource rent adjusted for subsidies, as follows:

$$R = LV - (C + S) \tag{5}$$

where LV represents the landed value of officially reported marine landings. The total variable cost of fishing is represented by C and subsidies are represented by S.

The computed resource rents for the six major Food and Agriculture Organization of the United Nations (FAO) regions (Africa; Asia; Europe; North America; Oceania; South and Central America plus the Caribbean) are summarized in Table 2.

We compute the gains from rebuilding (P_{gains}) as the value of the rebuilt resource rent ($R_{rebuilt}$) minus the value of current resource rent ($R_{current}$):

$$P_{gains_t} = R_{rebuilt_t} - R_{current_t} \tag{6}$$

where t represents time. We assume that globally, rebuilt fisheries will be successful in avoiding subsequent unsustainable increases in effort.

We calculate the present value of net gains from rebuilding global fisheries as follows:

$$PV = \sum_{t=0}^{49} \frac{P_{gains}}{(1+r)^t} \tag{7}$$

where PV is the present value of the net gain in resource rent, r is the prevailing rate of discount and t represents time from present $(t \in [0,49])$. In our analysis, we assume a fixed discount rate of $r=0.03$ (i.e., 3%) and compute the present value of net gains in resource rent for 50 years after rebuilding. We use this discount rate because many environmental economists have argued for and applied lower-than-market rates due to the central role of environmental resources in ensuring sustainable economies through time [33]–[35] Changes in the value of fisheries landings, costs, subsidies and resource rent through the transition time period are summarized in Figure 3.

Table 3. Wages, normal profit, resource rent and increase in rent from rebuilt fisheries.

Region	Wages	Earnings/Normal Profit*	Resource rent	Increase in rent
	(US$ billions)			
Africa	0.76	0.30	0.85	3.48
Asia	7.46	1.56	30.82	35.54
Europe	3.38	0.39	8.80	13.12
North America	2.36	0.56	7.98	6.86
Oceania	0.69	0.08	2.73	3.31
South America	0.83	0.15	2.51	4.31
World Total	15.48	3.04	53.69	66.61

*Profit is defined here as the return to capital or normal profit, i.e., payments to owners of capital.

4. Estimating the Cost of Rebuilding Global Fisheries

In addition to differences between current resource rent and that which is captured during the period of rebuilding, we estimate the costs necessary to reduce fishing capacity to levels required to allow fish stocks to rebuild. These costs are estimated based on the cost of effort reductions described earlier in the methods. We estimate wages, profits, resource rent and increase in resource rent from rebuilding for the six major FAO regions (Africa; Asia; Europe; North America; Oceania; South and Central America plus the Caribbean) in Table 3.

Since the real cost of rebuilding fisheries is foregone resource rent that may occur as fishing effort is reduced initially, we estimate the cost of rebuilding global fisheries through the transition to rebuilt fisheries as the difference between current fisheries resource rent and that which is realized through the period of transition. We hold the assumption that all harmful capacity-enhancing and ambiguous subsidies (Table 4) must be cut immediately or re-directed to make them beneficial subsidies, e.g., by investing in managing the rebuilding process.

Table 4. Annual global fisheries subsidies by category [17].

Category	Subsidies (US$ billions)
Beneficial[a]	8
Harmful[b]	16
Ambiguous[c]	3
Total	27

[a]Lead to 'investment' in the natural capital of fishery resources. They enhance the growth of fish stocks through conservation programs, and control and surveillance measures.
[b]Lead to 'disinvestments' in the natural capital of the fishery resources, including all forms of capital inputs and infrastructure investments from public sources that reduce cost or enhance revenue.
[c]Have the potential to lead to either 'investment' or 'disinvestment' in the fishery resources, and lead to resource enhancement or to resource overexploitation.

5. Calculating the Unit Cost of Reducing Fishing Effort

Policy makers generally prefer to minimize the employment impact of rebuilding fisheries. It would therefore be attractive to target effort reductions on large-scale vessels only, as they employ less people per unit of fish landed [36]. While the goal of matching global fishing capacity with the productive capacity of the resource by cutting only large-scale vessels seems theoretically possible, it would be ineffective in areas that are overfished but dominated by small-scale vessels. Available gear-related data [36], [37] reveal that the split between large- and small-scale vessels in the developed world is about 50:50, while it is 25:75 in developing countries. Our analysis of fishing effort cuts show that permanently removing around 213,000 large-scale and 1.2 million small-scale vessels (60% and 30% reductions, respectively), would halve the world's fishing capacity. This weighting between large- and small-scale fishing capacity is supported by evidence that large-scale vessels currently land roughly two-thirds of the world's annual reported marine landings [28].

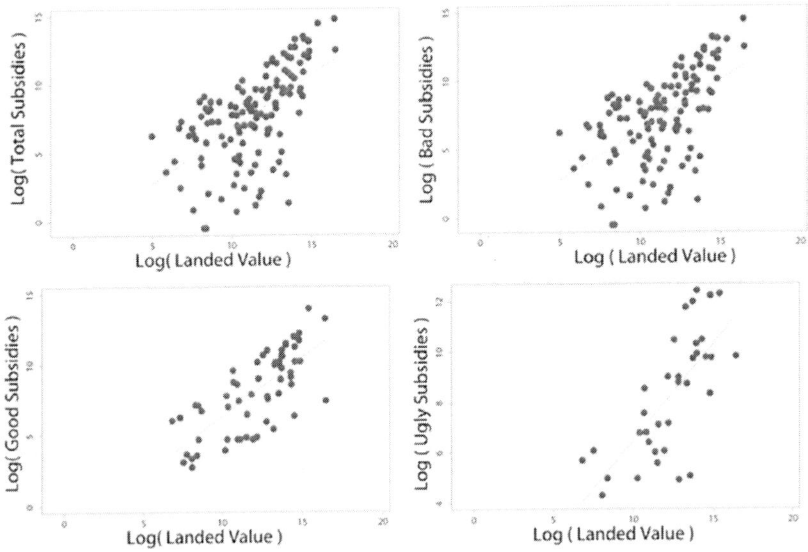

Figure 8. Correlation between reported subsidies [17] and landed-value [16].

Cost data [18] reveal that crew from large- and small-scale fisheries earn, on average, wages of US$20,000 and US$10,000 per year, respectively. Furthermore, vessels in large- and small-scale fisheries pay, on average, US$11,000 and US$2,500 per year for capital. Based on vessel and crew data from the European Union [38], we estimate that the average cost of a vessel buyback is roughly equal to the average interest payments on a vessel for five years, and the average cost of crew retraining is estimated at 1.5 times the average annual crew wages. Therefore, the average cost of decommissioning large- and small-scale fishing vessels would be US$55,000 and US$12,500, respectively. Likewise, payout/retraining costs for large- and small-scale fishers to leave fishing permanently would be US$30,000 and US$15,000 per person, respectively. Clearly, decommissioning costs for the extremely large industrial vessels with global roaming abilities would be higher than the above vessel averages.

Table 5. Global cost (mean +/−95% CI) of fishing (Year 2005 US$ per t of catch), separated into variable and fixed cost component [18].

	Lower 95% CI	Mean	Upper 95% CI
Cost category	**(US$ per t)**		
Variable	639	928	1,413
Fixed	123	192	164
Total	**762**	**1,120**	**1,477**

6. Data and Databases

We utilize four interrelated global databases of fisheries statistics, namely, databases of fisheries landings, ex-vessel fish prices, subsidies, and fishing costs. Each database represents the work of an international team of fisheries scientists and economists, and collectively represents the world's most comprehensive collection of truly global fisheries (economic) data.

Our catch data, the main source of which is the FAO global capture production database supplemented by several more detailed regional catch data sources, allocates the reported fish landings to a global system of 30-minute latitude by 30-minute longitude cells (just under 180,000 marine cells globally) using the intersection of statistical reporting areas, biological taxon distributions of reported taxa, general habitat preferences, global fishing access agreements and fishing patterns of reporting countries. Details of the methods and procedures of this spatial allocation are described in Watson et al. [15], and country-specific data by FAO region are presented in Tables S1, S2, S3, S4, S5 and S6.

The global ex-vessel fish price database used here, described by Sumaila et al. [16], covers annual average ex-vessel prices for all marine fish taxa by country reported as caught from 1950 to 2006. Through their extensive search of publicly available, but widely scattered and incompatible, national and regional statistical reports and grey literature, Sumaila et al. [16] accumulated over 31,000 records of observed ex-vessel prices in 35 countries, representing about 20 percent of the global landings over the 60 year period. In order to 'fill the gaps' in the database, a series of rules were developed whereby all catches with no reported prices were inferred to have an estimated price computed from the reported prices from related taxa, similar markets or years. Since the database was first presented, new reported prices have been included from various additional sources, and rules as to how prices relate across taxa, markets or years have been modified to improve the quality of the estimated prices. The time series of landed values of the world's marine fisheries, computed through the combination of the spatially allocated catch data with the ex-vessel price database (country-specific landed values by FAO region) are presented in Tables S1, S2, S3, S4, S5 and S6, and have been used in various analyses, such as the estimation of global subsidies [17] and costs of marine protected areas [39].

Table 6. Sensitivity of present value of rebuilding costs to parameter assumptions.

Effort reduction to achieve MSY (%)	Adjustment costs (US$ billions)*		
	−20%	Mean	+20%
40	130	162	195
50	162	203	243
60	195	243	292

*Adjustment costs include the cost of vessel buybacks and payouts for fishers to ease the transition to alternate employment.

The fisheries subsidies database defines subsidies as financial transfers, directly or indirectly, from government to the fishing industry [17]. This database is the most comprehensive collection of publicly available data on fisheries subsidies at the global level, spanning the years 1990 to 2006. Each record in the database represents expenditure in one of twenty-six identified subsidy categories for a given country and year combination. Where qualitative information indicates the presence of a subsidy program, yet quantitative data are not available, the database records the expenditure data as 'missing' for later estimation.

Table 7. Sensitivity of present value potential gain in resource rent from rebuilding to parameter assumptions.

Effort reductions to achieve MSY (%)	Potential gains (US$ billions)*		
	Lower 95% CI	Mean	Upper 95% CI
40	976	1,128	1,428
50	821	972	1,272
60	665	816	1,117

*Gains are denoted as the present value of increased resource rent from rebuilt fisheries.

Estimation of 'missing' subsidy data follows the method of Sumaila et al. [17] who utilize the strong relationship between fisheries subsidies and landed value (Figure 8) to estimate subsidy expenditure in cases where programs are documented without quantitative information. We use this procedure to estimate existing but unquantified fisheries subsidies for any of the twenty-six subsidy categories in any of the 144 maritime countries of the world, and summarize these globally by three general categories (Table 4): 'beneficial' (lead to 'investment' in the natural capital of fishery resources, thus enhancing growth of fish stocks through conservation programs, and control and surveillance measures), 'harmful' (lead to 'disinvestments' in the natural capital of the fishery resources, including all forms of capital inputs and infrastructure investments from public sources that reduce cost or enhance revenue) and 'ambiguous' (have the potential to lead to either 'investment' or 'disinvestment' in the fishery resources, and lead to resource enhancement or to resource overexploitation). Fuel- and non-fuel subsidies by country within FAO regions are summarized in Tables S1, S2, S3, S4, S5 and S6.

Lam et al. [18] developed a global database of fishing costs by country and gear types, capturing two types of fishing cost, variable (operating) and fixed costs in 144 maritime countries, representing approximately 98% of global landings in 2005. Results from this database are summarized in Table 5. Each record in the database represents a country and gear type combination. The gear types included in the database are based on the gear categorization system of the *Sea Around Us* project [40].

Fishing cost data were collected from secondary sources in major fishing countries in each of the six FAO regions. In order to include as many data of observed cost as possible, Lam et al. [18] accessed all available sources, irrespective of publication year, thus extending their efforts in collecting cost data from 1950 to the most recent year for which data were available. The data were then converted to 2005 real values using the consumer price index (CPI) for each country obtained from the World Bank. To make the comparison of fishing cost among different regions and countries possible, they converted all fishing costs from local currencies to US dollars by using currency exchange rates provided by the World Bank, and standardized the original cost to annual cost in US$ per tonne of catch.

A process of progressive refinement [16], [18] was then used to estimate the cost of all gear types in each fishing country from the observed, collected cost. Therefore, Lam et al. [18] ensured that all gear types in each maritime country of the world were assigned a cost, either the observed value where available, or an appropriate average estimated cost. Variable fishing costs by country within FAO region are summarized in Tables S1, S2, S3, S4, S5 and S6.

7. Sensitivity Analysis

We test the sensitivity of our results both in terms of the benefits and costs of rebuilding. An estimate of the potential contribution of rebuilt fisheries from Srinivasan et al. [19] is 89 million t per year (50% prediction interval: 83 million to 99 million t per year) and US$100 billion per year in landed value (50% prediction interval: US$93 billion to US$116 billion per year). This represents an increase over the current value of fisheries landings of US$13 billion per year (range: US$5 billion to US$29 billion per year). We also test the sensitivity of our estimates to changes in the cost of fishing by allowing the needed effort reductions to attain maximum sustainable yield to be within a range of 40–60% of current fishing effort, with a mean value of 50% of current effort levels.

Table 6 presents the results of the sensitivity analysis with respect to the costs of adjusting factors of fisheries production to maximum sustainable yield. The top-left and bottom-right cells of this table display the best- and worst-case scenarios for our estimates of adjustments costs. Table 7 presents the results of our sensitivity analysis on the estimates for rebuilding benefits in a three-by-three matrix, thus showing potential best- and worst-case scenarios of costs of rebuilding fisheries in the bottom-right and top-left cells of the table, respectively. It can readily be seen from these two tables that, even under drastic changes in our estimates, the combination of the worst-case scenarios result in large net present value of resource rent from rebuilt global fisheries.

ACKNOWLEDGMENTS

This is a contribution from the Global Ocean Economics Project and the *Sea Around Us* Project, both of which are collaborations between the University of British Columbia and the Pew Environment Group. This contribution benefited from work carried out by contributors to the United Nations Environmental Programme Green Economy Chapter on fisheries. We also thank Jennifer Jacquet for reviewing an earlier version of the paper.

REFERENCES

1. Pontecorvo G, Wilkinson M, Anderson R, Holdowsky M (1980) Contribution of the ocean sector to the United States economy. Science 208: 1000–1006.G. PontecorvoM. WilkinsonR. AndersonM. Holdowsky1980Contribution of the ocean sector to the United States economy.Science20810001006
2. Dyck AJ, Sumaila UR (2010) Economic impact of ocean fish populations in the global fishery. Journal of Bioeconomics 12: 227–243.AJ DyckUR Sumaila2010Economic impact of ocean fish populations in the global fishery.Journal of Bioeconomics12227243
3. FAO (2010) The State of World Fisheries and Aquaculture (SOFIA) 2010. Rome, Italy: Food and Agriculture Organization. 197 p. FAO2010The State of World Fisheries and Aquaculture (SOFIA) 2010. Rome, Italy: Food and Agriculture Organization.197 p
4. Zeller D, Booth S, Pauly D (2006) Fisheries contributions to GDP: Underestimating small-scale fisheries in the Pacific. Marine Resource Economics 21: 355–374.D. ZellerS. BoothD. Pauly2006Fisheries contributions to GDP: Underestimating small-scale fisheries in the Pacific.Marine Resource Economics21355374
5. Pauly D, Christensen V, Guénette S, Pitcher TJ, Sumaila UR, et al. (2002) Towards sustainability in world fisheries. Nature 418: 689–695.D. PaulyV. ChristensenS. GuénetteTJ PitcherUR Sumaila2002Towards sustainability in world fisheries.Nature418689695
6. Worm B, Hilborn R, Baum JK, Branch T, Collie JS, et al. (2009) Rebuilding global fisheries. Science 325: 578–585.B. WormR. HilbornJK BaumT. BranchJS Collie2009Rebuilding global fisheries.Science325578585
7. Cheung WWL, Close C, Lam V, Watson R, Pauly D (2008) Application of macroecological theory to predict effects of climate change on global fisheries potential. Marine Ecology Progress Series 365: 187–197.WWL CheungC. CloseV. LamR. WatsonD. Pauly2008Application of macroecological theory to predict effects of climate change on global fisheries potential.Marine Ecology Progress Series365187197
8. Sumaila UR, Cheung WWL, Lam V, Pauly D, Herrick S (2011) Climate change impacts on the biophysics and economics of world fisheries. Nature Climate Change 1: 449–456.UR SumailaWWL CheungV. LamD. PaulyS. Herrick2011Climate change impacts on the biophysics and economics of world fisheries.Nature Climate Change1449456
9. Merino G, Barange M, Blanchard JL, Harle J, Holmes R, et al. (2012) Can marine fisheries and aquaculture meet fish demand from a growing human population in a changing climate? Global Environmental Change in press. G. MerinoM. BarangeJL BlanchardJ. HarleR. Holmes2012Can marine fisheries and aquaculture meet fish demand from a growing human population in a changing climate?Global

Environmental Change in pressAvailable: http://dx.doi.org/1 0.1016/j.gloenvcha.2012.1003.1003

10. Mason F (2002) The Newfoundland cod stock collapse: a review and analysis of social factors. Electronic Green Journal 1. F. Mason2002The Newfoundland cod stock collapse: a review and analysis of social factors.Electronic Green Journal 1
11. SCFO (2005) Northern cod: a failure of Canadian fisheries management. Ottawa, Canada: Report of the Standing Committee on Fisheries and Oceans. SCFO2005Northern cod: a failure of Canadian fisheries management.Ottawa, Canada: Report of the Standing Committee on Fisheries and Oceans
12. Clark CW (1990) Mathematical Bioeconomics: The Optimal Management of Renewable Resources. New York: Wiley. CW Clark1990Mathematical Bioeconomics: The Optimal Management of Renewable Resources.New York: Wiley
13. Béné C, Hersoug B, Allison EH (2010) Not by rent alone: analysing the pro-poor functions of small-scale fisheries in developing countries. Development Policy Review 28: 325–358.C. BénéB. HersougEH Allison2010Not by rent alone: analysing the pro-poor functions of small-scale fisheries in developing countries.Development Policy Review28325358
14. Heal G (2000) Valuing ecosystem services. Ecosystems 3: 24–30.G. Heal2000Valuing ecosystem services.Ecosystems32430
15. Watson R, Kitchingman A, Gelchu A, Pauly D (2004) Mapping global fisheries: sharpening our focus. Fish and Fisheries 5: 168–177.R. WatsonA. KitchingmanA. GelchuD. Pauly2004Mapping global fisheries: sharpening our focus.Fish and Fisheries5168177
16. Sumaila UR, Marsden AD, Watson R, Pauly D (2007) A global ex-vessel fish price database: construction and applications. Journal of Bioeconomics 9: 39–51.UR SumailaAD MarsdenR. WatsonD. Pauly2007A global ex-vessel fish price database: construction and applications.Journal of Bioeconomics93951
17. Sumaila UR, Khan A, Dyck A, Watson R, Munro R, et al. (2010) A bottom-up re-estimation of global fisheries subsidies. Journal of Bioeconomics 12: 201–225.UR SumailaA. KhanA. DyckR. WatsonR. Munro2010A bottom-up re-estimation of global fisheries subsidies.Journal of Bioeconomics12201225
18. Lam V, Sumaila UR, Dyck A, Pauly D, Watson R (2011) Construction and potential applications of a global cost of fishing database. ICES Journal of Marine Science 68: 1996–2004.V. LamUR SumailaA. DyckD. PaulyR. Watson2011Construction and potential applications of a global cost of fishing database.ICES Journal of Marine Science6819962004
19. Srinivasan UT, Cheung WWL, Watson R, Sumaila UR (2010) Food security implications of global marine catch losses due to overfishing. Journal of Bioeconomics 12: 183–200.UT SrinivasanWWL CheungR. WatsonUR Sumaila2010Food security implications of global marine catch losses due to overfishing.Journal of Bioeconomics12183200
20. World Bank, FAO (2009) Sunken Billions: The Economic Justification for Fisheries Reform: Case Study Summaries. Bank WorldFAO2009Sunken Billions: The Economic Justification for Fisheries Reform: Case Study Summaries.Available: http://go.worldbank.org/MGUTHSY7U0.Accessed 2009 Apr 15. Accessed 2009 Apr 15
21. Porter G (1998) Estimating overcapacity in the global fishing fleets. Washington, D.C.: WWF. G. Porter1998Estimating overcapacity in the global fishing fleets.Washington, D.C.: WWF
22. Daw TM, Cinner JE, McClanahan TR, Brown K, Stead SM, et al. (2012) To fish or not to fish: Factors at multiple scales affecting artisanal fishers' readiness to exit a declining fishery. PLoS ONE 7: e31460.TM DawJE CinnerTR McClanahanK. BrownSM Stead2012To fish or not to fish: Factors at multiple scales affecting artisanal fishers' readiness to exit a declining fishery.PLoS ONE7e31460
23. Teh L, Cheung WWL, Cornish A, Chu C, Sumaila UR (2008) A survey of alternative livelihood options for Hong Kong's fishers. International Journal of Social Economics

35: 380–395.L. TehWWL CheungA. CornishC. ChuUR Sumaila2008A survey of alternative livelihood options for Hong Kong's fishers.International Journal of Social Economics35380395

24. Plimsoll (2011) The Commercial Fishing (Global) Industry: A comprehensive financial analysis of the top 1000 companies. England: Plimsoll Publishing Limited. 1256 p. Plimsoll2011The Commercial Fishing (Global) Industry: A comprehensive financial analysis of the top 1000 companies. England: Plimsoll Publishing Limited.1256 p
25. Delgado CL, Courbois C (1997) Changing fish trade and demand patterns in developing countries and their significance for policy research. MTID Discussion Paper 18, International Food Policy Research Institute. CL DelgadoC. Courbois1997Changing fish trade and demand patterns in developing countries and their significance for policy research.MTID Discussion Paper 18, International Food Policy Research Institute(IFPRI). (IFPRI)
26. Costello C, Kinlan BP, Lester SE, Gaines SD (2012) The economic value of rebuilding fisheries, OECD Food, Agriculture and Fisheries Working Papers, No. 55. Paris: OECD Publishing. 68 p. C. CostelloBP KinlanSE LesterSD Gaines2012The economic value of rebuilding fisheries, OECD Food, Agriculture and Fisheries Working Papers, No. 55. Paris: OECD Publishing.68 p
27. Gordon HS (1954) The economic theory of a common property resource: the fishery. Journal of Political Economics 62: 124–142.HS Gordon1954The economic theory of a common property resource: the fishery.Journal of Political Economics62124142
28. Pauly D (2006) Major trends in small-scale marine fisheries, with emphasis on developing countries, and some implications for the social sciences. Maritime Studies (MAST) 4: 7–22.D. Pauly2006Major trends in small-scale marine fisheries, with emphasis on developing countries, and some implications for the social sciences.Maritime Studies (MAST)4722
29. Froese R, Zeller D, Kleisner K, Pauly D (2012) What catch data can tell us about the status of global fisheries. Marine Biology (Berlin) 159(6): 1283–1292.R. FroeseD. ZellerK. KleisnerD. Pauly2012What catch data can tell us about the status of global fisheries.Marine Biology (Berlin)159(6)12831292
30. Branch T, Jensen O, Ricard D, Ye Y, Hilborn R (2011) Contrasting global trends in marine fishery status obtained from catches and from stock assessments. Conservation Biology 25: 777–786.T. BranchO. JensenD. RicardY. YeR. Hilborn2011Contrasting global trends in marine fishery status obtained from catches and from stock assessments.Conservation Biology25777786
31. FAO (2009) The State of World Fisheries and Aquaculture (SOFIA) 2008. Rome, Italy: Food and Agriculture Organization. 176 p. FAO2009The State of World Fisheries and Aquaculture (SOFIA) 2008. Rome, Italy: Food and Agriculture Organization.176 p
32. UNEP (2011) Towards a Green Economy: Pathways to Sustainable Development and Poverty Eradication. Nairobi: UNEP. 631 p. UNEP2011Towards a Green Economy: Pathways to Sustainable Development and Poverty Eradication. Nairobi: UNEP.631 pAvailable: www.unep.org/greeneconomy.
Available: www.unep.org/greeneconomy
33. Weitzman ML (2001) Gamma discounting American Economic Review 91: 260–271.ML Weitzman2001Gamma discounting American Economic Review91260271
34. Sumaila UR (2004) Intergenerational cost benefit analysis and marine ecosystem restoration. Fish and Fisheries 5: 329–343.UR Sumaila2004Intergenerational cost benefit analysis and marine ecosystem restoration.Fish and Fisheries5329343
35. Stern R (2006) Stern Review on the Economics of Climate Change. London: Her Majesty's Treasury. ix +579 p. R. Stern2006Stern Review on the Economics of Climate Change. London: Her Majesty's Treasury.ix +579 p
36. Gabriel O, Lange K, Dahm E, Wendt T, editors (2005) Surrey, England: Fishing News Books. O. GabrielK. LangeE. DahmT. Wendteditors2005Surrey, England: Fishing News Books.Von Brandt's Fish Catching Methods of the World, 4th ed.521 p. 521 p

37. Anticamara JA, Watson R, Gelchu A, Pauly D (2011) Global fishing effort (1950–2010): trends, gaps and implications. Fisheries Research 107: 131–136.JA AnticamaraR. WatsonA. GelchuD. Pauly2011Global fishing effort (1950–2010): trends, gaps and implications.Fisheries Research107131136
38. European Commission (EC) (2006) Economic performance of selected European fishing fleets. 2005 Annual Report. European Commission (EC)2006Economic performance of selected European fishing fleets.2005 Annual Report
39. Cullis-Suzuki S, Pauly D (2008) Marine Protected Area costs as 'beneficial' fisheries subsidies: a global evaluation. Coastal Management 38: 113–121.S. Cullis-SuzukiD. Pauly2008Marine Protected Area costs as 'beneficial' fisheries subsidies: a global evaluation.Coastal Management38113121
40. Watson R, Revenga C, Kura Y (2006) Fishing gear associated with global marine catches I. Database development. Fisheries Research 79: 97–102.R. WatsonC. RevengaY. Kura2006Fishing gear associated with global marine catches I. Database development.Fisheries Research7997102

CHAPTER 12

Fishery Fee and Tax Rate in an Oligopoly Industry with Entry and Exit*

Yu Shao, Kai Shen, Weiru Zhang, Shuntian Yao#

Division of Economics, Nanyang Technological University, Nanyang Ave, Singapore

ABSTRACT

Over the last two decades, fisheries sector has been playing an important role in the global economy as a vibrant sector. By looking into the interactions among natural resources, human beings and government, this paper first briefly examines the issue of complicated conflicts in fisheries together with its influence on fisheries management. Under the assumption that the fishing resources will not be exhausted, this paper further reveals the particular conflicts between economic performance and community welfare by modelling the strategic interaction amongst the government and fishing firms based on the two stage non-cooperative game theory. With a dynamic model enclosed, this paper demonstrates with how the government establishes the fishing fee rate to achieve its primary goal and how the fishing firms react to the policy in each scenario. In the end, it concludes with some brief suggestions regarding policy-making for future work priorities. Although fishery industry is used as an example in this article, our main results, however, can be applied to any industries of oligopoly competitions with entry and exit.

Keywords: Fisheries Management; Fishing Fee; Non-Cooperative Game Theory; Entry and Exit; Optimal Tax Rate; Social Welfare

1. INTRODUCTION

As a sunrise sector of our economy, the fishery industry in China has emerged from a subsistence traditional activity into a government-controlled commercial enterprise. However, fishing conflicts are among the persistent problems due to its complex and dynamic bio-socio-economic system, with its many interactions among natural resources, human beings and government. Conflicts are broadly defined as a situation of non-cooperation between parties and contradictory objectives (FAO 1998) [1]. Conflicts in fisheries usually arise among stakeholders with differing economic and social motivations regarding the allocation and access rights to the limited natural resources.

Why are there conflicts in fisheries? There is a set of three fisheries "world views" which reflect the philosophical basis of fisheries conflicts. This framework was specified by Charles (1992) [2] to analyse the conflicts by introducing a trio of fishery paradigms, i.e. the conservation, rationalization, and social/community paradigms. Together the three paradigms form the corners of a "paradigm triangle" within which differing approaches to fisheries conflicts and policy problems can be analysed (see Figure 1).

Figure 1 features three philosophical paradigms and their unique policy objectives, which are:

- The conservation paradigm operates with a policy objective aimed at resource maintenance.
- The rationalization paradigm focuses on the pursuit of economic performance and productivity. That is, the government should centre on fishery fee maximization.
- The social paradigm emphasizes community welfare, equity of distribution, and other social and cultural fishery benefits, which takes into account of the total welfare of government plus the fishery firms.

Charles (1992) [2] enclosed three cases in his study to illustrate the framework of analysing conflicts. However, the hotspot discussed in academic field is about the conflicts between the conservation paradigm and the rationalization paradigm. Most of the coastal countries' policies are focusing on enhancing the community welfare while ensuring the sustainability of fishery resources.

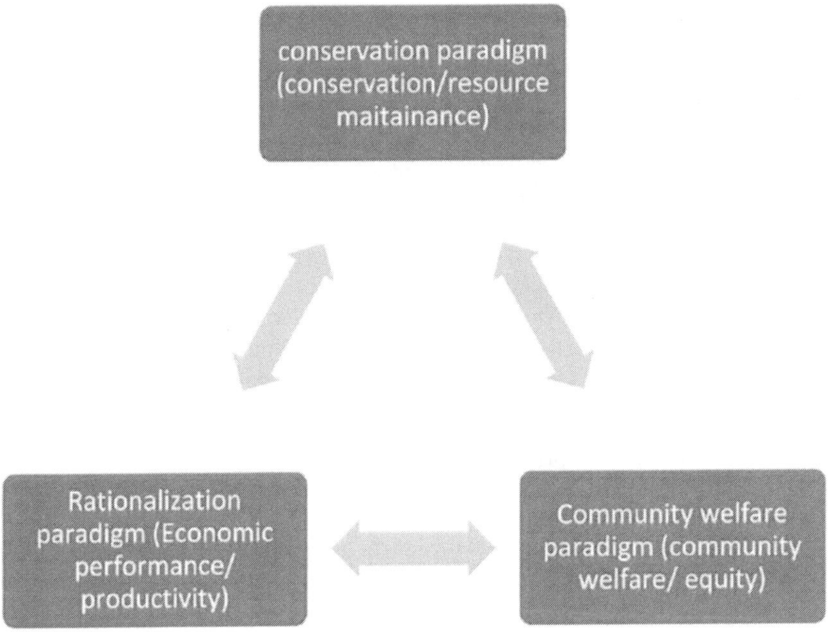

Figure 1. Framework for understanding and resolving conflicts in fisheries.

Those researches propose policy makers to adjust and control the fisheries by targeting on the total harvest amount. For example, Mainland Fisheries Authority of China imposed a two-month fishing moratorium in the South China Sea during Jun and July 1999 to preserve fishery resources (AFCD 1999) [3]; there are established systems of individual transferable quota (ITQ), which "gives fishers, vessels and/or producers dedicated access privileges to land a specified portion of total allowable catch (TAC)" and hundreds of fisheries globally is managed by this TAC-ITQ system (Chu, 2009) [4]. We therefore raise the interest of building a mathematical model to study the conflicts between the bottom two factors in Figure 1, namely the economic performance and community welfare.

In most of studies, conflicts are modelled by game theory. Essentially, it is a strategy that uses mathematics to describe player strategies in sources of conflict and common interest, and predicts what rational players should do, and not what they actually do (Luce & Raiffa, 1957) [5].

We begin our review with the early principal-agent model developed by Clarke and Munro (1987) [6], in which a coastal nation imposes a fishing fee on a single distant water fishing nation (DWFN) which has the same rate of discount as it does. They (1991) [7] further generalize their first model by considering the possibility that the coastal nation and DWFN have different discount rates on future return from fishery. They find that the coastal nation can increase its discounted net return from the fishery by simultaneously using a tax on harvest and a tax on effort rather than only using one tax.

Raissi (2001) [8] examines a model in which uses a dual tax system with an inferior fishing technology on two fishing firms—a domestic firm and a foreign firm. Raissi demonstrates that if there is no regulation on fisheries then the foreign firm will eliminate domestic competition by exerting the maximum effort. However, it may be optimal if the coastal nation encourages the two firms to converge to the equilibrium at which they are both exploiting the fishery by using taxes. Neither Raissi (2001) [8] or Clarke and Munro (1987, 1991) [6,7] allow the coastal nation to choose the number of firms that operate in its fishery.

Additionally, some other authors use non-cooperative game theory to model the tactical interactions between firms in an international, unregulated fishery. Among those studies, Levhari and Mirman (1980) [9] and Fischer and Mirman (1996) [10] uses a utility maximization model while Dockner et al. (1989) [11] and Ruseski (1998) [12] uses a profit maximization model. They all conclude that the firms will over-fish the stock. Ruseski [12] shows that the respective government of each firm that competes over the fishery has the incentive to license more vessels than that at socially optimal level. The received revenue is used to subsidize its own firm's effort, which in turn exacerbates over-fishing.

Yoav Wachsman (2002) [13] develops a model to simulate interactions between coastal nation (principal) and foreign fishing firms (agents) by involving n firms into a 2-stage game. Besides, the study further reveals the conflicts between economic performance and social welfare, which means when government chooses to maximize its own profit it has to sacrifice the maximum level of total return of the principal-agents system as a whole. One of the significant contributions of this study is that it allows the number of firms to be a variable in the coastal nation's objective function. However, this solution is possible only based on the strong assumptions of identical firms and zero fixed costs so that the size of firm is allowed to be extremely tiny when number of firms goes to infinity. Nevertheless, this study is a unique approach to evolve previous 2-players or 2 groups-players game theory model.

Our paper is to provide a thoughtful advice on the fishery policy by simulating the dynamic process of market interactions and considering the full aspect of fishery paradigm triangle. Resource sustainability is set as a pre-requirement, and a proper tax rate is obtained by a compromise between economic performance and social welfare. Furthermore, the model is made to be more realistic by taking into account of the fixed cost of each individual firm. While we analyse the fishery industry as an example, our results actually applies to any oligopoly industries with entry and exit.

This paper is structured as follows: model description is presented first; followed by the mathematical analysis. Next, we simulate a simple case where firms have the same marginal cost of efforts and fixed costs, and supplemented by further discussion on social welfare prospect. Finally we conclude the paper with policy reviews for the governments.

2. MODEL DESCRIPTION

In most of previous studies concerning resource constraint, Schaefer's Model (1957) [14] was used. In his model, harvest of each firm from the fishery can be calculated from a "catch per unit of effort (CPUE) product function". According to this function, each firm harvests a constant portion of the stock per unit of effort, e_i. Define q_i as the quantity (harvest) for firm i. Consider the situation when the total harvest of n existing firms Q is less than the stock size x, i.e., $Q = \sum_{i=1}^{n} q_i < x$, each firm's harvest can be calculated as:

$$q_i = lxe_i$$

where l is the catching ability coefficient, $0 < l < 1$. It was assumed that l is independent from the stock size and it is identical for all the firms.

In a working paper written by Wachsman (2002) [13], the author proposed a fishery model based on Schaefer's Model. In Wachsman's Model [13], no fixed cost is assumed, and potentially it was assumed that the number of firms could tend to infinity.

The advantage of Wachsman's Model is that the constraint of the resource is explicitly modeled and discussed. The weakness of Wachsman's Model is that, the assumption of zero fixed costs does not really capture the reality of the fishery industry, and under such an assumption, entry into or exit from the industry cannot be discussed. Additionally, when the number of firms, n, becomes larger and larger, and tends to infinity eventually, the capacities of the firms must become smaller and smaller given the fixed amount of natural resource, this looks quite absurd.

Comparing to the Wachsman's study, this paper releases the zero-fixed-cost assumption and gives a snapshot of the dynamic process in the fishery industry. In this dynamic process of interaction, each individual firm would decide whether to continue operating under a specific tax rate imposed. A firm will quit once the after-tax profit turns negative.

In our model there are totally N firms. Each firm i has a marginal cost of c_i and a fixed cost of f_i. The market inverse demand for the produce is given by

$$P = a - Q \tag{1}$$

where Q is the total quantity supplied to the market and P is the market clearing price. In addition to the production marginal costs as mentioned above, we also consider the tax rate r against revenue imposed by the government.

The model is derived by backward induction. We start with solving the level of effort that each firm will select in the second stage. Subsequently we solve the optimal fishing tax rate r, which will maximize the government's tax revenue after computing the equilibrium quantity as a function of r.

We will not consider the resource constraint in the first place. Instead we will have a section to discuss it later.

3. MATHEMATICAL ANALYSIS

Throughout our discussion, we will assume that any incumbent firm will not exit from the industry so long as its equilibrium profit is greater than or equal to zero, and we will also assume that any potential entrant will enter into the industry if it can get a nonnegative equilibrium profit after entry.

In our model firm i's profit function is expressed as:

$$\pi_i = PQ - VC - FC = (1-r) \times Pq_i - c_i q_i - f_i \tag{2}$$

where r is the tax rate that the government set, and c_i, f_i are the marginal cost of quantity and fixed cost respectively for firm i. Therefore,

$$\pi_i = (1-r)(a - \sum_{i=1}^{n} q_i)q_i - c_i q_i - f_i \tag{3}$$

Notice that firm i will continue to operate only if its profit is greater than or equal to zero.

For firm i, its objective function is

$$\max_{q_i} \pi_i = (1-r)\left(a - q_i - \sum_{j \neq i}^{n} q_j\right)q_i - c_i q_i - f_i \tag{4}$$

Rearrange firm i's profit function:

$$\pi_i = (1-r)\left[-q_i^2 + \left(a - \sum_{j \neq i}^{n} q_j\right) q_i\right] - c_i q_i - f_i \tag{5}$$

Mathematically, when the first-order condition (F.O.C) with respect to q_i of the profit function is satisfied, i.e. $\partial\pi_i / \partial q_i = 0$, the corresponding quantity maximizes firm *i*'s profit.

The first-order-condition of Equation (5) is

$$2(1-r)q_i = (1-r)\left(a - \sum_{j \neq i} q_j\right) - c_i \tag{6}$$

Subtracting $(1 - r)q_i$ from both sides:

$$(1-r)q_i = (1-r)(a-Q) - c_i \tag{7}$$

Adding cross all firms:

$$(1-r)Q = (1-r)na - (1-r)nQ - \sum_{i=1}^{n} c_i \tag{8}$$

therefore, we have:

$$Q = \frac{(1-r)na - \sum_{i=1}^{n} c_i}{(1-r)(n+1)} \tag{9}$$

Substituting Equation (9) to Equation (7) and solve for q_i:

$$q_i = \frac{a}{n+1} + \frac{\sum_{j \neq i}^{n} c_j - nc_i}{(1-r)(n+1)} \tag{10}$$

For our model, we assume that any firms stay in as far as the after-tax profit is greater than or equal to zero. With a process of a continuously increasing tax rate *r*, we assume that firms exits one by one. Let $n(r)$ be the number of firms remained in business for a given *r*.

In the first stage, government chooses a tax rate, *r*, to maximize the total amount of tax revenue received, *T*:

$$\max_r T(r) = rPQ \tag{11}$$

Notice that there exists a supreme for all feasible *r*, noted by r_{sup}. It is the largest tax rate government could impose due to the limited fisheries stock. That is to say, any r high than r_{sup} is not feasible.

The following method can be used as a good approach for finding an optimal *r*.

Lemma 1: *Assume that a group S of existing firms can survive in the industry under a tax rate r (i.e. each has a nonnegative equilibrium profit) when no entry occurs, then when the tax rate is reduced to* $r' < r$*, these same group of firms can also survive.*

Proof: We only need to show that, under the smaller tax rate, the new equilibrium profit for each firm increases.

Originally, the profit function for firm *i* is

$$\pi_i = (1-r)\times(a-Q)q_i - c_i q_i - f_i \tag{12}$$

Substitute Equations (9) and (10) into Equation (12) and simplify:

$$\pi_i = \frac{1}{(n+1)^2}\left[(1-r)a^2 + 2ab + \frac{b^2}{1-r}\right] - f_i \tag{13}$$

Where $b = \sum_{j\neq i}^{n} c_j - nc_i$, approximately equals c_i. At the critical point that firm *i* could survive, its profit equals to zero. Set Equation (13) equals to zero, we will get two solutions of *r*. In this case, we choose the smaller *r* because beyond this *r* firm *i* has already decided to quit the industry and the larger *r* is not feasible.

Take first-order derivative of Equation (13), we have:

$$\frac{\partial \pi_i}{\partial r} = -\frac{a^2}{(n+1)^2} + \frac{b^2}{(n+1)^2(1-r)^2} \tag{14}$$

Equivalently:

$$\frac{\partial \pi_i}{\partial r} = -\frac{\left[a(1-r)-b\right]\left[a(1-r)+b\right]}{(n+1)^2(1-r)^2} \tag{15}$$

In this function, a denotes the maximum price determined by supply-demand mechanism and *b* is approximately equals to firm *i*'s marginal cost. Since firm *i* can survive at given *r*, its unit after-tax revenue (*1* – *r*)a should be able to at least cover its marginal cost, *b*, and fixed cost per unit. Explicitly, (*1* – *r*)*a* – *b* > *0* Thus $\partial\pi_i/\partial r < 0$ at *r*. In addition, from Equation (14) we know that when *r* decreases, $\partial\pi_i/\partial r$ also decreases. Therefore the derivative is negative for a given n on $[0,r)$. By First Derivative Test, π_i is decreasing on $[0,r)$. Therefore, under the smaller tax rate, r', the new equilibrium profit for each firm will be larger. To be exact, the same group of firms will remain survivable.

Let N be the number of all potential firms in fishery industry. Now let us consider all possible nonempty subset of the set of N firms. There are $2^N - 1$ of these subsets.

Definition: *A* subset *S* of firms is said to be coexistible if there exists a tax rate r ³ 0 under which each member in *S* earns a nonnegative profit and no entry will occur in the industry.

Now let us discuss the situations for coexistable subsets. Obviously the number of coexistible subsets of *N* is finite.

Lemma 2: For any coexistiblesubset *S* of firms, there must exist a maximal tax rate $r_h(S) \geq 0$ such that under this tax rate *S* remains coexistible and at least one member in S has 0 equilibrium profit.

Proof: Because *S* is coexistible, there is a tax rate $r \geq 0$ under which each member in *S* earns nonnegative profit and no entry occurs in the industry. When the tax rate increases, by Lemma 1 no entry can occur, and at the same time the profit of each of the existing firm is reduced gradually. Note that each existing firm's profit depends on *r* continuously, and as a result the smallest profit among them also continuously depends on *r*. The smallest profit is negative when *r* is increased to 1, by continuity there must exist some tax rate $r_h(S) \geq 0$ making this smallest profit exactly equal to 0.

Lemma 3: For any coexistible subset S of firms, either there exist some $r_l(S) \geq 0$ such that entry into the industry will start to occur under this tax rate, or entry never occurs even the tax rate is reduced to 0.

Proof: The argument is trivial.

Let *S* be a coexistible subset of firms. Now we consider two different cases in Lemma 3.

Case 1: Suppose entry will occur when tax rate is reduced to $r_l(S)$.

The tax revenue $T(S,r)$ for the government from these firms is continuously dependent on *r* in the interval on $(r_l(S), r_h(S)]$. In this interval, there is a least upper bound on government revenue. If it can be attained at some *r*-

value $r^*(S)$ within the half open interval, we will denote it $T^*(S)$; if it is cannot attained within this interval, we define:

$$\tilde{T}(S) = \lim_{r \to r_l(S)+} T(S,r) = T\left(S, r_l(S)^+\right)$$

Note that $\tilde{T}(S)$ cannot be precisely achieved, but can be approximately achieved as r is reduced to sufficiently closed to but not equal to $r_l(S)$.

Case 2: entry will never occur When r varies from 0 to $r_h(S)$, the firms in S are always coexistible. Again we use $T^*(S)$.

Denote $T^-(S)$ to be either $T^*(S)$ or $\tilde{T}(S)$. As there are at most $2^N - 1$ coexistible subsets, which is finite, $\{T^-(S) : S \subseteq N\}$ has finite elements. Therefore, a maximum value among all the elements T^M exists as either a maximum value achievable or a supremum.

To sum up we have Proposition 1: For each subset S of coexistible firms, let $r^-(S)$ be $r^*(S)$ or $r_l(S)$, dependent on whether or not the maximum of $T(S)$ can be achieved within $(r_l(S), r_h(S)]$. Let $T^-(S)$ to be either $T^*(S)$ or $\tilde{T}(S)$. Let $T^M = \max_s \{T^-(S)\}$. Then T^M is either the maximum revenue achievable, or is the least upper bound.

4. NUMERICAL EXAMPLES

To better illustrate the optimization process on government's decisions, we give a simple example with all firms to be identical.

Suppose there are 5 potential firms in total in the industry, i.e. $N = 5$.

We randomly assign the values for parameters as follows:

As all firms are identical, all subsets of n firms, denoted by S_n, are exactly the same combination as long as nis fixed. We can simplify Equation (13) into

a	c	f	N
7	1	2	5

$$\pi_i = \frac{1}{(n+1)^2}\left[(1-r)a^2 - 2ac + \frac{c^2}{1-r}\right] - f \tag{16}$$

Substitute the values into above equation:

$$\pi_i = \frac{1}{(n+1)^2}\left[49(1-r) - 14 + \frac{1}{1-r}\right] - 2 \tag{17}$$

when $r = 0$, $n = 5$, we get $\pi_i = -1 < 0$; i.e. the subset consisting all 5 firms is non-coexistible.

When $r = 0$, $n = 4$, we get $\pi_i = -0.56 < 0$; i.e. the subset consisting 4 firms is non-coexistible.

When $r = 0$, $n = 3$, we get $\pi_i = 0.25 > 0$; i.e. the subset consisting 3 or less firms are coexistible For subset S_3 consist of 3 firms operating in the industry, to find $r_h(S_3)$, we set $\pi_i = 0$ for Equation (17). Solving the equation gives us $r_h(S_3) = 8.35\%$. When r decreases from 8.35% to 0, no entry will occur.

Government's Revenue Function is:

$$T_r = rPQ = r(a-Q)Q \tag{18}$$

Substitute Equation (9) into Equation (18):

$$T(r) = \frac{nr}{(n+1)^2}\left(a + \frac{nc}{1-r}\right)\left(a - \frac{c}{1-r}\right)$$
$$= \frac{3r}{16}\left(7 + \frac{3}{1-r}\right)\left(7 - \frac{1}{1-r}\right) \tag{19}$$

Take the first-order derivative of Equation (19):

$$\frac{\partial T}{\partial r} = \frac{147}{16} + \frac{33-51r}{16(1-r)^3} \tag{20}$$

Take the second-order derivative of Equation (19):

$$\frac{\partial^2 T}{\partial r^2} = \frac{768 - 1632r}{256(1-r)^4} > 0 \quad \text{for } r \in [0, 0.0835] \tag{21}$$

By first-order derivative test, $\partial T/\partial r$ is increasing for $r \in [0, 0.0835]$. Thus $\partial T/\partial r > \partial T/\partial r|_{r=0} = 18$ for $r \in [0, 0.0835]$. Therefore T is an increasing function on $r \in [0, 0.0835]$. So we have $T^*(S_3) = T(0.0835) = 0.950$.

We continue to solve the cases with subset S_2 and S_1 by similar methods and the results are shown in Table 1:

We plot $T(S_n)$ against r for n = 1, 2 and 3, as shown in Figure 2.

The result shows that government can achieve the maximum revenue of 6.426 by imposing a revenue tax at the rate of 60.23% and when only one firm remaining in the industry.

5. SOCIAL WELFARE PERSPECTIVE

The total profit earned by the firms can be obtained by summing Equation (12) across the n existing firms, which is $(1-r)(a-Q)\sum_{i=1}^{n} q_i - \sum_{i=1}^{n} c_i q_i - \sum_{i=1}^{n} f_i$, where Q is the total harvest of all existing firms. Then the government's revenue is $r(a-Q)\sum_{i=1}^{n} q_i$. Define W as the net return from the entire fishery industry that is the summation of total profit from all the firms and the government's revenue, i.e.:

$$W(r) = (a-Q)\sum_{i=1}^{n} q_i - \sum_{i=1}^{n} c_i q_i - \sum_{i=1}^{n} f_i \tag{22}$$

Table 1. Calculations results 1.

	$r_l(S)$	$r_h(S)$	$r^*(S)$	$T^*(S)$	$\tilde{T}(S)$
S5			Non-coexistible		
S4			Non-coexistible		
S3	0	8.35%	8.35%	0.950	-
S2	8.35%	37.98%	37.98%	4.649	-
S1	37.98%	60.23%	60.23%	6.426	-

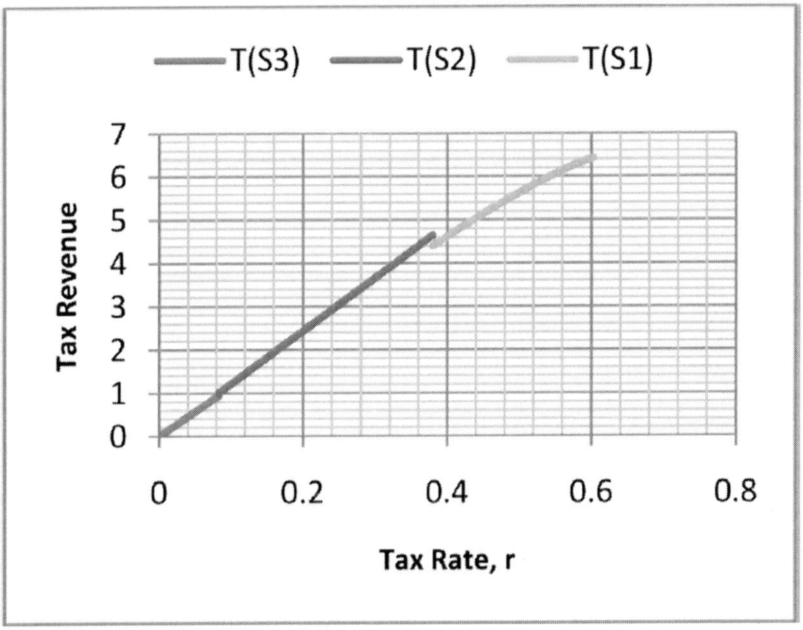

Figure 2. Tax Revenue against Tax Rate.

Instead of achieving maximum revenue, government can also set tax rate *r* targeting to maximizing social welfare. The process of determining this *r* can follow the same method of finding the tax revenue maximization tax rate.

For each *S* of coexistible firms, consider welfare *W*(*r*) when *r* varies in $(r_l(S), r_h(S)]$. Either the maximum $W^*(S)$ is attained within the interval at some $r_W^*(S)$, or the supremum $\tilde{W}^*(S)$ is achieved as *r* tends to $r_l(S)$ from the right. Very much similar to the arguments for Proposition 1, we have Proposition 2: For each subset *S* of coexistible firms, let $r^\wedge(S)$ be $r_W^*(S)$ or $r_l(S)$, dependent on whether or not the maximum social welfare of *W*(*S*)can be achieved within $(r_l(S), r_h(S)]$. Let $W^-(S)$ to be either $W^*(S)$ or $\tilde{W}(S)$. Let $W^M = \max_s \{W^-(S)\}$. Then W^M is either the maximum social welfare achievable, or is the least upper bound.

In the example of identical firms above, substitute Equations (9) and (10) into Equation (22), we get the welfare function when all firms are identical:

$$W(r)=\frac{n}{(n+1)^2}\left[(1-r)a^2-2ac+\frac{c^2}{(1-r)}+r\left(a+\frac{nc}{1-r}\right)\left(a-\frac{c}{1-r}\right)\right]-nf \tag{23}$$

Plot $W(S_n)$ against r for n = 1, 2 and 3.

Figure 3 shows that when government imposes a tax rate larger than but extremely close to $r_l(S_3)$ = 37.98%, the social welfare can be ultimately close the maximum value. Yet this maximum value can never be exactly achieved. However, this preferred tax rate is clearly different with the tax rate that achieving T^M, which is 53.18%.

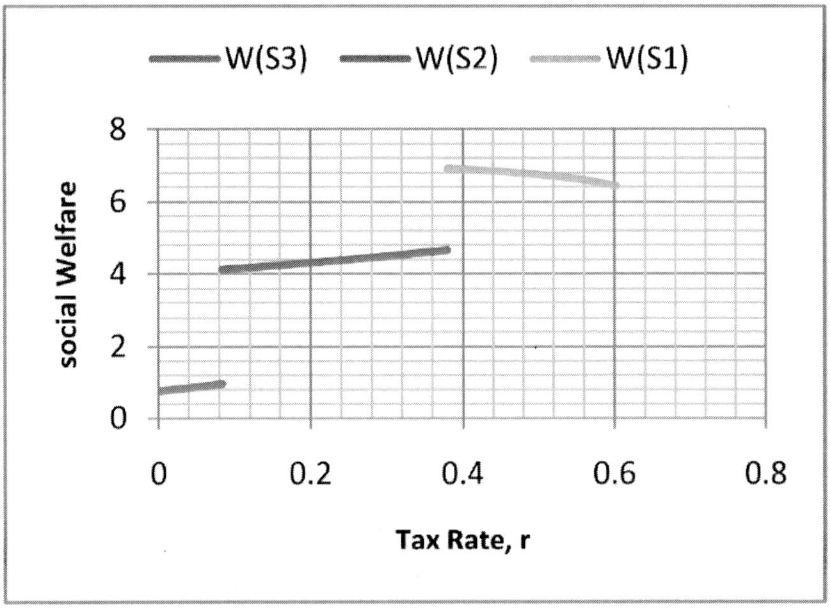

Figure 3. Social Welfare against Tax Rate.

In above example, we notice that social welfare is maximized when the firm is operating as a monopolist. This may due to a relative large fixed cost compare to the demand. Explicitly speaking, when firm quits with an increasing tax rate, the increase in demand is less than the decrease in total fixed costs; and hence the total return from the entire industry increases. Next we reduce the fixed cost of each firm to 0.7 to see whether social welfare will still be maximized in monopoly market.

In a new example, the parameters have following values:

a	c	f	N
7	1	0.7	5

Calculation results are listed in Table 2 below:

Figure 4 plots both tax revenue and social welfare against rate.

In this example, social welfare is maximized in duopoly market with a tax rate of 60.01%; while the maximum tax revenue, 7.56, is achieved when the tax rate is 64.28% and 2 firms remain in the industry. Again, the two goals are achieved at different tax rate.

Generally speaking, government can target to either maximize its revenue T or social welfare W but hardly to achieve both of them simultaneously.

6. TAXATION AS A MEASURE FOR PROTECTION OF RESOURCE

In the above discussion so far we did not directly consider the resource constraint, and when social welfare was considered, it's the one period welfare. In the long term an economy may also need to consider resource protection. Fortunately in our model, the government could use taxation to control the total amount of resource being consumed.

Actually it is easy to see that, when the tax rate is increased, the total amount Q harvested is reduced, as is stated in.

Table 2. Calculation results 23.

	$r_1(S)$	$r_2(S)$	$r^*(S)$	$T^*(S)$	$\hat{T}(S)$	$r_W^*(S)$	$W^*(S)$	$\hat{W}(S)$
S5	0	22.61%	22.61%	2.413	-	22.61%	2.415	-
S4	22.61%	39.06%	39.06%	4.543	-	39.06%	4.543	-
S3	39.06%	52.90%	52.90%	6.467	-	52.90%	6.467	-
S2	52.90%	64.28%	64.28%	7.560	-	60.01%	7.600	-
S1	64.28%	73.37%	67.59%	6.671	-	64.28%	-	7.49

[3] $r_T^*(S)$ denotes the tax rate that maximizes tax revenue and $r_W^*(S)$ represents the tax rate that maximizes social welfare. $T^*(S)$ (or $W^*(S)$) is the maximum tax revenue (or social welfare) when it can be exactly achieved and $\hat{R}(S)$ (or $\hat{W}(S)$) is the supreme of tax revenue (or social welfare) when it cannot be exactly achieve.

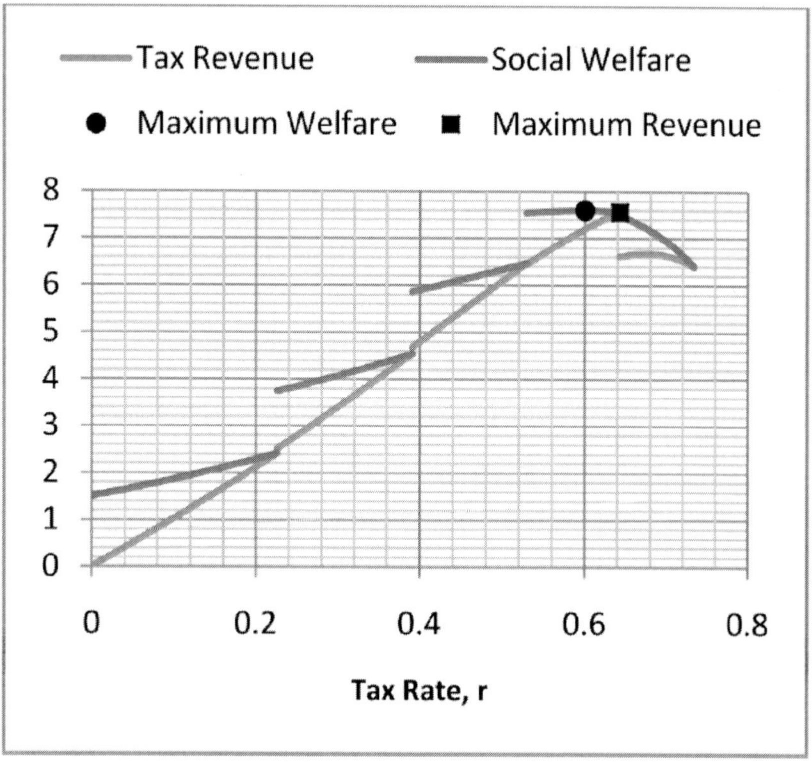

Figure 4. Revenue and Social Welfare against Tax Rate.

Proposition 3: For any coexistible subset S of firms, as the tax rate r increases, the total quantity harvested Q is reduced.

Proof: Rewriting Equation (9):

$$Q = \frac{na}{(n+1)} - \frac{\sum_{i=1}^{n} c_i}{(1-r)(n+1)} \tag{24}$$

Note that, when r increases, the second term in the right-hand side has a larger absolute value, and as a result Q is smaller.

Remark: While the proof of Proposition 3 is easy, the result is not completely trivial. From equation (10), it is interesting to note that, when the firms are asymmetric, the firm with the smallest marginal cost might increase its output when the tax rate r increases, yet the industry total output is always reduced. This is an important feature for an oligopoly industry.

It can be better illustrated with the following example.

To be consistent, we again assume that there are 5 firms in the market and keep a = 7. However, now the firms are asymmetric, their marginal costs and fixed costs differ with each other. They are listed as follows:

i	1	2	3	4	5
c_i	0.5	1.2	1	0.9	1.1
f_i	2	0.4	0.6	0.7	0.5

Note that firm 1 is the leading firm in the market, which has the smallest marginal cost.

Set tax rate r as 8%. We calculate each firm's harvest according to Equation (10):

q_1	q_2	q_3	q_4	q_5	Total (Q)
1.47	0.71	0.93	1.04	0.82	4.98

Now we assume the tax rate increases to 10%. The harvest of each firm becomes:

q_1	q_2	q_3	q_4	q_5	Total (Q)
1.48	0.70	0.93	1.04	0.81	4.96

In this example, we can see that with tax rate increasing from 8% to 10%, although the total production dropped by 0.02, firm 1's individual harvest raised from 1.47 to 1.48. This is a noteworthy feature of an oligopoly market with the presence of a leading firm. In fact we can state a general result:

Proposition 4: In an oligopoly industry with n firms, as the corporate tax rate r increases, any member with its marginal cost less than $n/(n+1)$ of the average marginal cost will increase its output.

Proof: Let c be the average marginal cost:

$c = \frac{1}{n}\sum_{j=1}^{n} c_j$. Suppose for firm i it holds that $c_i < \frac{nc}{n+1} = \sum_{j=1}^{n}\frac{c_j}{n+1}$. This is equivalent to $(n+1)c_i < \sum_{j=1}^{n} c_j$, or $nc_i < \sum_{j \neq i} c_j$.

Recall Equation (10):

$$q_i = \frac{a}{n+1} + \frac{\sum_{j \neq i}^{n} c_j - nc_i}{(1-r)(n+1)}$$

Now the second term in the right-hand side is positive, and as r increases its value increases.

In view of Proposition 3, for resource protection the government could always choose a tax rate making sure the total resource consumed does not exceed some upper bound. With this additional constraint the government could redo the calculations, discovering an optimal tax rate to maximize the per-period social welfare and at the same time protecting the resource for long term consumption. In view of Proposition 4, increasing tax rate in the long run will force firms with backward technology (with higher MC) to exist and encourage firms with more advanced technology (with lower MC) to enter.

7. CONCLUSIONS

Throughout the paper, non-cooperative game theory is used to model the strategic interactions between fishery firms and the government.

We construct a dynamic model in which different marginal costs and fixed costs are applied to different firms respectively. In this dynamic model, with a specified tax rate, firms may earn negative profits due to the presence of fixed costs. Hence it may choose to retire from fishery. Government will revise the tax rate according to its goal because of the changes in the number of firms and the existing firms' strategies. New tax rate will induce firms' new decisions. The iteration will go on until equilibrium achieved when government accomplishes its goal and no existing firms chooses to quit. The dynamics in this model makes the government revenue and net return are not always continuous in r (tax rate). However, we can determine the optimal r by a discrete optimization process. We assert that government can set a tax rate either to achieve (or to be infinitely closed to) the maximum revenue or to achieve (or to be infinitely closed to) the social welfare maxima. It is hard to find a single tax rate that can achieve both goals simultaneously. Moreover, taxation on fishing firms' revenues can be used as a tool for protecting fishing resources. When government increases tax rate, the total production will always reduce. One interesting point to note here is that, even though the total production reduces, the leading firm with the smallest marginal cost in an oligopoly market might still increase its production.

Yu Shao, Kai Shen, Weiru Zhang, Shuntian Yao# Based on the model constructed, the existence of fishery conflicts paradigm is revealed automatically. Nowadays, because of the upsurge discussion on environmental sustainability among the public, policy makers pay most of their attention to resource maintenance. However, conflicts are prevalent because of the poor implementation and enforcement of most fishery laws and regulations. Thus, it is necessary to involve all stakeholders in the fishery industry and related sectors as well as the policy makers and fisheries managers in a thorough and periodic review of policies and institutions. This paper suggests that future policy making should take into account of thorough argumentation in favor of the necessity not only to maintain the natural resources but also to increase yields and social welfare in future.

REFERENCES

1. Food and Agriculture Organization of the United Nations, "Integrated Coastal Area Management and Agriculture, Forestry and Fisheries," FAO Guidelines, Rome, 1998.
2. T. Charles, "Fishery Conflicts: A United Framework," Marine Policy, Vol. 16, No. 5, 1992, pp. 379-393. doi:10.1016/0308-597X(92)90006-B
3. Agriculture, Fisheries and Conservation Department, "Hong Kong Annual Report 1999," 1999. http://www.yearbook.gov.hk/1999/eng/08/08_00.htm
4. C. Chu, "Thirty Years Later: The Global Growth of ITQs and Their Influence on Stock Status in Marine Fisheries," Fish and Fisheries, Vol. 10, No. 2, 2009, pp. 217-230. doi:10.1111/j.1467-2979.2008.00313.x
5. D. R. Luce and H. Raiffa, "Games and Decisions: Introduction and Critical Survey," Wiley, New York, 1957.
6. F. H. Clarke and G. R. Munro, "Coastal States, Distant Water Fishing Nations and Extended Jurisdiction: A Principal-Agent Analysis," Natural Resource Modeling, Vol. 2, No. 1, 1987, pp. 81-107.
7. F. H. Clarke and G. R. Munro, "Coastal States and Distant Water Fishing Nations: Conflicting Views of the Future," Natural Resource Modeling, Vol. 5, No. 3, 1991, pp. 534-569.
8. N. Raissi, "Features of Bioeconomics Models for the Optimal Management of a Fishery Exploited by Two Different Fleets," Natural Resource Modeling, Vol. 14, No. 2, 2001, pp. 287-310. doi:10.1111/j.1939-7445.2001.tb00060.x
9. D. Levhari and L. J. Mirman, "The Great Fish War: An Example Using a Dynamic Cournot-Nash Solution," Bell Journal of Economics, Vol. 11, No. 1, 1980, pp. 322-334. doi:10.2307/3003416
10. R. D. Fischer and L. J. Mirman, "The Compleat Fish Wars: Biological and Dynamic Interactions," Journal of Environmental Economics and Management, Vol. 30. No. 1, 1996, pp. 34-42. doi:10.1006/jeem.1996.0003
11. E. Dockner, G. Feichtinger and A. Mehlmann, "NonCooperative Solutions for a Differential Game Model of Fishery," Journal of Economic Dynamics and Control, Vol. 13, No. 1, 1989, pp. 1-20. doi:10.1016/0165-1889(89)90008-0
12. G. Ruseski, "International Fish Wars: The Strategic Roles for Fleet Licensing and Effort Subsidies," Journal of Environmental Economics and Management, Vol. 36, No. 1, 1998, pp. 70-88. doi:10.1006/jeem.1998.1038

13. Y. Wachsman, "A Model of Fishing Conflicts in Foreign Fisheries," Working Paper No.02-16, 2002.

14. M. B. Schaefer, "Some Considerations of Population Dynamics and Economics in Relation to the Management of Marine Fisheries," Journal of Fisheries Research Board of Canada, Vol. 14, 1957, pp. 669-681. doi:10.1139/f57-025

CHAPTER 13

Desirability of a Standard Notation for Fisheries Assessment

Sergio Ragonese, Sergio Vitale*

Institute for Coastal Marine Environment—IAMC, Italian National Research Council—CNR, Mazara del Vallo, Italy;

ABSTRACT

The worldwide increase of the publications concerning the assessment of marine renewable living resources is highlighting long-standing problems with symbols and annotations. Starting from the symbols presented within the classic fisheries masterpieces produced, mainly in the fifty of the last century, a first "Milestone" list was organised. Thereafter, the pertinent literature was (not exhaustively) browsed in order to integrate this Milestone list on the base of a set of decisional criteria. The present contribution consists in using the Latin letters as well established symbols for the corresponding parameters, leaving free to specific use (with few historical exceptions) the Greek letters in view to open a discussion among all the fisheries scientists and bodies in order to move towards a common language and better communication standards.

Keywords: Fisheries; Assessment; Symbols; Standard

1. INTRODUCTION

The intuition to separate out the effects of births, deaths and growth on fish populations in order to estimate the surplus production of an exploitable stock has been developed since the first decades of the twentieth century [1-4]. The origin of fisheries science, as an integrated and structured discipline, however, might be generally placed at the second half of fifties and first half of sixty, when the milestones of Beverton and Holt [5-7], Gulland [8-11], and Ricker [12,13] were published. At that time, the main goals of such discipline were assessing and managing the living renewable aquatic resources under a theoretical based exploitation pattern. Just little after the Beverton and Holt and Ricker works, it was evident the opportunity to be clear about the definition of physical quantities and naming conventions came [14- 16]. Almost half a century later, notwithstanding the growing importance of assessments to promote credible and effective rebuilding or managing plans for the highly depleted fishable resources in the world, on our knowledge no general agreement exists about a proper and unambiguous annotations and symbols. That notwithstanding a big effort has been realised regarding the definitions [17-23]. In the present contribution, which recalls the Holt's heading of a section in one of his reports presented at the Biarritz symposium in 1956, the most common pitfall, ambiguities and lack of consistency arising from the literature in the field were analysed and a scheme of symbols usage was proposed in order to encourage fisheries scientists and bodies to move towards the establishment of a common language and better communication standards in assessments terminology.

2. MATERIAL AND METHODS

2.1. Organising the Symbols Lists

The pertinent fisheries literature was browsed to highlight the different symbols uses and attributions within a "classic" definition of assessment. In fact, it is worth noting that different assessment interpretations can be found in both grey [24] and "official" publications [25,26] according to a more or less broad application of the word. In particular, within the "broader" interpretations can be quoted 1) the part of fisheries science that studies the status of a fish stock as well as the possible outcomes of different management alternatives; it tells us if the abundance of a stock is below or above a given target point and by doing so lets us know whether the stock is overexploited or not; it also tells us if a catch level will maintain or change the abundance of the stock [27]; and 2) the application of statistical and mathematical tools to relevant data in order to obtain a quantitative understanding of the status of the stock as needed to make quantitative predictions of the stocks reactions to alternative future regimes [22].

On the contrary, the goals of the classic and narrower assessment definitions can be identified in a) for any given fish populations 1) what are the present quantities that will be available for capture in one or more years and what factors are determining the quantity of the fish and how are they changing [28]; 2) providing a means of expressing population properties by a relatively few parameters [26] and of codifying the effects of fishing on stocks [29]; 3) using stock demographic parameters to determine the total catch and how the catch and the catch per unit of effort varies with changes in the pattern of fishing [30]; and 4) the study of population structure, dynamics, and past exploitation of a single population [31] and its reaction to the dominant influence of fishing pressure [32].

Considering the classic definitions, the first step consisted in establishing a "Milestone" list with the historical symbols on the base of the fisheries science masterpieces produced in the fifties and sixties, and integrated with the successive contributes of the same Authors. In particular, the following contributes were consulted: Beverton [33], Beverton and Holt [5-7], Beverton and Parrish [34], Holt [15,35-37], Holt et al. [16], Gulland [8-11,14,24,30,38-45], Gulland and Holt [46], ICNAF [29,47-50], Kesteven and Holt [51], Ricker [12,13,52-58], and Ricker and Foerster [59]. For convenience, these Authors were abbreviated within the Milestone list as BH (Beverton, Holt and Beverton and Holt); I (ICNAF reports), G (Gulland) and R (Ricker).

Hence, the literature (both official and grey), books, manuals and programmes dealing with fisheries assessments were (not exhaustively) consulted in order to compare and integrate the basic scheme previously established. The final step was proposing a standard set of symbols, preferably (or whenever possible) according to the following Decalogue of criteria and conventions [14-16]. In particular, the symbols:

1) should be referred to the most relevant and studies items in "classic" fishery assessment;
2) should have a unique correspondence for each quantity, at least in base of their position (prefix, suffix, subscripts, superscript);
3) should be found within a standard commonly and easily available (in the specific case, the symbols lists in word Microsoft) avoiding other difficult, already existing, to find symbols [60];

4) should avoid special characters, such as the circumflex accent or the " " symbol, which was employed for example by Gulland [11,30] with the meaning of "therefore" or "consequently" [60; page 415];
5) should be different from those symbols traditionally well established in other related discipline (such as statistics);
6) in the masterpieces proposals or traditionally established should be maintained;
7) should consist of one to three "components" (never more than four letters), considering that in many instances many subscripts could be necessary [61];
8) should represent the initial words of the considered variable;
9) should help in expressing the relevant formula in a simple and compact form, which is easy to write, type and print;
10) similar should indicate similar measures.

In both Milestone and proposed list, the symbol § and §§ denote some remarks relative to the proposed symbol definition and alternative meanings of the same symbols, respectively. At the beginning of the proposed list, the X symbol was employed to represent some generalisations. Finally, the following abbreviations were adopted: coef. for coefficient, cons. for constant, n° for number, par. for parameter, prop. for proportion, and VBGF for von Bertalanffy Growth Function.

2.2. The "Milestone" List

a—BH_ Coef. of linear equations relating growth to density; unit haul swept area; intersect in regression line; const.; time interval (for example, in length increment— average length regression); average (strictly speaking) median length in Tauti's expression. G_ Probability (in age estimation); slope in the linear density dependence regression between asymptotic length and abundance; coef.; hook catch probability; area or volume under the influence of a gear; prop. of full fishing mortality rate; age at recruitment; proportionality coef. (usually 0.5) among yield, natural mortality and stock biomass (Y = aMB) in potential yield computation. I_ Mesh size. R_ Annual (or seasonal) total mortality rate; annual expectation; actual mortality; the first (multiplier) coef. in the (individual) length-weight functional relationship; coef. in Ricker's recruitment curve (when stock size and recruitment are in the same units); competition coef.; intercept in regression; sex ratio as males over females; Brody's coef.; intercept in yield per effort against effort; initial size; cons.

α (alfa)—BH_ Derived coef. of pre recruited mortality in the eggs-recruitment relationship; angle definition; cons. G_ Vulnerability; proportionality coef. R_ Par. in the Ricker (dimensionless) and Beverton and Holt R/S curves.

A—BH_ The area occupied by the fish population; Russell's recruitment. G_ Sum of squares; coef. in ageing error; area of fished region; Russell's recruitment; (Heincke's) annual death rate; cons. in stock recruitment curves. I_ Fish breadth/fish depth ratio. R_ Average population in successive years; annual (also Heincke) or seasonal mortality rate; A' annual or seasonal rate of disappearance of fish; maturity categories n° in Murphy's method; par. in other growth model than VBGF; par. in the Beverton and Holt R/S curve when S and R are in the same units; $A_{0.95}$ upper age limit according to Taylor's approximation [62].

b—BH_ Selection factor; ratio of length at 50% point of selection ogive (or mean selection length) to mesh size; coef. of linear equations relating growth to density; Huxley's coef. in fish length-weight relationship ($w = bl^k$); coef. (scale factor) in the Richards curve; coef. in length increment—average length regression; oldest age in Tauti's expression. G_ Probability (in age estimation); generic coef.; selection factor; density dependence in recruitment; generic slope in linear regression; b_n expanding term in the analytical computation. R_ The exponent in the allometric ($b \neq 3$) and isometric ($b = 3$) length-weight (individual) relationship; the slope of any line; proportionality coef. in recruits parental relationships; Brody's coef.; annual catch in Baranov's food biomass relationship; intercept in yield-effort (Y/f) against f; cons.

β (beta)—BH_ Derived coef. of pre recruited mortality in the eggs-recruitment relationship; cons. R_ N° of marked fish; incomplete beta function; par. in the Ricker and Beverton and Holt R/S curves.

B—G_ Stock biomass (size in weight); sum of squares; coef. in ageing error; biomass, B' in exploited phase; B_∞ at maximum population (carrying capacity); B_P predators; cons. in stock recruitment curves. I_ Mesh length/mesh width ratio. R_ N° of natural death; biomass of a group of fish or an entire stock; maturity categories n° in Murphy's method; par. in other growth model than VBGF; prop. of new recruits in the catch (Allen's method); B_∞ and B_S maximum stock size (the environment will support) and "optimum" stock size corresponding to maximum Y at equilibrium in Graham surplus production curve.

c—BH_ Cons. given by the ratio of fishing mortality and "effective overall fishing intensity" (effort); catch per net. G_ Cons.; c' proportionality coef. relating catch per unit of effort to density of fish (provided that the availability is cons.); reciprocal of vulnerability—aggregation product; coef.; ratio of length at capture and maximum length (in potential yield computations); mean selection or entry to the catch or first capture; Y/Y_{max} ratio in marginal yield analysis; raising factor (from haul catch to stock size). I_ Capture related general par.; X_c at first capture; first liable to capture by the fishing gear in use; t_c age at entry to the exploited phase or first liability to capture; selection factor. R_ The catch up to any time; Widrig's catchability; the fraction of the whole stock captured in a single unit of effort; Brody's coef.; X_c compensatory component in natural mortality; par. in growth models different from (or previous than) VBGF.

C—BH_ Catch in n°; local density (concentration) of fish; C' catch per effort. G_ Catch in n°; sum of squares; total catch. $C_{1 \ldots c}$ species competing with target species; cons. in stock recruitment curves. I_ N° of fish in the catch (catch in n°); head girth/head breadth ratio. R_ Cons. of integration; catch in n° (usually in 1 year); n° of fish examined for tags or marks; maturity categories n° in Murphy's method; average minimum age limit of usable stock.

χ (chi)—BH_ Fecundity (/g) coef. G_ χ^2 chi square statistic.

d—BH_ Average distance of fish in random movement. G_ Catch per unit of effort; sample catch; density of fish as derived in a given haul catch; mean depth. R_ Annual increase in length in Baranov's Yield method; absolute increase in length.

D—BH_ Dispersion coef.; unconditional natural mortality rate; expectation of death by natural causes; average density of fish. G_ Density of fish on fishing grounds; n° dying of disease; cons. in stock recruitment curves. I_ Expectation of death by capture (unconditional natural mortality rate); expectation of death by natural causes. R_ Total deaths; maturity categories n° in Murphy's method.

Δ (delta)—BH_ R_ Interval; variation; change. G_ One or unit operation.

e—G_ Sample effort. R_ Base of the natural (or Napierian) logarithms.

ε (epsilon)—BH_ Coef. of food utilisation for growth and maintenance; efficiency of utilization of food for growth. G_ Coef. in mortality estimation; random variable in fishing effort analysis.

η (eta)—BH_ Suffix denoting reference to spawning or (first) maturity; marked change in growth; cons. in the differential Richards.

E—BH_ G_ I_ R_ Exploitation rate [$F/Z(1 - \exp^{-Z})$]. BH_ Egg production; X_E equilibrium; coef. of anabolism; unconditional fishing mortality rate; expectation of death by capture; par. in the differential length based VBGF; Taylor's KL_∞ product. G_ X_E expected value; probability of ultimate capture (often exploitation ratio in steady state condition). I_ Unconditional fishing mortality rate; expectation of death by capture. R_ Escapement; (absolute) n° of eggs; X_E equilibrium (steady state); cumulative fishing effort.

E.F.—I_ Escape factor as length/mesh size.

f—BH_ Fishing activity; effort; intensity; power; overall; weighed mean fishing effort per unit area. G_ Fishing intensity; fishing effort (in some suitable units); f' adjusted; f(X) function. I_ "Japanese" mortality rate; effective overall fishing intensity. R_ Fishing effort (n° of units of gear in use); Widrig's effective fishing effort adjusted, when necessary; f_S corresponding to MSY (optimum f).

♀—R_ Females (reproducing the Petersen 1892 table). I_ ♀♀.

φ (phi) —BH_ Dummy (time) variable or general funcion.

ϕ (phi)—BH_ Dummy (time) variable or general function; ϕ' and ϕ'' function relating cost and revenues to F; ratio (generic).

Φ (phi)—BH_ Total n° of age groups into which recruitment occurs.

F—BH_ G_ I_ R_ Instantaneous fishing mortality coef. BH_ 'F/K. G_ F_1... F_f species on which the target species feed. R_ Size of a progeny in the recruits parental relationships.

g—BH_ Grazing mortality coef. (individual); total fishing effort; fishing effort as recorded; standardized fishing effort. G_ Fishing effort; F/K ratio; rate of growth in short time interval. I_ Fishing effort as collected (uncorrected) or "crude". R_ Instantaneous rate of growth in models different from VBGF.

γ (gamma)—BH_ Annual egg-production per recruit.

G—B_ Grazing mortality coef.; Russell's population growth; standardized total fishing effort. G_ exponential growth rate; net (and long term) gain from a change in gear selectivity or area closure. I_ Girth factor; weight of a fish. R_ Instantaneous growth rate (general); Russell's population growth.

G.F.—I_ Girth factor as fish length/max. girth; girth/length ratio.

Γ (gamma)—BH_ Index of competition (force of concurrence).

h—BH_ N° of hour fished; par. in the egg-production per recruit computation, h_1 and h_2 cons. in dependence growth on food. G_ Cons. in growth equation. I_ Degree of precision. R_ Annual (or seasonal) relative individual growth rate; weight increment/initial weight ratio; annual growth rate.

H—BH_ Coef. of synthesis in the differential VBGF; par. in M at age variation; par. in the egg-production per recruit computation.

i—G_ X_i generic group identification. I_ X_i year class. R_ Widrig's instantaneous rate of (total) mortality of a stock; X_{th} period; X_i density-independent component of natural mortality.

I—BH_ Index of fishing intensity; par. in M at age variation.

j—BH_ Exponent related to maintenance requirements (usually 2/3). I_ Fishing intensity; X_j age group. R_ X_{th} period of recovering.

k—BH_ Coef. of catabolism in the differential VBGF; growth equation coef.; Huxley's coef. in $w = bl^k$; k_2 and k_0 Baranov's fishing and natural mortality coef.; k_1 coef. in linear approximation to a selection ogive; coef. G_ Total n° of fish in unit weight; n° of strata; sub-areas proportionality coef.; coef. in the Ricker exponential and VBGF; slope in the linear density dependence regression between mortality and abundance; cons. in growth equations; average fecundity. R_ Factor of proportionality; growth coef. in model different than VBGF; Ford's growth coef.; a rate used in various connections; instanttaneous rate of increase in Graham surplus production curve (V of Graham); instantaneous growth rate of a stock.

κ (cappa)—BH_ Cons. relating F to "destruction" mortality.

K—BH_ I_ One of the two main par. of the VBGF, proportional only to the catabolism coef. (hence, more sensible to temperature variation); par. expressing the relative rate of approach to asymptotic size; coef. Defining the sampling efficiency of a gear (in fish dispersion); X_K an alternative gear to be compared. G_ Coef. Proportional to the rate at which the fish completes its growth; the rate at which the limiting length is reached in the VBGF; selection factor in gill-nets; a new gear. I_ Coef. in the allometric relationship $W = KL^n$; head depth/head breadth ratio. R_ A rate used in various connections; rate of change in length increment in VBGF; Brody coef.; any rate; generic cumulative catch; integration cons.; n° of degree days; Ivlev's growth efficiency coef.

l—BH_, I_, R_ Length. BH_ l' some arbitrary length above which all fish are considered recruited to the gear. G_ Length; mean n° per unit weight for a length group; l_c at first capture; l_d gill net deselection; l_m gill net maximum efficiency; l_p partial recruitment (discards); l_r recruitment; $l_s$minimum size.

λ (lambda)—BH_ Fishable life span; maximum age a fish can attain; end of (fishable) life span; the age above which older fish contribute to fisheries can be considered practically negligible (in most cases considered $\lambda = \infty$). G_ True mean n°; true survival rate. R_ The last period (the greatest age) considered in which an appreciable catch is made; the end of life span or maximum age attained; Halliday's (1972) "maximum age of significant contribute to fisheries"; λ_1 the probability of capture between competing species; difference between maximum and recruitment ages.

L—BH_ Length of individual fish (particular); loss rate of marks; L_∞ upper asymptote of length; $X_L$age at which the fish do not appear in the catch (for different reasons). G_ Mean n° per unit weight for a length group; total life span in the fishery; L_∞ maximum length, especially in the VBGF; immediate losses after gear selectivity change or area closure. I_ Fishable life span; length. R_ (Mean) length at recruitment (in Baranov's yield method); fork length.

Λ (lambda)—G_ Average n° of landed fish.

m—BH_ Apparent mortality coef.; m_0 and m_1 in linear regression; coef. in M at density; instantaneous mortality (in fish dispersion); slope coef. in regression line; exponent in the differential Richards; X_m size at maturity. G_ Mesh size; average n° of fish; exponent in general production model; cons. in Richard's growth curve; M/K ratio; haul catch. I_ Mesh size. R_ The fraction of the initial population which has been caught up to time t for which the term "fishing mortality" will be reserved; m_1 the fishing mortality up to the season end (also rate of exploitation); annual or seasonal/(fishing) mortality rate (if no other causes operate; Widrig's m); conditional fishing mortality; sample size; variable exponent in Pella and Tomlinson surplus production curve; c_m maximum recruitment; slope of the Richards curve at the inflexion.

♂—R_ Males (reproducing Petersen 1892 table). I_ ♂♂.

μ (mi)—BH_ Density independent and interaction components of mortality; μ_1 and μ_2 linear coef. in M at density. G_ Apparent survival rate. I_ Coef. defining the relationship between mesh size and fish girth. R_ (Annual) expectation of death by capture; rate of exploitation.

M—BH_ G_ I_ R_ Instantaneous natural mortality coef. BH_ M'; M/K; M_1 and M_2 density independent and dependent. G_ M' limiting value approached by biggest fish; apparent n° in a year class; X_M maximum. I_ Mesh; mesh size. R_ M' n° of fish marked; mean abundance of predators; average age of first recruitment.

M.I.—I_ Mesh index.

n—BH_ N° of marked fish recaptured; generic n° of items. G_ Generic n° (ships, years, sampling days, hooks in a long-line); generic exponent in the VBGF in weight. I_ X_n age group; exponent in the allometric relationship $W = KL^n$. R_ Annual or seasonal (natural) mortality rate (if no other causes operate), Widrig's n; conditional fishing mortality; sample size; exponent in growth related models (Richards, Ursin and growth—temperature models); any generic n°.

υ (ni)—BH_ Nutritional factor. R_ (Annual) expectaion of natural death.

N—BH_ G_ R_ N° of fish in a homogeneous group (N_X year class). BH_ Total n° in stock. G_ N° of fish measured; real n° in a year class; abundance; n° in the stock; sampling days; N_R and N_k released and retained after a change in gear selectivity. I_ Total n° of fish in the exploitable phase of the stock. R_ N° of fish in a year class or populations; N_0 at the beginning.

o—G_ Subscript denoting observed values.

ω (omega)—BH_ Average weight of individual food organisms during their "grazeable" life span.

O—G_ N° of fish dying of other causes than fishing.

Ω (omega)—BH_ Expanding term (summation cons.) in year class weight computation.

p—BH_ Standing crop; prop. of fish caught in unit haul swept area; probability of capture; fish of a given length present in the swept area; shape related coef. G_ Fraction; n° (or prop.) of fish (also retained) or items (hooks) of a given type; p_i relative fishing power; X_p production of plant; positive cons. in the incomplete Beta function. R_ Population at the start of the fishing season; Widrig's instantaneous rate of fishing mortality; total mortality multiplied by the ratio of fishing deaths to all deaths; complement of catchability; par. in Allen's recruitment method; generic coef. in regressions.

π (pi)—G_ Prop. in the whole sample.

P—BH_ Generic (mean) abundance or population size in weight (P_W) or n° (P_N); (annual) production; total weight in stock; P_m fishing power; P_l n° of fish of length l which is liable to capture by the gear. G_ Probability level; year period; population size; n° of fish dying from predation; P_H production of herbivorous; P_1 ... P_p predators eating the target species; selective gill-net; fishing power; generic par. I_ Total weight (biomass) of fish in the exploitable phase of the stock; probability. R_ Size (n°, weight, egg production etc) of parental stock; level of statistical probability; par. in Jones yield computation.

Π (pi)—BH_ Population size.

ψ (psi)—BH_ Dummy time variable.

Ψ (psi)—BH_ Dummy time variable; par. in M at age variation. I_ Girth related probability of escaping.

q—G_ I_ R_ Catchability coef. relating F to f. BH_ Weight-length coef. in $W = ql^3$; prop. completion ($q = 1 - p$); fish of a given length retained in the cod-end; generic n° of years. G_ Availability. I_ The ratio between the best index of effective overall fishing intensity and the resulting instantaneous fishing mortality coef. R_ Widrig's instantaneous rate of natural mortality; total mortality multiplied by the ratio of fishing deaths to all deaths; Baranov's integration factor; generic coef. in regressions.

Q—BH_ Physiological-temperature coef.; expansion term; maintenance energy coef. (also per unit physiological surface); Q_{10} physiological-temperature coef. (Arrhenius) rule. G_ Gross long term increase in catch following change in selectivity or area closure; prop.; coef. in non linear catch per unit effort—effort relationship. I_ Initial slope of the eumetric catch curve (Holt's responsiveness of the stock); fishing intensity/fishing mortality ratio. R_ The yearly n° of fish which reaches the minimum reference age (X_Q) used in yield computations; the cons. which appears in the integration of Baranov's yield computation; par. in Jones yield computation; par. in Allen's recruitment method.

r—BH_ Annual recruitment to a food population; r_{th} period; amount of food consumed per unit time (unit ration); generic cons.; fish escaping from parts of the net other than the cod-end. G_ Replacement; radius (or influence) of a fishing gear; maximum rate of population growth; time intervals; prop. recruited to the gear; age at first capture. I_ X_r at recruitment to fishable stock; t_r age at entry to the exploitable phase. R_ The fraction of the population remaining at time t; Widrig's availability fraction; the fraction of the stock susceptible to fishing; rate of accession (analogous to survival rates); greatest age involved; difference between recruitment and initial ages; food ration; correlation coef..

ρ (*ro*)—BH_ Pre exploitation phase; recruitment at the area where the fishing is in progress; *ρ'* entry to exploited phase (first retained). R_ Rate of fishing.

R—BH_ G_ I_ R_ Recruitment related par. B_ N° of fish recruited annually to the exploited area; n° entering the exploitable phase in a given period; R' n° entering exploited phase each year at age $t_{\rho'}$; maximal ration; n° of recaptured marks. G_ R' n° at the age of first capture; R_1 ... $R_R$species affecting the recruitment of target species; raising factor. I_ N° of recruits entering the exploitable phase. R_ (Absolute) n° of recruits to the vulnerable (catchable) stock whatever by movements in to the region fished or by change in size or behaviour; n° of recaptured marks; multiple correlation coef.; feeding ration.

s—BH_ Sex ratio as % of mature females in total mature population; physiological surface area; mean survival rate; fish which would have been caught in a large cover applied to the body of the gear; ratio of the catch obtained in a haul to the saturation catch. G_ Surface area of a fish; raising factor. I_ Observed annual fraction surviving; selection factor. R_ Rate of survival; standard deviation; X_s condition of maximum sustainable yield.

σ (sigma)—BH_ Standard deviation; σ^2 variance; generic cons. G_ σ^2 population variance (var for the sampling variance).

S—BH_ Grazing efficiency; Russell's stock size in weight; variance (in recruitment). G_ (Annual) rate of survival; Russell's stock size; abundance of spawning stock; prop. of retained fish; $X_S$minimum limit; standard deviation. I_ Annual fraction surviving (survival rate); selection factor; the girth at which the fish is meshed. R_ Rate of survival; S' apparent.

S.F.—I_ Selection factor as fifty percent (retention or escaping) point/mesh size.

Σ (sigma)—BH_ R_ Summation sign.

t—BH_ Age of fish; t' and t'' coef. in linear approximation to a selection ogive; t_0 scale cons. in the VBGF (or theoretical age at which the size is zero); t_r age at recruitment; t_c age at which fish are liable to be retained by the gear; t_L mean age of the oldest fish. G_ Time period; t_0 some previous time and cons. in VBGF; t_c at first capture; t_L maximum age in the fishery; t_p partial length recruitment; t_r at recruitment. I_ Time or age. R_ Time or age; time required for growth in growth—temperature models.

τ (tau)—BH_ Recapture period in marking; calendar date.

θ (theta)—BH_ Age group n°; X_θ prop. of females of age-group θ that are mature; angle definition; the youngest age group free from the influence of recruitment of gear selectivity.

T—BH_ Transport coef. (rate of interchange of fish between adjacent areas); mean age in catch or exploited phase; T_{max} maximum age in the sample; gross tonnage; fraction surviving; total. G_ Upper bound of a given time interval; time; non selective gear; gross tonnage. I_ Transport coef. R_ Interval of time; successive intervals in the life of the fish (not necessarily of equal length); weighed summation of age groups n° (also in Chapman and Robson); totals; temperature (in Celsius); metabolism.

u—BH_ Yield per recruit contributed by fish; ratio of grazing mortality coef.; weighting coef.; co-ordinate defining sub-areas. R_ The fraction by n° of fish caught by men; rate of exploitation (annual expectation); ratio of recovery to marked fish released; generic ratio; u_E equilibrium rate of exploitation (as captures divided by recruits).

υ (upsilon)—BH_ Age group.

U—G_ R_ Catch per unit of effort in n° (U_C) or weight (U_Y). BH_ Sum of square residuals; cons. Summation in yield analytical computation. G_ U_0 U_3 cons. in the expression for yield in weight. R_ Instantaneous rate of "other loss" (also emigration and shedding of tags).

v—B_ Co-ordinate defining sub-areas; fraction of a year. R_ Expectations of natural death; v' apparent.

V—G_ R_ Virtual population and cohort analysis par. BH_ Effective velocity of (random) movements; whole n° of years; vulnerability. G_ Value of individual fish; fish surviving the n_{th}year of life (in cohort analysis). R_ Utilized stock; variance.

w—BH_ R_ Weight of individual fish. BH_ w_c weight corresponding to the (theoretical) greatest steady catch obtainable by catching all fish at once (F

= ∞). G_ Mean weight in a group; weight sampled. R_ N° of spawners divided by the replacement n° of spawners.

W—BH_ Weight of individual fish (particular) and at stock level; W_∞ one of the two main par. of the VBGF proportional to the cube of the ratio of the coef. of anabolism and catabolism (hence, less sensible to temperature variation); W_∞ upper asymptote of weight. G_ Weight of landings; mean weight of a age group; individual fish; weight; W_∞ maximum (limiting) especially in VBGF; W_c at mean selection length; average larger than mean selection length; W_k of retained catch. I_ W^∞ par. of the VBGF in weight; fish weight. R_ Reproductive stock; weight of a group of fish (year-class, stock); W_0 initial weight of a stock; W_∞ theoretical maximum stock weight in unfished condition; prop. of recruits in Allen's recruitment method; n° of spawners divided by the replacement n° of spawners.

x—BH_ Mid point of the length interval. G_ Independent variable in regression generic coef.; $X_x$value in a particular year. I_ X_x year class. R_ The ratio of two initial populations; any (mainly dependent) variable (in regression); fractional representation of each age in the catch.

ξ (xsi)—BH_ Annual food consumption (individual).

X—BH_ Denotes a particular year, usually as suffix; "other loss than fishing" coef. in marking theory; fishing effort (in case of no ambiguity). G_ Effort in surplus production; fishing power; other loss rate (in tagging). I_ Mean girth; effective overall fishing intensity ("Japanese" fishing intensity); to be used only in case of no ambiguity (otherwise f should be employed). R_ Different kind of fishing effort; classification of stock composition; par. in Jones yield computation.

Ξ (xsi)—BH_ Annual food consumption of a fish population.

y—BH_ Growth increment (and increment per unit time) in length increment—average length analysis. G_ Dependent variable in regression; generic coef.; instanttaneous rate of capture of hooks. R_ Instantaneous rate of emigration; ratio in Baranov's food biomass relationship.

Y—BH_ G_ I_ R_ Yield (catch in weight). BH_ Yield in weight (Y_W) or n° (Y_N); total weight in fish catch; Y* expected post-regulation catch; maximum sustained yield or potential yield. G_ Catch in n°; (yield) in weight; ultimate yield after (Y_2) a change in size limit (Y_K); performance. I_ Weight of fish in the catch (catch in weight). R_ Catch by weight; Y_S maximum sustainable yield; Y' surplus production; different kind of fishing effort; classification of stock composition; dependent variable in regression; individual n° of eggs.

z—BH_ Ratio of fished to unfished areas (z = ∞ when the whole area is fished); cons. in year to year recruits variation. R_ Newcomers;

instantaneous rate of recruitment or immigration; n° of recruits divided by the replacement n° of spawners (and recruits).

Z (zeta)—BH_ Maintenance food coef..

Z—R_ G_ I_ Instantaneous total mortality coef. BH_ Correction term (in recruitment-egg relations); 'Z/K. R_ Recruitment; recruits to a stock divided by the "replacement" n° of recruits; instantaneous rate of disappearance (F + M + U).

3. RESULTS AND DISCUSSION

3.1. The "Proposed" List

The *x*, *X*—Individual—(lower case) observation and stock- (capital letter) level par., respectively. § Individual "fish" refers to any generic fish, shellfish or other organism exploited or exploitable [22], if not otherwise specified.

**X*—The asterisk, as left superscript, denotes that the X symbol results already well established (and maintained), but with a different meaning. §§ As right subscript (or superscript) denotes equilibrium quantity [63] or condition (also in unfished equilibrium), or "critical" par. [25]. As right superscript, it also specifies better an almost equivalent (or homologous) par.; for example, N and N* as n° and biomass of recruits at the start of year [64], R and R* as recruitment in n° and weight [65] or L* as maximum observed length (or largest specimen) in a sample [66]. Other diffuse uses are: "true" value [67]; generic critical or optimum (optimal in [67,68]) the average size of a fish of a given cohort when the instantaneous rate of natural mortality equals the growth. It also characterises symbols not used in Ricker's textbook; a par. estimated with methodology different from the usual. Further as right superscript, it has been used to identify a statistic or estimate of a par. [69] or a value at maximum or at stable condition [70].

'*X*'—A special variant of the X par. As left superscript, a conceptual close par. (for example see 'U). As right superscript, a special case of the same given par. (for example, see A'). Another example is the Y' defined as total yield as a fraction of the RW_∞ product or X' as an adjusted value[67]. §§ As superscript, a statistic or estimate of a par. [69]. Pristine level or condition [70].

X^+—Cumulated par., integrated beyond an age, size or time limit. ${}_tX^+$ plus (terminal) group. ${}_{(x,y)}X^+$ accumulation over the considered range, for example, all previous ages [70]. § C_t terminal catch in VPA; $C(L_{1,\infty})$ [71]. C^+ cumulated catches [72]. X_+ total; overall [73]. The sum of all catches of a year-class in subsequent years starting at age A [65]. §§ F^{+} at extinction [70].

${}_{c}X$—Constrained estimation (to be specified).

$X_{(\bullet)}$—Estimated via invariants or empirical equations rather than a true estimate. § Estimates uncertain [48]. Preliminary estimate [71]. To not be confounded with a proxy estimate [21].

${}_{/}X$—The dot at the left subscript denotes a variant (to be specified) from the basic definition. For example, $._{}K_{j}$ denotes the juveniles K in the biphasic or double VBGF.

${}_{\approx}X$—Non equilibrium condition. § Not cons. or random variation without trend par. condition (alternative to steady state).

${}_{\updownarrow}X$—Array of values. § X_{array} [74].

X_{Δ}—Par. referring to a finite interval. For example in Chen and Watanabe [75] procedure to estimate M.

${}_{\cap}X$—The Maximum value of a par. § The "maximum" has been often interpreted [72] as "optimum" [6, page 389] or "critical" [56]. For example, a_{cri} age at maximum cohort weight t_{cri} in [76]; c* and t* critical size and age as the value where biomass of a cohort is maximum in the absence of fishing and c_{opt} the value which maximizes yield when fishing mortality rate is fixed at the $F_{0.1}$[67]; l_{cri} the length above which all fish are vulnerable [25] or optimum age at exploitation); t_y in [77]. The most famous example is the Holt's optimum age (valid only in case of no dependence relationships) at which an (unexploited) cohort reaches its maximum abundance or reflecting the catch obtainable allowing a year class to reach its greatest total weight and hence catch it at once[6, page 374]. Other cases are: the Gulland's approximation ($F_{opt} \sim M$); the maximum value of an equilibrium curve as function of fishing intensity [78; 6, page 389]; the L_{opt} as the length at maximum yield-per-recruit [70]; the X_{SUP} [76] and X_m maximum or optimum [74]. Strictly speaking, "optimum" (i.e. any point which might be considered the "best" within the ultimate objective of fishery regulation) should be restricted to economic criteria [6, page 390] or management, but not to assessment. "Optimum" should not be confounded with the optimum sustainable yield, OSY, or Roedl's optimum yield OY [19]. "Critical" should not be confounded with the Hjort's critical period in which the numeric consistency of a cohort is determined; it is classically the larval or plankton stage (Hjort's hypothesis) or (more recent) the juvenile stage [79,80].

X_{obs}—Observation [25]. Observed value.

${}_{\square}X$—Inflexion. For example, $W_a = 0.29W_{\infty}$ in the isometric (cubic) VBGF.

X.—25, 50, 75 the par. corresponding to 25%, 50% and 75% of a logistic (anti symmetric) curve (or ogive) mostly in selection studies [81]. § X50 or X_{50} are often used to indicate size at (first) maturity or size at gear retention. The 50% has been defined "fifty percent (retention or escaping) point [49]. The 75% - 25% difference has been denoted as "selection span" [49]; nowadays, it is often denoted as "range" [66]. Reference [49] attributed a different (and apparently more appropriated) meaning to "selection range" as the range in fish length over which a fishing gear exercise selection.

$_{c}X$—Conventional par. For example, the Taylor's 95% maximum length approximation as index of longevity [62,71].

a—Age [25,61,64,65,67,82-84]. (Absolute) individual age [85,70]. $_{r}a$ relative. § t, T. Population age [85]. Age class; arbitrary reference age [86]. Age at maturity or first reproduction [87]. Generic age [68]. §§ Σa all individuals (new stock) transferring from the non-catchable to the catchable category [88]. Total mortality fraction and available population (in fished area) [69]. Annual rate of total mortality [89]. Instantaneous rate of loss of bait [90]. Cons. in the Putter expression of the VBGF and exponent relating length to surface in the generalisation of the VBGF [91]. a^2 as mean square coefficient of dispersion [92]. Coefficient of intraspecific competition [71]. Production/rate of recruitment ratio [67]. Vulnerability-selectivity age related par.[86]. Par. in Schnute model [93]. Mortality related par. in Ecopath [82]. Par. in different models [76]. Search rate; generic coefficient; n° of marked released [65]. Par. in different models; par. in the $w = al^b$ or non exponent coefficient in the length/ weight relationship [25]. Par. in different stock-recruitment relationships; effective search rate in Ecosim; [70]. Y-intercept in (AM or GM) linear regression, or multiplicative term in a length/weight relationship [66]. Positive cons. in von Bertalanffy (isometric) weight function [63].

**a*—Area fished (interested) by the gear by unit of effort [94;65]. Reference unit area interested by a fishing or experimental unit. a_s swept [74] or covered unit area. a_f over which the fishing effort is distributed. § a [25,71,74,76].

α—Free, to be specified. §§ Season [95]. Cons. per unit of effort [96]. Level of significance of a test [97]. Cons. in the generalisation of the VBGF [91]. Par. in total metabolism (T) and weight (W) according to $T = \alpha W^{\gamma}$ [98]. Coefficients in Jones and Johnston [99] growth, reproduction and mortality analysis. Age [100]. Par. in different models, especially SRR i.e. stock recruitment relationship [76]. Par. of the "asymptotic (overall) yield model" [71]. Par. in the recruitment function [64]. Par. (also scaled) of the Ricker stock-recruitment relationship; special values related to $F_{0.1}$, F_{ey} and MSY [67]. Par. in age-length growth model [101]. Subscript indexing length frequency data in MULTIFAN [102]. Instantaneous tagging mortality (dying immediately after tagging); par. in catching power and vessel standardization; intercept in the Ford-Walford plot [25]. Biological lower critical value [103]. Model par.; density independent coefficient (also slope at the origin) in SRR; coefficient (condition factor) in both isometric and allometric weight-length models; coefficient in Brody model; significance level; prop. of fish of length l; tag loss [65]. Generic par., for example, governing stock production especially at small size (such as intrinsic rate of natural increase) or specifying the state variable in Individual Based Models; slope at the origin in the general form of the SRR also defining the compensation capability of a stock; age at maturity; α_i Manly-Chesson food preference index [70]. Cons. in SRR; limit of R/S when $S \to 0$ [104].

A—Age [67]. Absolute age at stock level [70]. $_rA$ relative. § t, T. Total n° of age classes [86]. Mean age [84]. A_0 frequency of age a fish in a random sample A [65]. §§ Recruitment in weight [88,105]. Total n° of hooks [90]. Mouth area of the net [106]. Smaller mesh size in gill net selection experiment [71]. Par. relating L_∞, L_1, and L_2 with M/K [72]. Annual (or seasonal) total mortality rate [107]. N° of tagged fish alive at time t; abundance; area; par. in Jones' length based cohort analysis [25]. Height of the growth production function [87]. N° of fish dead after a given time; relative abundance [108]. Adult; A_j "total age" of predators [82]. Aspect ratio of the fish caudal fin [109]. Attrition rate [74]. Annual death fraction; maximum age of reproduction; the last age group; the n° of age groups [65]. Area [68]. Average [61]. Attribute vector in Individual Based Models; threshold value of viability in economics of fisheries [70]. Cons. (intercept) of the simple linear model [104].

A_c—Age of entry to capture. Age at which 50% of fish enters the exploited phase. A^c knife edge (the probability of capture becomes suddenly finite at this point). § $t_{\rho'}$, t_c. Age at which fish are first retained; age at first capture; 50% of selection age for the mesh in question.

A_{ch}—Age of the cohort; all the fish born in a discrete time interval of a given year. For example: A_{1st80}. § Usually coincident with the age class in case of continuous recruitment or one discrete recruitment per year; in case of multiple recruitment pulses, "micro cohort" or "stock let" [110] should be used and specified.

A_{cl}—Age of the year class; age class. All the fish born in the same year. For example, A_{1980}.

A_g—Age at generation. Average age of the parents when their offspring are born. § t_g.

A_m - Age at 50% onset of sexual maturity, based on the present gonadic activity. A^m knife edge. .A_m other to be specified; for example, mean age of spawners. § t_m, T_m. Age at first maturity; mean age at maturity; massive maturation [66]. A50 [22]. Fish from A_m onwards are usually considered adults [22].

A_{mean}—Mean age of the stock. Ratio between the integrals of the weighed by age consistency of the different cohorts (numerator) and the consistency of the different cohorts. § T, t, t_{media}.

A_{mx}—Massive age at maturity, the minimum size above which all the fish are able to reproduce independently from the present activity or the production pattern (discrete, continuous, intermittent, batch ecc.). § Almost never implemented; often confused with A_m.

A_M—Age at end of the reproductive span. Age above which the contribute to spawning of a cohort is negligible. § t_M. [75].

$A_\cap$—Age at maximum. Age at which an unexploited or exploited ($.A_\cap$) cohort reaches its maximum living biomass corresponding to the balance between growth rate and natural mortality. § Originally referring to the unexploited condition as t_{cri} or t*; critical age [65]. In case of isometric VBGF, the critical age is empirically related to maximum age [$A_\cap/A_{max} \approx 0.38$; 11].

A_l—Age of ultimate significant contribute to the fishery. The greatest age for which adequate data usable for fisheries assessment are available. Maximum age above which scanty and not statistical representative samples can be gathered from the stock given the reduction in n° as a consequence of fishing mortality (arbitrary threshold: 5% of caught or sampled specimens). § t_l, L, A_L. Fishing or ecological longevity; maximum exploited age. A par. variable according to the fishing pattern and true longevity of the stock often confused with life span after the classic Jones' approximation $t_l \sim \infty$.

A'—Age of fully capture. The youngest age that is fully represented in the gross catch sample. A'' the age immediately successive to A' (which should be preferred in computations). § t' [74]. To not be confounded with the age at entry to capture.

A_R—Age at which the 50% of fish enter the area where the fishing is in progress and becomes liable to encounter with the gear. § t_ρ, t_r. Recruitment at the fishing grounds; age at which the fish become present in the exploited area and susceptible to the capture with the given gear.

A_{dR}—Age at de-recruitment from the fishery. Age at which the fish are still present in the exploited area, but become no more susceptible to the capture with the given gear (fish will no longer be vulnerable or accessible to the gear for a given fishing pattern). § A_d, D50%, R, t_{rif}. Deselection (length) [74]. Gill net de-selection; age at de-selection; at reform; right-end de-selection [66].

A_L—Longevity. True life span in unexploited condition. Estimated or observed theoretical (true) longevity (maximum age). Average age of the specimens in the upper tail (95th quantile) [111,112] as estimated from a representative (not biased) sample extracted from an unexploited (or lightly exploited) stock sampled from its natural environment. § T_{max} [113]. T_L experimental wild longevity. T_m [74]. t_{max} [114].

A_{Lx}—Present longevity. The maximum age recorded for the present investigated stock or (A_{Lx}) species by aging just the largest few fish at hand [as proxy of A_L; 112]. A_{mx} estimated according to a method to be specified (mean of nth extremes, extreme values theory etc). § T_{max}, a_{max} [70]. t_{max} [114].

A_{Lxe}—Ever observed longevity. The maximum age ever recorded for the investigated stock or (A_{Lxe}) species in nature (*A_{Lxe} from captivity data). § T_{max}.

${}_{cx}A_L$—Conventional longevity. Age at which the cohort has been reduced to $_{0.05}$ (x = 5) or $_{0.01}$ (x = 1) of the initial reference abundance. ${}_{c95}A_L$ age at 95% of the asymptotic length or Taylor's approximation, originally expressed as = (2.966/K) + t_0; often reported as ≈ 3/K).

A_x—Age group where x varies from I, II, III IV etc [8].

A—Age at inflexion. Age at which a discontinuity occurs (to be specified). § t_η.

A_0—Age at theoretical zero size in the VBGF and allies [70]. The theoretical (hypothetical, artificial, arbitrary) "age" at which the fish would have been zero length/weight if it had always grown according to the VBGF (hence, it can be either positive or negative). Location par [65]. A_0 any initial or starting age (such as length at birth in sharks); scale par. to be specified. § a_0 [64]. $t_0$theoretical age at which the weight/length is zero; cons. which simply moves the curves along the abscissa and can be interpreted as the time measured from 0 at which the animal would had zero length if it had followed the same growth curve all its life [115]. Adjusting par. [22]. Almost always it takes non-zero negative values [108], and does not usually express "prenatal growth" [66]. Strictly speaking, the par. should be referred to isometric (cubic) weight-length relationship. Often it is omitted in the general treatment or assessment based on the VBGF, but it must be considered in real computations [113].

*A—Amount of area occupied by the population or stock [94,76,65]. Reference area of stock distribution. *A_s study area. *A_f over which the fishing effort is distributed. § A [67,71].

AA'—The eumetric line joining the (locus) of maxima of yield-mortality curves in the yield-isopleths diagram.

ASP—Annual surplus production [65].

b—The exponent in the length-length (.b) and lengthweight relationships[25,66,70,74,76,116,117] according to = *kL^b. b_e and b_{10} after ln and log transformation. b = 1 and b = 3 conventionally denote an isometric (or isogonic) relationship in length-length and weight-length relationship, respectively. ≠1 and ≠3 denote a positive or negative allometry (heterogonic or disharmonic relaionship) [116]. § The original Huxley and Teissier formula was expressed as $y = bx^k$. Strictly speaking, the isometry and allometry would be used for length-length relationship. n [118,119]. §§ Instantaneous rate of hooking fish [90]. Life span [120]. Exponent relating length to "effective metabolic rate" in the generalisation of the VBGF [91]. Par. in different models [76]. Par. in the autoregressive time series model [64]. Selectivity age related par. [86]. Shape par. in Schnute model [93]. Par. in different models; recruitment per spawner [25]. Yearly clutch size or scale par. [87]. Juveniles produced/unit adult biomass/time [82]. Selectivity coefficient [107]. Generic coefficient [65]. Par. specifying the adaptive trait in Individual Based Models; mass of prey [70]. Slope in linear regression [66]. Positive cons. in von Bertalanffy (isometric) weight function [63]. (Scaling) par. relating some initial length (L_0) such as at settlement [121] or at birth [27] to L_∞.

b'—Slope in the relationship between trophic level and body weight § b [70].

β—Free. §§ Season indices of adult stock [95]. Exponent in the length weight relationship [122]. Unit value of the catch [96]. Probability to accept a null hypothesis (H_0), when in fact it is false [97]. Par. in different models, especially SRR [76]. Par. of the Ricker stock-recruitment relationship [67]. Par. in the recruitment function [64]. Par. in age-length growth model [101]. Tag reporting (or returning) rate; par. in catching power and vessel standardization [25]. Biological upper critical value [103]. Exponent in K1—W relationship; Bunsen coefficient [109]. Par. in effort standardisation; model par.; exponent in the allometric weight-length models; coefficient in Brody model [65] Shape par. in different SRR and yield —effort relationships; density-dependent mortality coefficient in the Ricker SRR [70].

B—Biomass [65,67,76,83,104,107,117]. Average biomass of the fishable stock at equilibrium [67]. Size in (living) weight [88,105], population [86], total [25] or stock biomass [70]. Fish biomass [63]. B_0 the pristine (unexploited, unfished, prior of any fishing) level [65]. B_∞theoretical asymptote biomass, the level to which an unexploited (or lightly exploited) stock tend in an undisturbed environment. B_s spawning stock. B_m at which

MSY occurs [65]. § B_F fecund biomass [76]. B as bioass of a cohort [65]. Generic Biomass [68]. B_0 virgin or unfished; B_∞asymptotic or "pristine", "virgin", "unfished" analogous to the logistic K; births [65]. S [70]. K carrying capacity (normally according to the logistic population model, also defined with L by [96] and A by Ivlev [123]. SSB [21]. B_{inf} or K carrying capacity or unexploited biomass [117]. B_0natural (no fishing) biomass curve; B** escapement [63]. §§ B n° of fish perished by natural causes [123]. Frequencies in equal length intervals [118,119]. Total n° of newborn individuals in Schaffer formula [121]. Larger mesh size in gill net selection experiment [71]. Bias [72]. B par. in Caddy [124] M asymptotic model. Variance component between length interval in LFA [65]. Cons. (slope) of the simple linear model [104].

BB'—The eumetric line (contours) joining the (locus) of maxima of yield-age at entry (mesh) curves F in the yield-isopleths diagram.

BI—Biomass index. Estimation of local abundance in weight of fish standardized to 1km^2. BI_h; in case of hour based standardization.

B/R—Biomass per recruit. B'/R, relative [66. § BPR [117].

c—Capture related general par. with the exception of the (commercial) catchability coefficient (q) relating fishing mortality to fishing effort. X_c of entry to fishery; at first liable to capture by the fishing gear in use. Fishable size [65,70]. c_s the fraction of fish captured in an experimental set $(0 < 1)$; $>$ 1 in case of herding effect [94]. § Σc sum of the weights of all fish caught during the year [88]. c = l/m relative releasing effect or ratio between the length at 50% of release and inner mesh size [125]. t_c at first capture. c, Q, overall gear efficiency, i.e. the prop. of fish which have been in touch of the gear and were at the end captured; exceptionally, fish can be attracted and actively enter inside the codend by the mesh [126]. Cons. of proportionality between F and f [127]. Proportionality coefficient relating to the efficiency of the gear [94]. §§ Instantaneous rate of loss of hooked fish [90]. Juvenile survival rate to maturity in Schaffer formula [121]. Index of ecological similarity [71]. Cons. in $P/B = bw^c$; CPUE [67]. Cons. [64]. Upper and lower limits for the mean asymptotic length [101]. Par. in different models; competition coefficient; shape par. in Deriso's SRR; n° of prey; capacity in multistage SRR [25]. Exponent in the height of the growth production function [87]. Food consumption rate per unit biomass [82]. c and c_y operating cost and cost per unit caught [68]. Coefficient in the allometric $W = cl^n$ [77] or $W = cl^m$ [Bayley's method; 70]. Par. in different contexts, for example, as correlation coefficient in the Shepherd's stock recruitment model; district-specific escapements vector in "run reconstruction" [65]. Competition [70]. Positive cons. in the von Bertalanffy (isometric) weight function [63].

**c*—Ratio of length at capture and maximum (asymptotic) length in potential yield computations. § c [67,76].

**c_1*c_2*—Hoenig and Lawing's coefficients; multipliers for estimating Z and its standard error using one of Hoenig's methods given the sample size from which the longevity estimation was derived. § c_1-c_2 [66]. §§ c_1-c_2 interaction coefficients in Lotka-Volterra model [70,71]. Coefficients in Beddington and Cooke potential yield computations [128]. To not be confounded with unit costs of measuring length and age [65].

cov—Covariance [65].

C—Catch in n° [64,67,86,89,95,104,117] related to the fishing activity [25,65,107,70]. C_g gross; C_b not target; C_r retained on board; C_l live; C_r rejected; C_u landed. § Catch in weight [88,129]. N° of tagged fish which will be caught [130]. C_W catch in weight [94]. C_n food intake [67]. Tags caught and returned; change in weight from tagging to recovery; total catch in weight units; catch per unit of time; aggregate productivity in multistage SRR [25]. From [66]: CLi∞ cumulative catch in n° from length i to L∞ and C_{im} cumulative catch in n° for mesh size m; t terminal catch in VPA. $C_{(t)}$ cumulative catch; n° of fish in age a group; capture of marked specimens [65]. §§ Cost of the fishing effort [96] Cons. in the Putter expression of the VBGF [91]. Average increment per day in larval life [131]. Cons. in trawl performance [106]. Multiplicative factor for debiasing recruitment estimates [71]. Generic cons. [67]. Efficiency par. in mammals [87]. Cons. in integration and matrix; total cost of measuring length and age; consumption rate in multispecies modelling [65]. C* prey consumption [68]. Consumption [61]. Amount of food consumed; cost per unit of effort; energy cost in handling or searching for prey; control; C_{24} daily ration [70]. Cons. of the SRR generalising the model [104]. Cost coefficient [63].

°C—Sea water temperature in Celsius degrees. °C_b at bottom; °C_s at surface.

**C*—Factor which expresses the amplitude (or magnitude) of the growth oscillation in the Pauly and Gatschuz's seasonal length growth VBGF [66,72]. § C [117]. It ranges between 0, no oscillation, up 1, growth stops at WP [109]. Values higher than 1 can be also obtained indicating a prolonged no growing phase (length shrinkage is a rare phenomenon in wild marine organisms [72].

C^2—Par. in Powell Z/K estimation [71].

CC—Catch curve, the ln of the abundance in n° (or index) at successive ages (CC_a) or sizes (CC_l).

CE—Coefficient of error [72].

CF—Condition factor [70,116], generally as [w/l^b] × 100. $_T$CF Tesch w/L^b. $_F$CF Fulton (in case of isometry). $_C$CF Clark. $_{rel}$CF Le Cren's relative

[116]. CF any other to be specified (for example, within age groups or a given age group). § c.f. [108]. q [74]. The symbol K is generally used for Fulton [56,70,116]. K_{mean} for Clark [116].

CH—Cohort (see A_{ch}).

CI—Confidence interval [117].

CL—Age class (see A_{cl}).

C/R—Catch per recruit in n°.

CR—Covered region [132,133]; the area included between the trawl doors.

CV—Coefficient of variation [65], as standard deviation/mean ratio [21]. .CV as standard error/mean ratio. As % if not otherwise specified. § C.V. [71]. CSC—Contact and selection curves; the relationships describing the probability of a fish to avoid, contact and escape after the contact/capture with a given gear. aCSC availability, rCSC contact-selection and sCSC selection [73]. § Selection curves [73].

χ—Free. §§ Arbitrary reference age in Francis' VBGF reparameterization [101]. Mid point of the i^{th}length frequency interval in MULTIFAN [102]. Sex ratio as fraction of spawning population that are mature females by weight; prop. of fertilised eggs that will result in females [65]. To not be confounded with the chi square statistic χ^2 [65,66].

d—Average distance of fish in random movement. Distance travelled [27]. §§ Increment in length in the compensatory growth analysis [91]. Time interval in capture-recapture [134]. Integration cons. [67]. d_1 and d_2 weighting factors in ELEFAN fit; pseudo-random n° [72]. Random variable and other deviation related par. [86]. Discount rate; temperature effect par. [25]. Power of weight to which anabolism is proportional; a cons. term [108]. d_1 and d_2 density independent and dependent recruitment effects, respectively; par. in different contexts, for example, correlation coefficient between weight at recruitment and n° of eggs; n° of deaths [65]. N° (or prop.) of prey in the diet; as subscript, deterministic [70]. Stage [27].

df—Degree of freedom. § DF [8]. d.f. [66,71,73].

δ (delta)—Free. §§ Par. in growth model derivation [122]. Recruitment par. for density dependence; effective discount rate [67]. Successive age increment [72]. Variance related par. in MULTIFAN [102]. Relative size at independence [87]. Death rate related to natural mortality [84]. Asymmetry (shape) par. in selection curves; probability of grid contact and allies in "grid" studies [73]. Discount rate [63,68]. Standard deviation and (δ^2) variance [66]. Variances ratio; par. in Schnute-Richards and seasonal growth models; δ_t additive and independent error in each year [65]. Natural mortality fraction (1-exp M); density dependent mortality component;

fraction of revenues to cover depreciation of the fleet; stochastic vector; derivate; zooplankton temporal "width" in the Cushing's match/mismatch hypothesis; isotopic "fingerprint"; stable isotope index [70]. Par. in Schnute-Richards' growth model [135]. D—Dispersion coefficient of fish [65]. § Usually referred to movements among adults distribution, spawning and juvenile concentration areas [136]. §§ Fishing (F) changes [127]. Par. (1-M/H) in the Allen's method [118,119]. Total n° of deaths [25,69,70, 104,107,123]. Finite time interval in capture-recapture [134]. A measure of the sensitivity of the output [71]. Jacobian matrix; n° of fish eaten [72]. Catch related par. [86]. Density of fish [107]. Density per km^2 of fish [25]. Door spread [137]. N° dying from natural mortality in VPA [74]. Duration (in days) required to reach any particular stage ([70]; for example, the larval stage [109]. From [68] density dependence, diversity indexes, egg stage duration, and density of fish. Fraunhofer diffraction function in Shepherd's method [66]. Coefficient in the $L_\infty = DK^{-h}$ relationship [87]. Natural death; population density; density of prey; decrease due to natural mortality; squared deviation; par. in delay difference model; square root of accumulated variance among ages or Lai's transformation; specified level of precision [65]. As D or subscript, X_D, it denotes discards; D as diet related par., for example, in Ecopath [70]. Deviance residuals; variance; variance matrix [73].

**D*—The shape par. in Pauly's generalised VBGF. § D (gill) surface factor [71,138].

DC—Diet composition [68,82].

DI—Density index. Estimation of local abundance in n° of fish standardized to 1km^2. DI_h in case of hour based standardization.

Δ—Any finite difference [71]. § Elapsed (time) and size increment in mark recapture studies [65]. §§ Taxonomic diversity related par. [68].

e—Base of the natural (or Napierian) logarithms; e = 2.71828... [65,66,107]. §§ Unit effort in tagging studies [139]. Effective [76]. Stochastic term [72]. Error term in model definition. Age specific fecundity; prey density [25]. $e_{detritus}$ instantaneous export rate of detritus [82]. Normal random (0 - 1) variable; ê rate of egg production per unit biomass and unit time; fishing effort in markrecapture experiments [65]. Elasticity (sensitivity) [27].

ε—Error term. §§ Mean of the logarithms of n° sampled [130]. Midpoint in a given class [122]. Efficiency of conversion of available energy into growth or gonad energy or converting surplus energy into body weight [99]. Sampling error [85]. Par. in catching power and vessel standardization [25]. Gross food conversion efficiency; ratio of growth increment/food ingested during a given period [109]. (Total) egg production [65]. Stochastic vector; variable (unrelated to abundance) mortality component [70].

ϵ—To not be used in order to avoid confusion with similar symbol. §§ Particle size conversion efficiency [67]. Error term [86]. Unexplained predictor error or residual [25]. Different error terms (process, normal, additive, multiplicative etc) [65].

η—Shape par. in different models (especially in growth curves). §§ Random variable in population equilibrium catch relationship [129]. Par. related to food [140]. Probability of being recaptured [141]. Random perturbation in the stochastic analysis of an exploited population [142]. Expected catch [73]. Par. in weigh at age models; residuals [65]. Random residuals [70].

E—Exploitation rate [67,74,76] or fraction [65]. Prop. of dying due to fishing [76]. The ratio of fish caught to total mortality when F and M take place concurrently and are unchanging or change proportionally. Overall E_∞ (t→∞) or E annual (t = 1) expectation of capture. § Fishing effort [96]. Exploitation fraction [86]. Defined as the F/Z ratio multiplied by 1-exp-Zt (the exploitation fraction, μ in [65] or, more properly, by 1-exp-Z(t_∞ - t_c), hence, as $t_\infty \rightarrow \infty$ E → F/Z. Rate of exploitation (u in [69,123]. E' probability of ultimate capture [118,119]; in case of cons. F/M ratio, it is equal to the exploitation rate. Exploitation pattern in Lleonart [22]. Heincke's fishing coefficient. §§ Total (cumulative) effort in tagging studies [139]. Expected estimate or value [65,70,72,73,122, 130]. Par. (=Hp/3q) in the length based derivation of VBGF [91]. Const. of locomotion [143]. Efficiency index in Tuna fishing effort analysis [144]. E'' and E emigrating rate [145,70]. True effort [67]. (Expected) mean value [72]. Fishing effort; size of fishing fleet; n° of prey eaten by a predator; total egg production; some environmental variable [25]. Reproductive effort [87]. Rate of fish encounter/escaping with/from the gear [137]. Fishing effort; expected par.; cumulative fishing effort in mark recapture experiments [65]. Instantaneous rate of gastric evacuation; effort rate; energy gained [70]. Reproduction rate; age or stage elasticity [27].

EE—Ecotrophic efficiency [68,82]. § Fraction of mortality not due to predation or fishing in Ecopath [70].

ER—Expected revenue [68]. § R net of operative costs [68].

f—(General) fishing effort [27,74,94,117,127]. f_c capacity; f_n fishing effort as collected (uncorrected; nominal); f_e overall (fleet and time); f_i intensity (by unit surface and time); f_t time; f_o effective overall intensity (weighted sum); f_{MSY} corresponding to MSY [21]. § F, f. N° of fishing efforts [123]. Full-recruitment fishing mortality or effective fishing effort coefficient [86]. Fishing mortality rate for fully vulnerable individuals; exploitation rate [25]. E as fishing effort [74]. The amount of "energy" (work, fishing boats, technique and technology) used to catch fish. f_{MSY} a.k.a. optimum f [21]. Fleet [70]. §§ Probability of tags recapture [139]. f* n° of eggs/ adults [145]. Feeding level [140]. Ration requirement coefficient [72]. f_y first year of fishery data [86]. Set of auxiliary factors; fecundity; $f_{(x)}$ net fecundity; average net fecundity; degree of freedom; females; par. in the (fixed allocation) age sample size determination; age [65]. Fraction of females spawning in a given time interval [68]. (Length) frequency [70]. Tag recovery rate [27].

♀—Females [72,87].

φ—Free. §§ Retention rate; function [76].

φ—To not be used in order to avoid confusion with similar symbol. §§ Generic par.; function [96]. Probability of survival in tagging experiments [146]. Par. related to food [140]. Translocation rate [145]. "Pseudovalue" in jackknife [71]. Fraction of population at some stage or condition [67]. Probability; Daan's food requirement [72]. Arbitrary reference age in Francis VBGF reparameterization [101]. Natural mortality hazard par. [86]. Coefficient in autoregressive process; general non linear term; par. in seasonal growth model [65]. Increase factor in R in surplus production modelling [70].

Φ—Pauly and Munro growth performance index in weight and Φ' in length; $\log K + 2/3 \log W_\infty$ and (in case of isometry) $\log K + 2 \log L_\infty$ [70]. §§ Angle [147]. Gear saturation par. [86]. Par. Vector [93]. The same symbol is employed to denote the Golden ratio.

F—Instantaneous coefficient of fishing mortality [63, 65,67,70,74,76,86,117,130]. If not specified (see below) it indicate the average (overall weighed) F over the range of age groups which can be considered fully represented in the samples. F_J juveniles. F_p parental (adults). F_{MAX} at maximum equilibrium yield. F_{max} corresponding to Y/R_{max} for a given entry to fishery. F_{MSY} at maximum sustainable yield [$F_{msy} \equiv F_m$ according to 65]. F_r ratio of fishing mortality on the oldest age group to the fishing mortality of the preceding age group, used in many tuned VPA assessments [21]. F_λ terminal (last year for which data are available for assessment; mainly in VPA). $F_\updownarrow$ array of values according to a model or equation to be specified. F' collateral mortality induced by fishing; for example, mortality due to discard [65]. § M_f [95]. F_{ey}, F_p where marginal yield per recruit is 10% or p-times the marginal equilibrium yield in a lightly exploited stock [67]. Fishing loss rate; fleet size [25]. $(F)_{max}$ force of fishing mortality. To remember that F_{MAX} is usually different than F_{MSY}. §§ Biomass flow up the size spectrum; scalar-valued function [67]. Variance within length interval in LFA [65]. Stomach content; starting area in fish migration [70].

F_c—Fecundity (general). N° of "mature" (hydrated) "eggs" (strictly speaking oocytes) produced on average by a female of a given size-age. aF_c absolute; pF_c potential (as the stock of eggs in the ovary before spawning; F_{pot} in [68];

fF_c free eggs released into water (F_{rea} realised fecundity, in [68]; rF_c relative (as function of size or age); mF_c life time fecundity (i.e. the progeny derived by a female during its life); dFc daily; $Fc_{/R}$ annual egg-production per recruit. § Often replaced by spawning biomass as a proxy. fec [76,86]. F_{atr} as fecundity and prop. of atresic eggs [76]. DEPM, daily eggs production model in Lleonart [22].

FP—Fishing power. Relative unit efficiency of capture of different vessels versus a standard vessel. § ρ, P, Q. Efficiency. [148,149].

FR—Fished region [150], the area covered by the wings of the gear. § Fished area; 50% of the head rope length according to the Baranov's approximation.

g—Individual general growth rate. g_a absolute; g_r relative; g_i instantaneous; g_f finite; g_s specific. § Σg sum of growth increments of all individuals surviving at the end of year [88]. Mean annual growth increment [101]. Net growth of adult population [65]. Coefficients of predator negative growth in Lotka-Volterra model [70]. Growth per unit time [27]. §§ Par. related to food [140]. g gonad [99]. F/K; global [76]. Probability density of fish length; ration requirement coefficient [72]. N° of groups [64]. Par. group or stratum index [86]. Par. combining mortality and growth effects; observation error; age specific vulnerability to fishing induced mortality [25]. Index identifying a group of fish tagged and released over a short period of time [151]. Gross food conversion efficiency [82]. Probability of a fish to be retained by a grid [73]. Coefficient in Fletcher's quadratic model; g_1 and g_2 density independent and dependent growth effects, respectively; F/K ratio in the allometric Y/R model; index of gear type [65]. The greatest true age group [70]. Gear type [27]. To be not confounded with gram [66]. γ—Free. §§ Par. in total metabolism (T) and weight (W) according to $T = \alpha W^{\gamma}$ [98]. Area successfully searched [143]. Shape par. in the Shepherd's SRR [76]. Par. in the recruitment function [64]. Par. in catching power and vessel standardization; tag shedding rate [25]. Par. in Fletcher's modification of the Pella-Tomlinson model; exponent in different stock recruitment models; coefficient in Brody model; par. in the Schnute and Schnute-Richards growth models; inflexion in the maturing model; vector of movement and population par. [65]. Transformation or shape par.; degree of compensation par. in Shepherd's general SRR relationship [70]. Fraction of individuals in a stage moving to the next stage; population growth rate [27]. Shape par. in Schnute-Richards' growth model [135].

G—General growth rate at stock level. ${}_aG$ absolute; ${}_rG$ relative; ${}_iG$ instantaneous; ${}_fG$ finite; ${}_sG$, specific; **.**G other to be specified. § Stock growth in weight [95]. Instantaneous growth rate [67,76]. Growth survival factor [25]. g in [70]. Specific growth rate. Increase due to growth of individuals [65]. Par. relating Z/K or Z/H to length and weight [118,119]. §§ Par.

related to food [140]. Scalarvalued function [67]. Cumulative length-distribution function [72]. Par. in the recruitment function [64]. Total n° of group or strata; true age composition [86]. Income for the fishing industry; fishing induced mortality [25]. Natural mortality factor in Pope's cohort analysis [74]. Sea area over which egg production is expressed [68]. Anabolic component [70]. Probability of moving form one stage to another stage [27].

**G*—General growth rate at stock level as whole population; for example, in surplus production [70].

Γ—Free. §§ Index of competition. Notation for the gamma distribution; environmental variable affecting recruitment in semelparous population modelling [65].

GI—Gonosomatic (or gonadosomatic) index; ratio between gonadic (ovaries or testis) and body weight. GI_0 whole body; GI_e eviscerated; **.**GI to be specified in case gonad weight includes also other reproductive annexes (for example, the ovary glands in cephalopods). § Usually it is considered an index of the state of maturity or of the level of sexual activity (especially in females).

Ger—Gastric evacuation rate. § E [70].

GML—Growth-maturity-longevity-plot [113].

GOF—Goodness-of-fit [70]. § G for the projection matrix method [70].

h—Hour [25,109]. §§ Time required to capture and consume [143]. Par. combining mortality and growth effects; annual harvest rate; time taken by a predator [25] Exponent in the $L_\infty = DK^{-h}$relationship [87]. Instantaeous longline fishing mortality rate [151]. From [109]: height of a fish's caudal fin; h^2 measure of genetic heritability; squared of caudal fin. Harvest rate related to fishing mortality [84]. Arbitrary increment; coefficient in Fletcher's quadratic model; derivative of the underlying deterministic growth curve; eigenvector [65]. N° of age groups; herbivore organisms [70]. Density dependence or steepness par. in B&H SRR [117]. *h—Haul. § h [73].

H—Loss rate of marks. §§ Weight synthesised per unit surface area in the derivation of the VBGF [91]. Instantaneous (exponential) growth rate [119]. Par. in surplus and Shepherd's SRR [76]. Distribution function [72]. Estimated age composition or percent of total catch [86]. Par. in the recruitment function [64]. Prop. of females mating [87]. Herding effect related par. [137]. Coefficient of anabolism used in the derivation of the VBGF [108]. Natural mortality factor in Jones' length-based cohort analysis [74]. Overall n° of hauls [73]. Probability of dying after being caught (and discarded); matrix related par. [65]. Evenness and diversity indexes or alternative hypothesis [68]. Harvest fraction; handling time for a prey [70]. General power function [104]. Hamiltonian in the conditional equation [63].

HI—Hepatosomatic index; ratio between liver and body weight. HI_0 whole body; HI_e eviscerated.

i—As subscript, generic index to designate stock or site [25] or group identification [61,65,66]. § Year [129]. For counting items [108]. §§ Index of total mortality [123]. Instantaneous total mortality [120]. Intercept in the generalisation of the VBGF [91]. Mesh opening [81].

ι (iota)—To not be used in order to avoid confusion with similar symbol.

I—Ingestion of food and related par. Food consumption in a given period. Coefficient of food utilisation for growth and maintenance. $_gI$ gross food conversion efficiency. I_S stock's feeding requirement (I) in [72]. I_{dr} daily ration i.e. the amount of food consumed by a fish of a given weight in one day, and often expressed as % of its own weight [109]; I in Ware [143]. § Growth efficiency, for example, the ratio between production and food consumption [70]. §§ Marked specimens [92]. Par. related to "optimal grouping"; index of cohort [72]. Money invested in a new boat [25]. As subscript, inflexion; groups deriving from cohort stratification; survey index in ADAPT approach; integral [65]. Index of relative abundance (68,70). Income [70].

**I*—Separation Index [74]. § SI [66].

∞—Infinite. The upper limit which can (probabilistic) or cannot (asymptote; integral) be touched by the considered par. In the asymptotic case, it might characterise the maximum size towards which a fish (or a stock; [70] would grow if it could assimilate energy at the maximum possible rate throughout its life.

IALK—Iterate age length key [65].

j—Juvenile. Fish which has not reached the maturity condition. j_a, always, which maintain non developed gonads in spite of a size larger than A_m [152] § $_j$ as subscript in Walters et al. [82]. Usually in Mediterranean stocks, they include the recruits or the young(est) of the year (YOY). Immature. §§ Location or area [94]. Time interval [86,97]. N° of par. [67]. For counting items [108]. Index for predators or consumers [82]. A given gear [73]. Different periods of life; as subscript, generic index or group identification [65].

J—Juvenile at stock level. §§ N° of recapture period [92]. Time intervals [97]. Jaws size in urchins [121]. Age at first capture; yield [67]. Jacobian matrix; n° of jobs in fishing industry [25]. An alternative to a given gear (73). N° of length interval; total value function [65].

k—The coefficient in the allometric length-weight relationship according to w = kl^b. k_e and k_{10} after ln and log transformation. § C in [118,119]. §§ Catchability in tagging studies [139]. Cons. in surplus model [96]. Terms which do not contain F or M in virtual population analysis [127]. Destruction per unit weight in the derivation of the VBGF [91]. Rate of deceleration in growth increment [77]. N° of discrete sampling occasion in Jones [92]. New age or length at capture related index [76]. Age at (of full) recruitment; discount rate; ratio of the size-specific mortality to the size-specific growth rate [67]. Youngest possible age at recruitment [64]. As k_a and k_t growth coefficient related to mean length at age and length increment (tagging) analysis [101]. Cohort index [86]. Carrying capacity; cost of a new boat [25]. Identifies size-at-release class of fish from a given group [151]. Coefficient of catabolism or n° of par. [108]. Time at recruitment to the adult stage [82]. Often employed to represent the growth coefficient in the VBGF [70,87]. Par. in effort standardisation; intrinsic growth par. analogous to r in surplus models [65]. Sampling effort factor; k1 and k2 cons. relating mode/spread with mesh size [73]. Catch per unit capital [68]. Cohort Index in Extended Survivors Method; coefficient of proportionality between Y and f in case of cons. density of fish; n° of prey categories [70]. Cons. in the SRR and (as carrying capacity) production model [104]. N° of par. estimates in a given procedure [66]. k_1, k_2,... k_n growth coefficients or rate in different compared models; n° of estimable par. in AIC computation [135].

κ—To not be used in order to avoid confusion with similar symbol. §§ Eigenvector; difference between intrinsic rate and exit rate in migration model [65]. Log of spawning level; curvature par. in different (growth and maturing) model; identifying the Ricker's conventional Brody coefficient in the VBGF; coefficient in Brody model [70].

K—Rate of curvature [113]. Curvature par. [66,72, 74,153] or coefficient in the monomolecular or inverse exponential equation VBGF [70,117,154] and its allies. It determines the rate at which the asymptotic-infinite size is approached. § von Bertalanffy growth par. [64] or coefficient [70]. Par. (=k/3) in the length based derivation of VBGF (91). Rate of change in length/weight increment; growth rate in the VBGF; Brody's (length) or Putter's (weight) coefficient [56,58] Stress factor [71,138]. Par. of the von Bertalanffy length equation [67].The Brody's denomination in the VBGF is also adopted by other Authors [25,65,68]. Growth completion rate [155]. Growth rate [70]§§ Total (cumulative) catch in tagging studies [139]. Efficiency of utilization of food [98] = j-1 in Jones [107], recalling the Robson and Chapman classification. Par. in gear saturation modelling [156]. Drag coeffcient [106]. Par. in surplus and Shepherd's SRR models [76]. Selectivity factor [126]. K_n gross production efficiency of food [67]. Oldest possible age at recruitment [64]. Index of food abundance in extended stock recruitment relationship [157]. Total n° of cohort [86]. Cumulative catch [in Leslie and De Lury model; 25] and fleet related par.s [68]. Growth-survival-catch analogous to the catch factor of closed populations [25]. Upper limit or carrying capacity; matrix related par. [65]. Par. in predator model to estimate natural mortality [61]. Size of the fishing fleet; threshold (biomass) level in the SRR relationship [70].

**K*—Threshold biomass in surplus production model [upper limit or carrying capacity; 65] and different stockrecruitment relationships (above which the relationship departs from linearity ([70]..l—Individual body length as effective or index of the whole fish extension [total in 11]$_e$l extreme (total with the caudal tips joined). $_n$l natural (total with tail tips in natural position). $_f$l fork. $_s$l standard. $_c$l carapace (crustaceans). $_d$l dorsal mantle (cepahalopods). $_h$l height. l recently killed. l.defrosted. l• other to be specified (see also age for other subscripts). § Length at 50% of release [125]. §§ l_x probability of living at age x [121]. l_t landing tax [25]. l_y last year of fisheries data [86]. Annual survival rate [65,70] after natural mortality [68]. Early life survival from the egg stage to recruitment; annual survival from natural sources; index for length interval [65].

•$_l$—Body component (organ, appendix etc) length.

l—Ultimate significant contribute to the fishery. § Maximum age [104]. §§ Par. relating variance and n° of observations [130]. Fraction of the catch inspected for tags [97]. Par. related to the growth in yield computation [120]. Length in the Putter expression of the VBGF [91]. Coefficients which takes account of the difference in summer and winter feeding [99]. Fecundity per recruit; cons. relating the resulting recruitment from a given spawning biomass [76]. Annual growth rate in Schaffer's formula [121]. Importance of residual [67]. Convenience notation to relate two growth expressions for tagging data [101]. Max par. in fish vision [109]. Probability that a tag is lost due to tag shedding or tag-induced mortality [151]. Coefficient of variation of length at age [153]. Poisson mean [73]. Term corresponding to F in stochastic model; attack rate; par. in Gompertz's differential growth derivation; eigenvalue; penalty weight; ratio of variance [65]. Intercept in linear weight at age plots; average per capita encounter rate [70]. Finite rate of population growth [27]. Theoretical initial growth rate in Gompertz's model [135]. Adjunct variable [63].

log—10 based logarithm [66,70] § $\log_{10}$, log, lg.

ln—e based logarithm [66] § $\log_e$.

L—Length (generic) at stock level [67,70]L_m at 50% sexual maturity [70,113]. L_{rt} at 50% of retention. L_{ml} minimum landing length. See l (individual length) and A (age) subscripts for other specifications. § N° of length class [73]. Mean length of a fish at age t [65] $L_{50\%}$, (l_{50} in 73) length at which 50% of fish are retained [70] MLS minimum landing size [22] §§ Catch or landings; production in weight [96]; limiting population [129]. t_L longevity [112]. Penalty weight [86]. Likelihood [25]. Depth [68]. Length of the search along the transect; cumulative survival; n° of individuals measured for length; observed and true length frequencies; likelihood; instantaneous rate of tags shedding [65]. As subscript, $_L$, also denoted fish caught and landed [70].

L∞—Asymptotic length. Length to which the curve approaches closer and closer as the independent par. Becomes extremely large (→∞) or extremely small (→ −∞) [30] $l_{\infty a}$ mean length of "old fish", where old means of an age beyond which the mean length at age does not increase appreciably [101]. Mean length of infinitely old fish [153]. Length at maximum age [t_∞ in65]. In the VBGF, the size at t = ∞]70] or that an average fish would achieve if it continued to grow indefinitely according to the VBGF [13], without touching it ($L^\infty > L_{mxe}$). § Usually reported as L_∞. The same symbol (L^∞) was recommended for VBGF by ICNAF [47]. Limiting size [11]. Maximum expected length [158]. Maximum possible length [154,72]. $l_{\infty\bullet t}$ maximum length achieved in a population where the subscript t refers to tag data analysis [101]. L_{inf} in Rosenberg and Beddington [115]. Infinite; the mean length of very old (strictly speaking, infinitely old) fish" [22]. Sometimes it is assumed [in Elasmobranchs VBGF modelling;159] or mistaken [21] for the largest observed size for the species (i.e. $L^\infty \equiv L_{max}$ approximation [76].

L∞—Pristine infinite. The mean length the fish of a given stock would reach if they were to grow forever [66, 72]. Mean size according to a probabilistic distribution function [160]. § Mean length of oldest specimens estimated in the pristine condition. In the VBGF, $L\infty < L_{mxe}$.

L_∞—Actual infinite. Mean size of oldest specimens esimated in the exploited condition. Mean size according to a probabilistic distribution function [160]. § L_n successsive (pseudo-) infinite lengths in seasonal VBGF [161]. In the VBGF, L_∞<<$l_{mxe.}$</l

Lmx—The present maximum size recorded for the investigated stock. Lmx ever recorded; Lmxestimated according to a method to be specified such as mean of n extremes, 95th percentile, extreme values theory etc LC50—Median lethal concentration [109].

LFA—Length frequency analysis [65].

Λ (lambda)—To not be used in order to avoid confusion with similar symbol. §§ Test statistic [67]. Diagonal matrix [65].

•*m*—Mesh [66,68,76,125,126] Mesh size [73]. The size of hole in a fishing net [22] and related par. dm diamond or qm squared; mb bar, ms stretched size. § Codend mesh remain open [148]. §§ Σm sum of the weights of all fish which have died from natural causes during the year [88]. Fraction of fish removed in 1 year [123] Bertalanffy anabolism exponent in the generalisation of the VBGF; slope in Ford-Walford plot [91]. Rate of change between metabolic rate and weight [154]. Prop. of larvae [145]. Shape par. in the Pella and Tomlinson's model; M/K (76). N° of marked specimens; cons. in the Lotka-Volterra equations [71]. M/K (Conditional fishing mortality rate [107). As subscript, marked specimens [93]. Generic slope in linear regression; arbitrary exponent in the Pella and Tomlinson's model; natural mortality rate; maximum age in multispecies analysis [25]. Maximum flow between preys and predators/consumers [82]. K/Z [74]. Median; maximum productivity in surplus models; estimated variance related par.; m(x), probability of maturity; males; M/K ratio; n° of marked fish recaptured; index of consumption of species I by species j [65] Maintenance and depreciation [68]. Conversion for prey mass to energy; fecundity [70]. As subscript, marking related par. [66]. To not be confounded with meter.

♂—Males [72,87]. § M, m.

μ—Free. §§ Mortality rate [139]. Population mean [130]. N° of tagged fish recaptured [123]. Par. in tagging experiments [92]. Prop. in length frequency analysis [121,162]. Cons. relating the spawning biomass resulting from a given recruitment [76]. Annual exploitation fraction [67]. Food size preference coefficient [72]. Fraction of the catch removed [64]. Mean length in MULTIFAN [102]. Term corresponding to M in stochastic model; annual exploitation fraction (C/N); mean; maximum growth rate [65]. Mesh size [70]. Exploitation rate [27].

mpa—Mean parental age, the average n° of progeny produced by a females during its life weighed by the different age class. § Most suitable for iteroparous species. As the mean fecundity, it is a par. not simple to estimate [163; pag. 128].

M—Instantaneous rate of natural mortality, where "natural" refers to all causes of mortality except fishing [61,63-65,67,74,76,86,117,130]. If not specified (see below), it indicates the average (overall weighed) M over the range of age groups which can be considered fully represented in the samples. M∞ infinite mortality (≈ 0), or mean of infinitely old and big fish [153] to which approach the M (assuming no senescence); it is analogous to Gulland [45] M' limiting value. Mjjuveniles. Mpa parental (adults). Mbur bursts, the high M values in early period such as eggs and larvae lifetime [164]. Masy the mathematic theoretical (lower) asymptotic natural mortality [Main163]. Masy the asymptotic upper biological mortality to which tend the oldest specimens of a stock. M↕ array of values according to a model or equation to be specified. MΔt phase mortality, defined as the product MΔt (in the presence of fishing, ZΔt) and represents the cumulative mortality which occurs in the time interval under consideration [165] M* the combination between M and the instantaneous rate in tags shedding [65]. M0 and M1, M2 etc., natural mortality excluding predators [82], and other forms of natural mortality, for example, in multispecies VPA [68,70]. § Mn [95]. Masy is a mathematical par. without any biological meaning (may also assume negative values). M% in % as recorded in aquaculture experiments [109]. M* due to predation by marine mammals [68]. M = + ∞ catastrophic mortality at the end of lifetime [63]. §§ Maximum population which the environment will support under average conditions and with no fishing [129]. Metabolism par. in the "surface law" or "two-thirds rule" or "Rubner's rule" [91]. tM age at reform; XM at MSY [76]. N° of fish in a sample [64]. N° of predators [25]. Projection matrix [65].). Mature specimens [61]. Metabolism term or catabolic losses [70].

MC—Migration coefficient [70].

MPA—Modal progression analysis.

**MPA*—Marine protected areas. § MPA [117].

MS—Mean square [72]. § MSE mean square error [72]. §§ It can be confounded with the Maximum Sustainable Economic Yield [MSE in 74].

MSY—Maximum sustainable yield [65,74,117] § At equilibrium [65].

MSEY—Maximum sustainable economic yield [74]. § MSE [74]. MEY [117].

n—Generic n° of specimens or items to be specified [68,70] § As superscript, indicates the order of a moment [64]. §§ Exponent in the allometric W = cln [77] Shape par. in Richard's growth curve [121]. Trophic level; year [67]. Conditional natural mortality rate [107]. N° of encounter; shape par. in Pella-Tomlinson model [65].

υ—To not be used in order to avoid confusion with similar symbol. §§ Fraction of the total mortality [64]. Par. in Schnute-Richards growth model; vector of par. (general and related to mark recovery) [65]. Selection par. related to haul [73]. Par. in Schnute-Richards' growth model [135].

N—N° [67] of fish [25,130] in a homogeneous group [61,68] N° of survivors of a cohort [70,76] Population in n° [63]. N0 total at the beginning of a given year. § Initial population [69] in n° [89] N° of fish of a given cohort alive at age a; n° marked and total recaptured [65]. §§ Par. related to Hill diversity [68] N° of schools; egg production [65]. Sample size in n° [70].

o—To not be used in order to avoid confusion with similar symbol. §§ Term in stochastic mortality model [65].

ω—Gallucci and Quinn's [166] growth performance index (KL∞). §§ Fraction of the total biomass due to newly recruited fish [64]. Weight at age ratio [65].

O—N° of dead fish in a given time interval. OZ dying of total causes. OF dying of fishing (C plus n° of fish dying for collateral effects of fishing). OM dying of other causes than fishing (all natural causes). OMp predators. OMd diseases. § Estimate of total fish taken [86]. §§ Coefficient related to the age at maturity of a cohort [76]. Objective function [65]. Mature ogive [61]

OI—Omnivore index [70].

Ω—Expanding term or factor [76]. Summation cons. in year class weight computation. §§ Yamanaka's yield function [120]. Period in seasonal growth models [65]. Region bounded by [63].

p—Prop. [70] of related, but not dependent variables. pr retained in the cod-end [81] pm mature. p♂ of males over females. p♀ females over males. p♀s females over sexed specimens. pm♀s mature females over mature sexed specimens (operational sex ratio; [70]§ SR sex ratio [76]. Prop. [67]. Prop. of ration (72). Probability [86]. Probability in selection curves [73]. Generic prop. [65] Prop. of mature at age [70] §§ Instantaneous rate of fishing mortality [89,120]. Annual mortality rate (1-exp-Z) [127]. Index of intensity of exploitation or mortality owing to catches; standard fishing effort [123]. Cons. of proportionality in the derivation of the VBGF [91]. Term for correcting injested to utilizable energy [98]. tp age at reruitment in the fishing ground [77]. Assimilation factor [143]. Duration of each recapture period in Jones [92]. Movement probability in diffusion model [141]. Price [67]. Weighting factor [72]. Landed price; probability of moving from an area to the other; movement rate; residence prop.; generic par. in GLM; productivity in multistage SRR; probability of encounter-chase-capture [25]. Probability that an individual is caught in stochastic mortality model; model par.; n° of tests; generic fraction; true prop. of fish by length interval; par. in the (proportional allocation) age sample size determination; probability of finding a marked member at time t; prop. Mature at age [65]. Relative fishing effort [73]. Coefficient related to food energy; index of predator type; predation [70]. Net price per unit weight [63]. To be not confounded with the probability level in statistical inference.

π—3.14 [66]. §§ πa prop. of each cohort which recruits at age a; fixed prop.; harvest rate [67]. Fraction of total recruitment (64). Fishing mortality hazard par. [86]. Fraction of total recruitment of a given cohort or at the start of a year (exploitation fraction) [65].

ψ (psi)—To not be used in order to avoid confusion with similar symbol. §§ Generic par. [96] N° of tidal cycle per day [145in key papers]. The mean direction of movement [92]. Arbitrary reference age in Francis VBGF reparameterization [101]. Moment-generating function; monotically decreasing function of r; instantaneous rate of movement from region i to region j [65]. Ratio between C and pristine biomass [70].

P—Production [67,70,82,83,167], the quantity of overall (dead and still alive) biomass produced during the interval under consideration. Pn net production, amount of living matter [22]. § Population of usable size [95]. Population size [56]. Biomass of adult stock [67]. Latent productivity; dynamic resource (or prey) term [65]. Productivity of the prey's food resource [70]. Cons. of the production model generalising the model [104]. §§ Prop. of tagged [139]. Survival rate [130]. Population size [96]. PW biomass in weight [77]. PW average population or standing stock in weight [94]. Pressure (m off the bottoms) in Harden Jones [146]. Pe n° of eggs spawned [145]. Rate of growth; price; probability [67]. Prop. [72]. Previously exploited fish in Schnute [64] model. Availability [86]. Points in a score (or objective) function; prop.; aggregate productivity in multistage SRR [25]. Probability of obtaining a catch distribution in stochastic mortality model; fishing power; ratio of catchability; average egg production per day; probability; prop. of individuals in size class [65]. Prey biomass and egg production in n° [68]. PM as prop. of mature [61]. Population; fish price; generic par. [70]. Probability of capture or of occurrence [66]. Probability of remaining in the same stage [27]. *P—Parental, the non juvenile component of a stock. *Pm "mature" fish which are almost mature or fully mature. *Ps spawners. *Pa adults, i.e. fish which have reached the capacity to reproduce independently from their present contribute to reproduction (i.e. including abortive fish).

Π—Free. § To not be confounded with "product".

Ψ—Free. §§ Dummy time variable. Function [76,96]. Natural survival rate; elasticity (sensitivity) [27].

P/B—Production biomass ratio [70,167]. § Turnover rate [71]. Equivalent to Z, under more circumstances [70,167].

PF—Power factor [cfr Fig. 1 in 34]. Fishing power. The relative performance of each vessel computed by applying a specific equation; for example, PF = cons. *(vessel length)n. § P, P.F., Piand Qj for power factor of the ith vessel and jth location [168].

PDF—Probability density function [65].

q—Catchability [65,67,70,74,76,86,94,97,107,117,169]. Coefficient of proportionality between F and f (specified for the f used). qf the stock fraction taken by 1 unit of effort (usually < 0.1). See "effort" for other speciications. § The fraction taken by 1 unit of fishing effort [122]. Efficiency of fishing; efficiency of each fishing boat [25]. Different catchability in Jennings et al. [68]. Availability [68]. q* in Lleonart [22]. §§ Instantaneous rate of natural mortality [89,120]. Total mortality (127]. Coefficient in the length weigh relationship [122]. Index of natural mortality [123]. Cons. of proportionality in the derivation of the VBGF [91]. Different par. in tagging experiments [92]. N° of par. [67]. Prop. of ration [72]. Essentially a cons. of proportionality [64]. Density independent mortality [25]. Complement to p in stochastic mortality model; model par.; time (years) range; calibretion coefficient in ADAPT [65]. Sub sampling fraction [73]. Recruitment ratio in [68]. Quarter or any time period in fish migration [70]. Q—Physiological-temperature related coefficients; mainly referring to the correction factor such as Q01 [98] or Q10 (Arrhenius) rule [70,103]. §§ Fishing effort efficiency [69]. N° of fish caught per moment [123]. Fraction of fish taken from a cohort [156]. Weight of prey consumed [71]. Gill-net catch per lift; harvest [67]. Fish group in Schnute [64] model. Q and Q/B amount of food consumed per unit weight of an age structured population of fish generally expressed on an annual base [109]. Par. (difference between maximum and minimum age considered) in Shepherd's lengthcomposition analysis [70,115]. Index of biomass concentration in the Csirke's SRR [107]. Probability or prop. in MULTIFAN [102]. Par. in the Charnov and Berrigan $K = Q(L\infty)-h$ invariant [113]. Quantity [25]. Total consumption rate [82]. One of the par. defining the REP (replication estimate of dispersion) par. [73]. Mean catchability; par. related to Fletcher's quadratic model [6]. Amount of food consumed [70,109]. Consumption and, as Q*, upper limit to fleet catching capacity [68]. Par. in surplus production modelling; catchability at fleet level [70]. Capturability coefficient [104]. In management designs, Quota, a share of total allowable catch in weight [22, 65,68,70].

QO2—Weight-specific oxygen uptake [109].

r—Product-moment (Pearson) and (r2) determination coefficient [66]. § R2 [117]. §§ Exponent in compound interest equation as applied to stock weights [105]. Year [127]. N° of tags recovered [27,97]. Distance [147]. Different par. in tagging experiments [92]. Accessibility [76]. Multipliers in yield per recruit equations [71]. Average of some rate; intrinsic rate [67]. Optimal length class interval; prop. of ration [72]. Age at recruitment; survival probability [86]. Relative recruitment rate; uniform random (0 - 1) n°; age at recruitment in multispecies analysis [25]. Coefficient in Jones and Johnston [99] growth, reproduction and mortality analysis. Instantaneous growth par.; probability that an individual is alive at time t in stochastic mortality model; region; model par.; intrinsic growth par. in surplus models; r1 and r2 density independent and dependent natural mortality effects, respectively; par. in Francis' growth model; (knife edge) recruitment age; sequential observation of biomass; logarithmic (median and average) growth rate of a population; the first age group; prop. of fish of length l and age a; par. in the age sample size determination; generation length in semelparous (once-breeding) population modelling [65]. Reciprocal of catchability [70].

**r*—sIntrinsic rate of initial population growth; the rate showed at very low abundance (assuming no depensation). § r intrinsic rate of growth [25]. rm intrinsic rate of growth of a stock [108]. Pollard's Malthusian par. in Quinn and Deriso [65]. rpop [70].

ρ—Ratio [70] of dependent variables. ρN ratio of surviving (absolute rate of surviving per year). ρZratio of complementary surviving (absolute rate of total mortality per year). ρGR part of the metabolic energy available for growth and reproduction. ρE part of the reproductive energy used for zygotes formation [167].§ Tags reported/ tags captured ratio [97]. S, A. ρ' exchange ratio per tidal cycle [145]. Recruit per spawner ratio [84]. §§ Probability of individual fish to be available [122]. Par. related to food [140]. Fluid density [106]. Slope in the Ford-Walord plot [25]. Ford's growth par. (coefficient in 67]; correlation factor; inverse of the capture probability [65].

R—Recruitment related par. [25,61,63-65,67,70,76,86, 95,107,127]. N° of recruits [89]. Raf to the area interested by fishery. Rgf to the gear used by fishery. RX effective recruitment (the total n° of fish recruited in one year from X year-classes. § Whatever recruitment by movements in to the region fished or by change in size or behaviour [56; page 40]. R as recruits abundance [68]. Recruitment to the exploitable phase [117]. §§ Net radius [147]. Food intake [98]. Food consumed by a species [140]. Standard vessel [76]. Rate of respiration per unit area [67]. Ration; rectangular distribution [72]. N° of tags released; n° of boat retired [25]. R0 net reproductive rate [87]. Rddaily ration [109]. Variance matrix [73]. Cons. ratio of change in population n° or Malthusian par.; life-time egg production; lifetime reproduction [65]. Current observation in Bayesian statistics [68]. Revenue from fishing; food ration; radius of an hard part in backcalculation; inherent riskiness in fish foraging and habitat choice [70]. RL—Reproductive load [145]. The ratio between the length at maturity (Lm) and L∞ [113].

RP—Recruitment pattern.

RV—Reproductive value [65].

RSS—Residual sum of square [65,67].

s—Sampling standard deviation and (s2) variance [25, 70,107] § s.d. and s.e. [108]. S.D. and S.E. [109]. SD [65]. τ in [73]. §§ Total survival fraction [69]. N° of tags reported [97]. Area of surfaces involved in anabolic process in the derivation of the VBGF [91]. s0 time when the organism begins to react to the net's approach [147]. Selectivity [76,86,107]. Oldest exploited age class [86]. Annual survival rate from natural mortality; survival rate; responsiveness of a fleet; speed of a vessel [25]. Surface area of a fish's caudal fin [109]. Gear selectivity; ratio between Ls, earliest age at spawning, and asymptotic length; gear selectivity; district-specific catches vector in "run reconstruction" [65]. Index for prey species in multispecies VPA; stage in Individual Based Models; as subscript, stochastic [70]. Selectivity; exploitation pattern [104]. To not be confounded with the unit time second.

σ—Stock standard deviation and σ2 variance [65,86] § SE2 [65]. §§ Fraction of population at some stage or condition [67]. Fraction of the fish population that survives mortality by natural causes [64]. Food size preference coefficient [72]. Annual survival rate [84]. Larval temporal "width" in the Cushing's match/mismatch hypothesis [70]. Stage [27].

S—Stock size in n°. S' unit stock. S0 the pristine level. S∞ asymptote size in n°, the level to which an unexploited stock tends. § Catchable stock in weight [88]. Escarpment biomass; adult abundance [67]. Spawning stock [65,157], usually as proxy of egg production [70]. N as stock size in n° [107]. Post harvest biomass in delay difference model; spawning biomass; n° of sets [25]. Stock size in n° or biomass [117]. (Standing) stock as an (homogeneous) group of fish which occupy a given area in a given period, which are subject to the same opportunities of growth and reproduction and to the same risks of natural and fishing mortality [25]. The unit stock represents a stock which can be kept unitary from the point of view of fishing when the assessments carried out on the basis of this assumption does not give contrasting results with the real situation [24,30,42]. Both terms are often synonym with assessment/management unit, even if there is migration of the same species to and from adjacent areas. "Stock let", S•, is used in case of short living species such as most cephalopods [110]. §§ Selection function [122]. Speed of the (fishing boat) ship [147]. Competition coefficient [131]. Partial recruitment factor [85]. Squared deviations [141]. Fishing mortality subdivisions; SR sex ratio [76]. (Generic individual) length or weight size [71,121]. Selectivity; recruitment; survival rate [67]. Index of samples or cruises; standard deviation of a distribution [72]. Survival rate [86,107]. N° of sets; natural survival rate [25]. Survival fraction over one year; S0 early life survival; average size of a school; size of any hard body structure uses for back calculation; escarpment of adults from the fishery [65]. From Jennings et al. [68]. species richness (n° of species), survivorship, scale radius, spawner abundance, and escarpment related par. Prop [61]. Swimming speed; fishing mortality on age group in the separable VPA; index for predator species in multispecies VPA; species-specific selection pattern; score; strategy vector in Individual Based Models; measure of suitability; coefficient in the logistic; for example, the ML50% maturity estimation [70].

**S*—Generic score, such as the S function in Shepherd's method [66]. § S [66,108]. §§ N° of surviving fish after a certain time period [66,108].

SF— Selection factor [66,74] The ratio between the 50% of entry to capture and mesh size. § S.F. [50]; s.f. [50,126].

SR—Selection range [73]. The size range in which the probability of capture varies. § sr [81]. §§ To not be confounded with the reversed logistic curve [SR in 74].

SS—Sum of squares [25]. § SQ [104].

ST—Steady state, situations in which all the demographic processes (recruitment, growth and mortality) are cons. along the time (deterministic models) or the same processes are randomly varying in time without no trend (stochastic models). § One of the more common hypotheses in classic fisheries science and the most criticised assumption nowadays [25].

Stc—Stomach contain in weight. § F [70].

SPR—Spawning stock-biomass per recruit [70].

SRR—Stock-recruitment relationships [11,70,117]. § Recruitment curve as a graph of recruits (Y ax) against spawners [56]. S/R [27,74]. S-R [104].

SSB—Spawning stock biomass [70,117] § SP [86]. SB spawning biomass [104].

Σ—Summation sign [65,71]

t—Time [63,76]. Time in month [61]. t0 cons. that adjust the time scale to an origin at the inflexion point of a curve [66] th at hatching. ts sets the beginning of sinusoidal growth oscillating with respect to t = 0 (72) in the seasonally version of the VBGF (also summertime; 74). As subscript, tagging related par. [151]. Time or phase shift according to Jennings et al. [68] and Quinn and Deriso [65], respectively. tz origin of the VBGF in calendar time expressed as fraction of a year (in Shepherd's method). § Age [67]. tD end of (finite) lifetime [63]. §§ Total population [69]. N° of tags liberated [97]. t° temperature [98]. Freedom period in tagging experiments [92]. ts as the age at the end of the first fishing season [72]. ts earliest age at spawning; time dimension in migration model; maximum age of the cohort [65]. Age. tL age when the year class leaves the fished area; t0 age at maximum biomass peak [63].

τ—Free. §§ Portion of the ration ultimately recovered as net energy [143]. Time [120] interval [72]. Fraction of population at some stage or condition [67]. Different par. related to survival from fishing and natural mortality [64]. Variance related par. in MULTIFAN [102]. Age at sex or sex ratio change [87]. Time interval; density independent mortality component [70]. Fraction of or whole year; time spent in searching and handling; reference ages in growth models; n° of years represented in the plus group [65] τP age at which 100(P)% of the population remains [114]. Mean length at first recruitment; as subscript, size at recapture [66].

θ—Z over K ratio (72). θB Beverton and Holt. θS Ssentongo and Larkin. θP Powell. θR regression. θW Wetherall. θJ Jones and van Zalinge [72]. § Z/K [119]. r [115]. §§ Angle [147]. Fraction of the total mortality [64]. Generic par. of interest; a priori estimate of a par. within the Bayesian-type approach; prop. of specimens which change sex; true prop. of fish by age; survival curve in length frequency analysis (LFA); ratio between the true but unknown n° of age a fish in a random sample and the overall fish population; annual rate of movement from region i to region j; migration prop. [65]. T—Transport coefficient or rate of interchange of fish between adjacent areas [140]. Instantaneous rate of transfer [65]. § T'' immigrating [145]. Destination area in fish migration [70]. §§ Tα period of marking; ratio in Allen's recruitment method. Total metabolism [98,143]. Different par. in tagging experiments [2]. Duration of pregnancy or time to hatching in Elasmobranchs VBGF modelling [159]. Time interval [76]. Total n° of marked and non marked specimen [71]. Time delay or horizon; temperature as degree days [67]. Th time at hatching; T0time of recruitment; test function; mean environmental [72] or annual habitat temperature in °C [66]. Age [93,101]. Tmax maximum age in absence of exploitation [113]. Tonnage of a vessel; balance index; ocean-surface temperature anomaly; fishing time [25]. Age at which transition to the next pool occurs [82]. Total death; transformation of original data; time periods; age of recruitment; nth moment; T0 mean generation length or mean age of reproduction; objective function; vector of tag release in different areas; threshold level [65]. Marked specimens [68]. Test function in Shepherd's length-composition analysis; vessel lifetime; age at maturity; biological zero time temperature; foraging time [70]. Time at which the final harvest is to be performed [63].

TL—Trophic level [70].

Θ (theta)—M over K ratio. Θ∞ as M∞/K [153]. § M/K invariant. §§ Vector of par. [93]. Mean; vector of unknown par. [73]. Set of par.; age-specific migration rate matrices and matrices in general [65]. Vector of par. [70].

u—Ratio of recovery to marked fish released. §§ Unavailable population (in unfished area) [69]. Ratio of exploitation [89]. Yearly increase in length [120]. Average fish velocity [147]. Normally distributed random variable [67]. Typical fish weight [64]. Age at maturity— longevity product [113]. Individual probability of capture; par. related to Fletcher quadratic model; normal random (0 - 1) variable [65]. Catch per unit of effort [70].

υ—To not be used in order to avoid confusion with similar symbol.

U—Catch per unit of effort [25,27,65,70,76,107,129]. UC in n° (pieces). UY in weight. UM per unit mortality (equivalent to CPUM by Quinn and Deriso [65]. § Catch-per-Unit-of-Effort [95]. c.p.u.e., C.P.U.E., CPUE, p.u.e, PUE, CUE [22,25,67,76,107]. C/f, Y/f [107]. C/f, Y/f, fishing success. Prop. harvested

[25]. The symbol "p" for "per" should not be used; also the c.p.u.e. and allies symbols should be no more employed. A term used in yield-per-recruit model [108]. §§ Energy introduced in the growth process in the "equation of continuity" [91]. Average net velocity [147]. Par. in tagging experiments [92]. Expansion factor [67]. Mean of a distribution; expansion factor [72]. Weight related par. in Schnute [64] model. Complement of Lc/L∞ as U= 1-Lc/L∞(74). Random variable; expansion factor in B&H Y/R model; penalty weight function in LFA; utility function [65]. Exploitation rate [27] U'—Catch per unit of area. § CPUA [74].

v—Co-ordinate defining sub-areas. §§ Natural death fraction [69]. Swimming speed [143]. Virgin or pristine; equilibrium [76]. Weight 1 year prior to recruitment; a particular weight of fish; normally distributed random variable [67]. Natural mortality—longevity product [113]. Age specific vulnerability [25]. Vulnerability [151]. Maximum mortality rate that a predator can exert on a prey [82]. Selection par. related to haul [73]. Par. related to Fletcher's quadratic model; residuals [65]. Vulnerability to the fishery [68]. Rates of behavioural exchange between vulnerable and invulnerable state in Ecosim; sets of observations [70].

**v*—Value of fish in the specified monetary unit. § v as average price [74]; for example, *v$. p as price or V as value of income [68]. v fish value [66].

V—Virtual population [127], cohort analysis [65] and, in general, age-structured related par. [86]. § VPA (74]. §§ The (logarithmic) rate of natural increase; growth rate of biomass; catch in weight at equilibrium [88,105]. Value [96]. Energy removed in the growth process in the "equation of continuity" [91]. Water volume [140]. Trawling speed and fluid velocity [106]. Weight at recruitment; a particular weight of fish; normally distributed random variable; value [67]. Variance [72]. Weight at recruitment to area and gear [64]. Amount of prey biomass or n° of species eaten by a predator; value; vessel effect; n° of vessels in the fishing fleet [25]. Reproductive value [87]. Biomass of prey I available to predator j [82]. Value in Thompson and Bell analysis [74]. Variance; V0 volume of eggs in the ovaries; penalty function for mortality and variance within length interval in LFA; accumulated variance among ages; variance-covariance matrix [65]. Overlap index in length frequency analyses; sustainable profit; value [70].

**V*—Swimming speed [146]. b in body lengths s-1 [99]. § Fish velocity [92].

VBGF—Von Bertalanffy Growth Function [70,71,82] or formula [71,113]. § VBGE, equation [153] LVB, Ludwig von Bertalanffy [65]. VBGM, model [135]. The (3 par.) von Bertalanffy equation [70].

VPA—Virtual population analysis [65,70]. § Age structured, sequential, integrated or synthetic analysis [65] MSVPA multispecies [70].

w—Individual [74,120] measured fish weight. Body mass [67] or weight [25,70]. Fish weight [63]. wr round (overall-total). wg gutted (eviscerated). wdry dehydrated organism after a standard time in an oven (incorporated in McGurk [168] model. § Typical fish weight and other related par. in Schnute [64] model. wk weight at recruitment; normally distributed random variable [25]. Round and gutted weight also in Lleonart [22]. §§ Par. in tagging experiments [92]. Non linear scaling par. [86]. Width of the length frequency intervals in MULTIFAN [102]. Relative (statistical) weight; width of a vessel action [25]. Statistical weight [for example, in the Extended Survivors Analysis;70], usually the inverse of variance, applied to a datum [65]. Akaike's weigh factor [135].

•*w*—Individual body component weight.

wd—Individual measured fish width; especially in rays and skates it indicates the distance from the tip of the left to the tip of the right "wing" [109].

W—Theoretical (parametric) fish weight at stock level [25,61,67,70,76,86]. See a for other specifications (such as Wc and WR). § Somatic weight [99]. Mean weight of a fish at age t [65]. Individual weight [117]. §§ Total weight of a stock [120]. Empirical weighting factor [86]. Net width [137]. Penalty term in LFA [65].

W∞—Asymptotic weight. See L∞ and allies for the analogous specifications.

WP—Winter Point in the seasonal VBGF; the period of the year (expressed as fraction of year) when growth is slowest; related to ts through WP = ts + 0.5 [66,72] § tw [72].

x—Independent variable generally in linear regression [66]. §§ N° of tagged specimens found in a sample [139]. Par. in tagging experiments [92]. Coefficient of proportionality between Gulland MSY and MB0 expression [128,173]. Label for categories of interest in gear selectivity; sampling units; position dimension in migration model [65]. Spatial coordinate in Individual Based Models; age [70]. xm age at maturity in Jensen's invariant [27].

ξ—Free. §§ Probability of being recaptured [141]. Variance related par. in MULTIFAN [102]. Par. in Deriso-Schnute's stock recruitment and mark recapture version of Schnute's growth model; fraction of fish not caught that do not die due to natural mortality [65].

Ξ (csi)—Free. §§ Annual food consumption of a fish population.

X—Mark (tagging) and recapture related par. NX n° of tagged specimens before the samplings [139]. rNX n° of tags (tagged fish) recovered [25,151] XD disappearance rate of marks [65]. XOLother loss rate (in tagging). XP probability of recapture. § m in Jones [92]. R for recaptures [68]. §§ Exploitation rate as F(1-exp-Z)/Z [127]. N° of baits undisturbed [90]. Annual instantaneous mortality other than fishing [92]. "Perturbed" and "unperturbed" outputs in sensitivity analysis [71]. Independent observation; population level; n° of groups [67]. Ancillary variable as l/L∞[72]. Average weight of the entire recruitment population and other related par. [64]. Coefficient of proportionality between M and FMSY [173]. Any "second" stock in multispecies analysis; generic independent variable [25]. Value of a environmental factor [103]. Multiplication factor of F in Thompson and Bell analysis [74]. Average year class size over a period of time; total sampling units; critical value; environmental variables; transformed variable; matrix related symbols; disappearance rate of marks [65]. Generic variable [70]. Proportional escapement [117]. XSA—Extended survivors analysis [70].

y—Dependent variable generally in linear regression [66]. §§ Par. in tagging experiments [92]. Total yield as a fraction of the N(t0)W∞ product [67]. Calendar year and year of commercial data [86]. Initial conditions [yl and y2; 93] in Schnute's growth model; position dimension in migration model [65].Year; spatial coordinate in Individual Based Models [70].

**y*—Year. § y [74;61;70]

Y—Yield [25,67,76,83,120]. (Total) catch in weight [70,82,105,117,122]. Total yield from a year-class [67]. (Total) catch in biomass units [65]. Y∞ total yield over the lifetime of a cohort. Y* at equilibrium [65]. Yg gross. Yby not target. Yr retained on board. Yl live. Yd landed. RY replacement yield [RY in 117]. § Net economic yield [96]. Cohort strength; actual catch [67]. YW average weight of fish and other related par. [64]. §§ N° of fish hooked whether or not they subsequent escape [90]. Average weight within the newly recruited population [64]. Total n° of year of fishery data [86]. Length, weight, or some other determinant of fish size [93]. Rate of a certain metabolic function [103]. Recruitment variable; generic individual size; n° of size classes [65]. Generic [dependent; 25] variable [70].

Y/R—Yield-per-recruit [76;65]; catch in weight per recruit. Y/Rmax maximum value (at Fmax). Y/RF∞ maximum possible value in the isopleth diagram. Y'/R relative. § (YW/R)max. (Y/R)' relative (74). YPR [70].

YEB—Yield exploitable biomass; the fraction of a stock which is considered economically usable. In case M is low then YEB ≈ MSY (on the opposite YEB > MSY) [170-172].

z—Ratio of fished to unfished areas (→ ∞ when the whole area is fished). §§ Function of the virtual population [127]. Cons. in the parabolic generalisation of the growth—metabolic process [91]. Standard metabolism [143]. Dummy variable [86]. Specific growth rate in Schnute's model; steepness of the SRR [25]. Par. (Nt – N* or Bt – B*) in delay difference models [65]. Spatial coordinate in Individual Based Models [70]. Shape par. in Pella-Tomlinson's model [117].

ζ—Free. §§ Maintenance food coefficient. Efficiency or correction factor [67].

Z—Instantaneous rate of stock total mortality from all causes [130,76,67,86,107,25,82,74,65,61,70]. ZH Heincke. ZR Robson and Chapman. Zr = 0 maximum level that a population can withstand [68]. Z*a area specified (as Zy,a,Ar,q in [70]. § Z* reference mortality in a variant of B&H expression relating mortality to length at maturity [128]. Sstock [70] See F ad M for other specifications. §§ Catch in term of n° of fish on the line [90]. Standardised normal variate [97]. Mean depth [67]. Average weight within the previously exploited population and other related par. [64]. Logarithm of relative rate of change in Schnute model [65].

3.2. Final Thoughts

A simply insight in both "Milestone" and "proposed" list allows the immediate perception that convenience and opportunity were the general criteria followed by the Authors in defining and using the symbols. Even in the same contribute or textbook, the same symbols were often employed with different meanings [68] and rarely according to any point of the decalogue, with some exception (for example, the 8th point in [25]. Overall, only an handful of symbols resulted to be used in a quite consistent and standard manner; in particular, those related to natural (M) and total (Z) instantaneous mortality rates, and the coefficient of proportionality (q as catchability) relating fishing mortality (F) to fishing effort (f).

Another interesting aspect consists in a sort of reluctance in using symbols in most of the contributes and books produced at the beginning of the fishery science. At the Symposium on Fish Populations, held in Toronto in 1947 [174], few specific symbols were used only in two of the nine contributes presented [59;95]; few symbols were also used in Ricker [53], a contribution likely prepared before his 1958 handbook). Apparently, the burst of the use of symbols in fisheries occurred with the Beverton and Holt contributes, especially those published in 1956, 1957 and 1959; in the 1957 book, in particular, the Authors were forced to adopt almost the whole Latin and Greek alphabet.

As previously stated, the opportunity of symbols standardisation was highlighted at the end of fifty [14-16], but thereafter, the feeling is that the Authors were more worried about the agreement on definitions instead of symbols.

The consequences of a lack of standardization have determined long-standing problems mainly due to confusing symbols, definitions and applications. One of the most interesting examples of such as ambiguity might be referred to the "asymptotic" length in the von Bertalanffy growth function. Beside the biological interpretation (where it has any), the main troubles arise from a) confounding the maximum individual length with maximum mean length at age [101,160]; b) assuming that the maximum ever length could be used as the theoretical asymptotic length [175] or as a proxy of it ($L\infty \approx L_{max}/0.95$; 71); and c) forgiving that present maximum length in a exploited stock might be quite lower than the maximum length in the pristine condition (even at small rate of fishing; [54]), resulting in lower asymptotic length even maintaining the same rate of approaching it [176]. Considering that $L\infty$ (more rarely L_{inf} or L_{asy}) is generally used for different defined asymptotic length estimations, which are on their turn used to compare data or to get invariants [177,178], it is evident of the possible confusing effects and the opportunity for a better correspondence between the symbol and the defined parameter.

Another example of misusing is represented by what Beverton and Holt (see the Milestone list) have defined as the "fishable life span" (denoted with λ), which represented an arbitrary upper bound to the computations, reflecting the progressive rarefaction of the oldest age classes in the samples. Convenience has induced the Authors to set $\lambda \approx \infty$, obtaining a simplification in the computations. Thereafter, the fishable life span was often associated to the longevity, which is evidently an error at least because the two parameters have a different definition.

Finally, present results support the Beverton and Holt's 1957 sentence about the discrepancy between the number of items employed in fisheries assessment and the number of possible not confusing symbols. The present proposal tends to offer an operative solution by associating all the Latin letters to key quantities/items relevant to fisheries assessment and leaving free (with few consolidated exceptions) the Greek letters to identify different parameters and items (Table 1 in Annex). Obviously, the scheme presented here should be intended as a consistent example which should be used as a starting point towards the establishment of an international communication standards in fisheries assessment.

Table 1. Synoptic table of correspondence between item/parameters and symbols proposed (PS) in the present contribute. In the examples column, a selection of other meanings attributed to the given symbol in fishery science are presented (see both the "Milestone" and "Proposed list" sections for abbreviations and other details).

Item/parameter	PS	Examples
Age	a, A	Age at recruitment: area occupied by the fish population; hook catch probability; mesh size. Total mortality fraction: instantaneous rate of loss of bait; recruitment in weight
Age group	A_x	-
Allometry (coefficient)	k	Growth coef. in model different than VBGF; total n° of fish in unit weight. Catchability in tagging studies; destruction per unit weight in the derivation of the VBGF
Annual surplus production	ASP	-
Area	$^*a$, $^*A$	-
Array	${}_tX$	-
Biomass	B	Mesh length/mesh width ratio: prop. of new recruits in the catch. N° of fish perished by natural causes; larger mesh size in gill net selection experiment
Biomass index	BI	-
Biomass per recruit	B/R	-
Biomass threshold (surplus production)	$^*K$	-
Catch curve	CC	-
Catch in number	C	Average minimum age limit of usable stock; head girth/head breadth ratio. Amount of food consumed: cost of the fishing effort; energy cost in handling or searching for prey
Catch per recruit	C/R	-
Catch per unit of area	U^a	-
Catch per unit of effort	U	Instantaneous rate of "other loss" (also emigration and shedding of tags). Energy introduced in the growth process in the "equation of continuity"; exploitation rate
Catchability	q	Widrig's instantaneous rate of natural mortality. Condition factor: instantaneous rate of natural mortality, recruitment ratio; total mortality
Capture related	X_c	Compensatory component in natural mortality; par. In growth models different from (or previous than) VBGF; selection factor. Instantaneous rate of loss of hooked fish; juvenile survival rate to maturity; food consumption rate
Coefficient of error	CE	-
Coefficient of variation	CV	-
Cohort	CH	-
Condition factor	CF	-
Confidence interval	CI	-
Contact/selection curves	CSC	-
Conventional	${}_pX$	-
Constrained	${}_cX$	-
Co-ordinate defining sub-areas	v	Expectations of natural death. Swimming speed; virgin or pristine; equilibrium
Covered region	CR	-
Cumulative	X^-	-
Curvature par. (VBGF)	K	Alternative gear to be compared; head depth/head breadth ratio. Fulton condition factor; total (cumulative) catch in tagging studies; total n° of cohort
Dead specimen (in number)	O	N° of fish dying of other causes than fishing. Coefficient related to the age at maturity of a cohort; mature ogive; objective function
Degree of freedom	df	-
Density index	DI	-
De-recruitment from the fishery	X_{dR}	-
Diet composition	DC	-

Difference (finite)	Δ	One or unit operation. Taxonomic diversity related par.
Dispersion coefficient	D	Average density of fish; n° dying of disease. Finite time interval in capture-recapture; total n° of deaths; n° dying from natural mortality in VPA
Distance (in random movement)	d	Catch per unit of effort; density of fish in a given haul catch. Increment in length in the compensatory growth analysis; time interval in capture-recapture
Ecotrophic efficiency	EE	-
Equilibrium	X^*	-
End phase	X_M	-
Error term	ε	Gross food conversion efficiency; midpoint in a given class, par. in catching power and vessel standardization
Eumetric line joining the locus of maxima of yield-mortality curves	AA'	-
Eumetric line joining the locus of maxima of yield-age at entry (mesh) curves	BB'	-
Expanding term or factor	Ω	Period in seasonal growth models; Yamanaka's yield function
Exploitation rate	E	Egg production; Taylor's KL_∞ product. Emigrating rate, total (cumulative) effort in tagging studies
Exponent (allometry)	b	Annual catch in Baranov's food biomass relationship; selection factor. Instantaneous rate of hooking fish; life span; selectivity coefficient
Fecundity	Fc	-
Females	♀	-
Fished to unfished areas ratio	z	Newcomers; instantaneous rate of recruitment or immigration. Function of the virtual population; shape par. In Pella-Tomlinson's model
Fishing effort	f	"Japanese" mortality rate. Feeding level; fecundity; probability of tags recapture
Fishing power	FP	-
Food (ingestion of)	I	Index of fishing intensity. Index of relative abundance; marked specimens; money invested in a new boat
Free, to be specified	α	Coef. of pre recruited mortality in the eggs-recruitment relationship. Age; season; generic parameter; instantaneous tagging mortality
Free, to be specified	β	N° of marked fish; incomplete beta function. Season indices of adult stock; exponent in the length weight relationship
Free, to be specified	χ	Fecundity coef. Arbitrary reference age, prop. of fertilised eggs that will result in females; sex ratio
Free, to be specified	δ	Effective discount rate; successive age increment; shape par. in selection curves
Free, to be specified	φ	Dummy (time) variable Retention rate; function
Free, to be specified	γ	Annual egg-production per recruit. Area successfully searched; fraction of individuals in a stage moving to the next stage; shape par. in the Shepherd's SRR
Free, to be specified	Γ	Index of competition. Environmental variable affecting recruitment in semelparous population modelling
Free, to be specified	μ	Coef. defining the relationship between mesh size and fish girth. Mortality rate; n° of tagged fish recaptured
Free, to be specified	Π	Population size. Product
Free, to be specified	Ψ	Girth related probability of escaping. Natural survival rate; elasticity (sensitivity)
Free, to be specified	ι	Recapture period in marking; calendar date. Fraction of population at some stage or condition; portion of the ration ultimately recovered as net energy
Free, to be specified	ξ	Annual food consumption (individual). Probability of being recaptured
Free, to be specified	Ξ	Annual food consumption of a fish population.
Free, to be specified	ζ	Maintenance food coefficient. Efficiency or correction factor
Gastric evacuation rate	Ger	-

Generation	X_g	-
Gonosomatic index	GI	-
Goodness-of-fit	GOF	-
Growth (intrinsic rate of initial population)	$^*r$	-
Growth-maturity-longevity-plot	GML	-
Growth oscillation (seasonal) factor	$^*C$	-
Growth performance index (Gallucci & Quinn)	ω	Average weight of individual food organisms during their "grazeable" life span. Fraction of the total biomass due to newly recruited fish; weight at age ratio
Growth performance index in weight (Pauly & Munro)	Φ	Total n° of age groups into which recruitment occurs. Fraction of the year before the start of fishery; gear saturation par.; golden ratio
Growth rate	g, G	Fishing effort as collected (uncorrected) or "crude"; net (and long term) gain from a change in gear selectivity or area closure. Age specific vulnerability to fishing induced mortality; F/K; par. related to food; income for the fishing industry; fishing induced mortality
Growth rate at stock level	$^*G$	-
Haul	$^*h$	-
Hepatosomatic index	HI	-
Hoenig & Lawing's coef.	$^*c1\,^*c2$	-
Hour	h	Annual (or seasonal) relative individual growth rate. Annual harvest rate; haul; instantaneous long-line fishing mortality; time required to capture and consume

Index (generic)	X_i	Widrig's instantaneous rate of (total) mortality of a stock. Index of total mortality; mesh opening
Individual level	x	-
Infinite	∞	-
Inflexion	X	-
Interval (finite)	X_Δ	-
Invariant	X_ω	Estimates uncertain
Juvenile	j, J	Age group; fishing intensity. Index for predators or consumers; location or area; time interval; jaws size in urchins; age at first capture; yield
Length	l, L	Immediate losses after gear selectivity change or area closure; fishable life span. Annual survival rate; early life survival from the egg stage to recruitment; probability of living, catch or landings; production in weight
Length at capture - maximum (asymptotic) length ratio	$^*c$	-
Length frequency analysis	LFA	-
Logarithm (e based)	ln	-
Logarithm (10 based)	log	-
Logarithm base (natural or Napierian)	e	Sample effort. Age specific fecundity; fishing effort in mark-recapture; prey density; unit effort in tagging studies
Longevity	A_L	-
Loss rate of marks	H	Par. in the egg-production per recruit computation. Instantaneous (exponential) growth rate; weight synthesised per unit surface area in the derivation of the VBGF
Males	♂	-
Mark (tagging)—recapture related parameter	X	Effort in surplus production; fishing power; other loss rate (in tagging). Annual instantaneous mortality other than fishing; exploitation rate; n° of baits undisturbed

Massive	X_{mx}	-
Maximum (as peak value)	${}_{n}X$	-
Maximum ever (as greatest value)	X_{mxe}	-
Maximum present (actual)	X_{mx}	-
Maximum sustainable economic yield	*MSEY*	-
Maximum sustainable yield	*MSY*	-
Mean	X_{mean}	-
Mean square error	*MS*	-
Median lethal concentration	*LC50*	-
Mesh	*m*	Apparent mortality coef.; instantaneous mortality (in fish dispersion). Fraction of fish removed in 1 year. M/K: shape par. In the Pella & Tomlinson's model
Migration coefficient	*MC*	-
Modal progression analysis	*MPA*	Marine protected areas
Marine protected areas	**MPA*	-
Mortality: fishing (instantaneous rate)	*F*	Size of a progeny in the recruits parental relationships. Biomass flow up the size spectrum: starting area in fish migration: stomach content
Mortality: natural (instantaneous rate)	*M*	Average age of first recruitment: mesh: mesh size. Maximum population which the environment will support under average conditions and with no fishing: mature specimens
Mortality (natural)—K ratio	Θ	Age-specific migration rate: matrices in general: vector of unknown par.
Mortality: total (instantaneous rate)	*Z*	Correction term (in recruitment-egg relations). Average weight within the previously exploited population: catch in term of n° of fish on the line
Mortality (total)—K ratio	θ	The youngest age group free from the influence of recruitment of gear selectivity. Fraction of the total mortality: survival curve in length frequency analysis
Non equilibrium	${}_{ne}X$	-
Number of specimens or items	*n*	Annual or seasonal (natural) mortality rate (if no other causes operate): generic exponent in the VBGF in weight. Exponent in the allometric $W = cl^n$
Number of specimens in a homogeneous group	*N*	Total n° in stock. Egg production: par. related to Hill diversity: sample size in n°
Observation	X_{obs}	-
Omnivore index	*OI*	-
Parental stock	**P*	-
Physiological-temperature related coefficient	*Q*	Gross long term increase in catch following change in selectivity or area closure. Amount of food consumed: fishing effort efficiency: fraction of fish taken from a cohort
Powell's Z/K parameter	C^2	-
Power factor	*PF*	-
Product-moment (Pearson)	*r*	Fish escaping from parts of the net other than the cod-end. Distance: multipliers in yield per recruit equations: n° of tags recovered
Production	*P*	Fishing power: n° of fish dying from predation. Population size: price: prop. of tagged: rate of growth
Production—biomass ratio	*P/B*	-
Proportion	*p*	Population at the start of the fishing season: standing crop. Annual mortality rate: assimilation factor. instantaneous rate of fishing mortality
Ratio of dependent variables	ρ	Recruitment at the area where the fishing is in progress. Probability of individual fish to be available: slope in the Ford-Walford plot
Ratio of recovery to marked fish released	*u*	Co-ordinate defining sub-areas. Ratio of exploitation: unavailable population (in unfished area)
Recruit: recruitment	*R*, X_0	Feeding ration: n° of recaptured marks. Food intake: net of operative costs: n° of tags released

Recruitment pattern	*RP*	-
Redundant (same symbol, different parameter)	**X*	Equilibrium quantity; "critical"; "true" value
Reproductive load	*RL*	-
Reproductive value	*RV*	-
Residual sum of square	*RSS*	-
Revenue (expected)	*ER*	-
Score (generic)	**S*	N° of surviving fish after a certain time period
Selection factor	*SF*	-
Selection range	*SR*	-
Separation Index	**I*	-
Sexual maturity	X_m	-
Shape par.	η	Suffix denoting reference to spawning or (first) maturity; marked change in growth. Probability of being recaptured; random variable in population equilibrium catch relationship
Shape par. (Pauly's generalised VBGF)	**D*	-
Spawning stock biomass	*SSB*	-
Spawning stock-biomass per recruit	*SPR*	-
Standard deviation (at sample level)	*s*	Sex ratio; surface area of a fish. N° of tags reported; total survival fraction
Standard deviation (at stock level)	σ	Food size preference coefficient, fraction of the fish population that survives mortality by natural causes
Stock-recruitment relationships	*SRR*	-
Steady state	*ST*	-
Stock level	*X*	Fishing effort (in case of no ambiguity); par. in Jones' yield computation
Stock size in number	*S*	Grazing efficiency; the girth at which the fish is meshed. Competition coefficient; selection function; survival rate; swimming speed
Stomach contain in weight	St_c	-
Sum of squares	*SS*	-
Summation	Σ	-
Swimming speed	**V*	-
Time	*t*	Age of fish; time required for growth in growth temperature models. Freedom period in tagging experiments; maximum age of the cohort; total population
To not be used in order to avoid confusion	ϵ	Random variable in fishing effort analysis. Particle size conversion efficiency
To not be used	φ	Dummy (time) variable. Probability of survival in tagging experiments
To not be used	ι	-
To not be used	κ	Cons. relating F to "destruction" mortality. Difference between intrinsic rate and exit rate in migration model
To not be used	Λ	Average n° of landed fish. Diagonal matrix; test statistic
To not be used	υ	Annual expectation of natural death; nutritional factor. Fraction of the total mortality; selection par. related to haul
To not be used	o	Subscript denoting observed values. Term in stochastic mortality model
To not be used	ψ	Dummy time variable. Arbitrary reference age; mean direction of movement; ratio between C and pristine biomass

To not be used	υ	Age group
Transport coefficient	T	Mean age in catch or exploited phase. Duration of pregnancy or time to hatching in Elasmobranchs VBGF modelling: total metabolism
Trophic level	TL	-
Ultimate (terminal) significant contribute to the fishery	λ	Probability of capture between competing species: true survival rate. Finite rate of population growth: fraction of the catch inspected for tags: length in the Putter's expression of the VBGF
Value of fish in the specified monetary unit	$^{*}v$	-
Variable (dependent)	y	Instantaneous rate of capture of hooks. Year: spatial coordinates in Individual Based Models
Variable (independent)	x	The ratio of two initial populations. N' of tagged specimens found in a sample
Variant (closed, but different parameters)	$:X''$	Pristine level: statistic: estimate
Variant from the basic definition of the same parameter	$_{i}X$	-
Virtual population	V	Effective velocity of (random) movements: vulnerability. Catch in weight at equilibrium: growth rate of biomass: trawling speed and fluid velocity
Virtual population analysis	VPA	-
Von Bertalanffy Growth Function	$VBGF$	-
Weight	w, W	N' of spawners divided by the replacement n' of spawners: reproductive stock: weight of landings. Width of the length frequency intervals in MULTIFAN: width of a vessel action: total weight of a stock
Weight-specific oxygen uptake	QO_2	-
Winter Point	WP	-
Year	$^{*}y$	-
Year class	X_{cl}	-
Yield	Y	Different kind of fishing effort. N' of fish hooked whether or not they subsequent escape
Yield exploitable biomass	YEB	-
Yield-per-recruit	Y/R	-

4. ACKNOWLEDGEMENTS

A special thank to Lady Alexis Pacey, Publications Manager of the North Atlantic Fisheries Organisation (NAFO), Dartmouth, NS, who has provided the original historical ICNAF reports and Dr. Adamo Giovanna for her genuine interest in this paper.

REFERENCES

1. Cushing, D.H. (1983) Key papers on fish populations. Irl Press, Oxford.
2. Smith, T.D. (1988) Stock assessment methods: The first fifty years. In: Gulland, J.A., Ed., Fish Population Dynamics: The Implications for Management, John Wiley and Sons Ltd., Chichester, 1-33.
3. Saetersdal, G. (1989) Fish resources research and fishery management: A review of nearly a century of experience in the Northeast Atlantic and some recent global perspectives. Journal du Conseil International pour l'Exploration de la Mer, 46, 5-15.
4. Angelini, R. and Moloney, C.L. (2007) Fisheries, ecology and modelling: An historical perspective. Pan-American Journal of Aquatic Sciences, 2, 75-85.

5. Beverton, R.J.H. and Holt, S.J. (1956) A review of methods for estimating mortality rates in exploited fish populations, with special reference to sources of bias in catch sampling. Rapports et Procès-Verbaux des RèunI ons Commission Internationale pour l'Exploration Scientifique de la Mer Méditerranée, 140, 67-83.
6. Beverton, R.J.H. and Holt, S.J. (1957) On the dynamics of exploited fish populations. Ministry of Agriculture, Fisheries and Food, Lowestoft.
7. Beverton, R.J.H. and Holt, S.J. (1959) A review of the lifespans and mortality rates of fishes in nature, and their relation to growth and other physiological characteristics. In: Wolstenholme, C.E.W. and O'Connor, M., Eds., Ciba Foundation Colloquia on Ageing, J and A Churchill Ltd., London, 142-179.
8. Gulland, J.A. (1955) On the estimation of growth and mortality in commercial fish population. Fishery Investigations, 18, 46.
9. Gulland, J.A. (1956) The study of fish populations by the analysis of commercial catches: A statistical review. Rapports Conseil Internationale pour l'Exploration Scientifique de la Mer Méditerranée, 140, 21-27.
10. Gulland, J.A. (1956) A note on the statistical distribution of trawl catches, in problems and methods of sampling fish populations. Rapports Conseil Internationale pour l'Exploration Scientifique de la Mer Méditerranée, 140, 28-29.
11. Gulland, J.A. (1962) The application of mathematical models to fish populations. In: Le Cren, E.D. and Holdgate, M.W., Eds., The Exploitation of Natural Animal Populations, Blackwell Scientific Publications, Oxford, 204-217.
12. Ricker, W.E. (1954) Stock and recruitment. Journal of the Fisheries Research Board of Canada, 11, 559-623. doi:10.1139/f54-039
13. Ricker, W.E. (1958) Handbook of computations for biological statistics of fish populations. Bulletin of the Fisheries Research Board of Canada, 1-300.
14. Gulland, J.A. (1958) Notations in fish population studies. ICNAF Special Publication, 1, 69-70.
15. Holt, S.J. (1958) Population dynamics: Devise means of reducing the time required to obtain data necessary for making assessment of stocks, especially those required as the basis for a program of conservation. ICNAF Special Publication, 1, 27-50.
16. Holt, S.J., Gulland, J.A., Taylor, C. and Kurita, S. (1959) A standard terminology and notation for fishery dynamics. Journal du Conseil International pour l'Exploration de la Mer, 24, 239-242. doi:10.1093/icesjms/24.2.239
17. IDWG-LUP (1994) Terminology for integrated resources planning and management. In: Choudhury, K. and Jansen, L.J.M., Eds., Integrated Resources Planning and Management, Food and Agriculture Organization, Rome.
18. Coad, B.W. and McAllister, D.E. (2008) Dictionary of ichthyology. http://www.briancoad.com/dictionary/references.htm
19. Garcia, S.M. (2009) Glossary. In: Cochrane, K.L. and Garcia, S.M., Eds., A Fishery Manager's Guidebook, 2nd Edition, Wiley-Blackwell, Hoboken.
20. NOAA (2010) Definition of fisheries technical term. http://www.nefsc.noaa.gov/techniques/tech_terms.html#ai
21. Restrepo, V.R. (2000) Glossary of fishery terms. ICCAT Standing Committee on Research and Statistics (SCRS). www.iccat.es
22. Lleonart, J. (2010) Draft glossary of scientific terms of interest for the SAC General Fisheries Commission for the Mediterranean. Thirty-fourth Session, Athens, 14-17 April 2010, 1-41.
23. FAO (2011) Glossary. www.fao.org/fi/glossary/default.asp
24. Gulland, J.A. (1983) Stock assessment: Why? FAO Fisheries Circular, 759.
25. Hilborn, R. and Walters, C.J. (1992) Quantitative fisheries stock assessment, choice, dynamics and uncertainty. Chapman and Hall, London. doi:10.1007/978-1-4615-3598-0
26. Kesteven, G.L. (1999) Stock assessments and the management of fishing activities. Fisheries Research, 44, 105- 112.
27. Musick, J.A. and Bonfil, R. (2005) Management techniques for elasmobranch fisheries. FAO Fisheries Technical Paper, 474, 1-251.

28. Huntsman, A.G. (1953) Fishery management and research. Journal du Conseil International pour l'Exploration de la Mer, 14, 44-55.
29. ICNAF (1960) The selectivity of fishing gear. Proceedings of the joint scientific meeting of ICNAF, ICES, and FAO on fishing effort, the effect of fishing on resources and the selectivity of fishing gear, Lisbon, 27 May-3 June 1957.
30. Gulland, J.A. (1969) Manual of methods for fish stock assessment. Part 1. Fish population analysis. FAO Manual Fisheries Science.
31. Clark, S.H. (1979) Application of bottom-trawl survey data to fish stock assessment. Fisheries, 4, 9-15. doi:10.1577/1548-8446(1979)004<0009:AOBSDT>2.0.CO;2
32. Caddy, J.F. (2009) Practical issues in choosing a framework for resource assessment and management of Mediterranean and Black Sea fisheries. Mediterranean Marine Science, 10, 83-119. doi:10.12681/mms.124
33. Beverton, R.J.H. (1953) Some observations on the principles of fishery regulation. Journal du Conseil Permanent International pour l'Exploration de la Mer, 19, 56- 68.
34. Beverton, R.J.H. and Parrish, B.B. (1956) Commercial statistics on fish population studies. Rapports et ProcèsVerbaux des Rèunions Commission Internationale pour l'Exploration Scientifique de la Mer Méditerranée, 140, 58-66.
35. Holt, S.J. (1958) The evaluation of fisheries resources by the dynamic analysis of stocks, and notes on the time factors involved. ICNAF Special Publication, 1, 77-95.
36. Holt, S.J. (1963) A method for determining gear selectivity and its application. ICNAF Special Publication, 5, 106-115.
37. Holt, S.J. (1965) A note on the relation between the mortality rate and the duration of life in an exploited fish population. ICNAF Research Bulletin, 2, 73-75.
38. Gulland, J.A. (1961) The estimation of the effect on catches of changes in gear selectivity. Journal du Conseil International pour l'Exploration de la Mer, 26, 204-214.
39. Gulland, J.A. (1964) A note on the interim effects on catches of changes in gear selectivity. Journal du Conseil International pour l'Exploration de la Mer, 29, 61-64.
40. Gulland, J.A. (1964) Catch per unit effort as a measure of abundance. Rapports et Procès-Verbaux des Rèunions Commission Internationale pour l'Exploration Scientifique de la Mer Méditerranée, 155, 8-14.
41. Gulland, J.A. (1964) Variations in selection factors and mesh differentials. Journal du Conseil International pour l'Exploration de la Mer, 29, 158-165.
42. Gulland, J.A. (1967) Further notes on the effect of possible regulatory measures on catches of Greenland cod. ICNAF Redbook, Part III: 34-47.
43. Gulland, J.A. (1977) Fish population dynamics: The implications for management. Wiley Interscience, New York.
44. Gulland, J.A. (1983) Fish stock assessment: A manual of basic methods. Wiley Interscience, Chichester.
45. Gulland, J.A. (1987) Natural mortality and size. Marine Ecology Progress Series, 39, 197-199.
46. Gulland, J.A. and Holt, S.L. (1959) Estimation of growth parameters for data at unequal time intervals. Journal du Conseil International pour l'Exploration de la Mer, 25, 47-49.
47. ICNAF (1958) Some problems for biological fishery survey, and techniques for their solution. ICNAF Research Bulletin, 1.
48. ICNAF (1962) Report of the working group of scientists on fishery assessment in relation to regulation problems. In: Beverton, R.J.H. and Hodder, V.M., Eds., A Proceedings of the International Commission of the NW Atlantic Fisheries, 11, 1-81.
49. ICNAF (1963) The selectivity of fishing gear. ICNAF Special Publication, 5.
50. ICNAF (1967) Selected papers from the 1966 annual meeting. Redbook Part III, Dartmouth.
51. Kesteven, G.K. and Holt, S.J. (1955) A note on the fisheries resources of the north west Atlantic. FAO Fisheries Paper, 7, 1-12.

52. Ricker, W.E. (1940) Relation of "catch per unit effort" to abundance and rate of exploitation. Journal of the Fisheries Research Board of Canada, 5, 43-70. doi:10.1139/f40-008
53. Ricker, W.E. (1958) Production, reproduction and yield. Veroffentlichungen Limnologie, 13, 84-100.
54. Ricker, W.E. (1963) Big effects from small causes: Two examples from fish population dynamics. Journal of the Fisheries Research Board of Canada, 20, 257-264. doi:10.1139/f63-022
55. Ricker, W.E. (1973) Critical statistics from two reproductive curves. Rapports et Procès-Verbaux des Rèunions Commission Internationale pour l'Exploration Scientifique de la Mer Méditerranée, 164, 333-340.
56. Ricker, W.E. (1975) Computation and interpretation of biological statistics of fish population. Bulletin of the Fisheries Research Board of Canada, 191.
57. Ricker, W.E. (1977) The historical development. In: Gulland, J.A., Ed., Fish Population Dynamics: The Implications for Management, Wiley Interscience, New York, 1- 26.
58. Ricker, W.E. (1979) Growth rates and models. In: Hoar, W.S., Randall, D.J. and Brett, J.R., Eds., Fish Physiology, III, Bioenergetics and Growth, Academic Press, New York, 677 743.
59. William, E. and Ricker, R.E. (1948) Computation of fish production. A Symposium on Fish Populations. Proceedings of the Meeting held at the Royal Ontario Museum of Zoology, Toronto, 10-11 January 1947, 173-211.
60. Liungman, C.G. (2004) Symbols: Encyclopaedia of western signs and ideograms. 3rd Edition, HME Publishing, Stockholm.
61. Gjøsæter, H., Bogstad, B. and Tjelmeland, S. (2002) Assessment methodology for Barents Sea capelin, Mallotus villosus (Müller). ICES Journal of Marine Sciences, 59, 1086-1095. doi:10.1006/jmsc.2002.1238
62. Taylor, C.C. (1958) Cod growth and temperature. Journal du Conseil International pour l'Exploration de la Mer, 23, 366-370.
63. Marutani, T. (2008) On the optimal path in the dynamic pool model for a fishery. Reviews in Fish Biology and Fisheries, 18, 133-141. doi:10.1007/s11160-007-9065-7
64. Schnute, J. (1987) A general fishery model for a sizestructured fish population. Canadian Journal of Fisheries and Aquatic Sciences, 44, 924-940. doi:10.1139/f87-111
65. Quinn, T.J. II and Deriso, R.B. (1999) Quantitative fish dynamics. Oxford University Press, Oxford.
66. Gayanilo Jr., F.C., Sparre, P. and Pauly, D. (2005) FAOICLARM stock assessment tools II (FiSAT II). User's guide. FAO Computerized Information Series (Fisheries), 8.
67. Christie, W.J. and Spangler, G.R. (1987) The international symposium on stock assessment and yield prediction. Canadian Journal of Fisheries and Aquatic Sciences, 44, 501.
68. Jennings, S., Kaiser, M.J. and Reynolds, J.D. (2001) Marine fisheries ecology. Blackwell Science, Oxford.
69. Widrig, T.M. (1954) Method of estimating fish populations with application to Pacific sardine. Fishery Bulletin of the United States of America, 56, 141-166.
70. Hart, P.J.B. and Reynolds. J.D. (2002) Handbook of fish biology and fisheries (1-2). Blackwell Science Ltd., Oxford.
71. Pauly, D. (1984) Fish populations dynamics in tropical waters. International Center for Living Aquatic Resources Management, Manila.
72. Pauly, D. and Morgan, G.R. (1987) Length-based methods in fisheries research. International Center for Living Aquatic Resources Management, Manila.
73. Millar, R.B. and Fryer, R.J. (1999) Estimating size-selection curves of trawls, traps, gillnets, and hooks. Reviews in Fish Biology and Fisheries, 9, 89-116. doi:10.1023/A:1008838220001
74. Sparre, P. and Venema, S.C. (1998) Introduction to tropical fish stock assessment, Part I: Manual. FAO Fisheries Technical Paper, 306, 1.

75. Chen, S. and Watanabe, S. (1989) Age dependence of natural mortality coefficient in fish population dynamics. Nippon Suisan Gakkaishi, 55, 205-208. doi:10.2331/suisan.55.205
76. Laurec, A. and Le Guen, J.C. (1981) Dynamique des population marines exploitées. Publication CNEXO Rapport Scientific et Technologic, 45, 1-111.
77. Kutty, M.K. and Qasim, S.Z. (1968) The estimation of optimum age of exploitation and potential yield in fish populations. Journal du Conseil International pour l'Exploration de la Mer, 32, 249-255. doi:10.1093/icesjms/32.2.249
78. Hjort, J., Jahn, G. and Ottestad, P. (1933) The optimum catch. Hvalradets Skrifter, 7, 92-107.
79. May, R.C. (1974) Larval mortality in marine fishes and the critical period concept. In: Blaxter, J.H.S., Ed., The Early Life History of Fish, Spring-Verlag, Berlin, 3-19. doi:10.1007/978-3-642-65852-5_1
80. Smith, P.E. (1985) Year-class strength and survival of 0 group clupeoids. Canadian Journal of Fisheries and Aquatic Sciences, 2, 69-82.
81. [82] Bethke, E. (2004) A simple general approach to codend selectivity of trawls and its application to the data of Fiorentino et al. (1998) for Hake (Merluccius merluccius). Fisheries Research, 70, 113-119. doi:10.1016/j.fishres.2004.05.012
82. Walters, C., Christensen, V. and Pauly, D. (1997) Structuring dynamic models of exploited ecosystems from trophic mass-balance assessments. Reviews in Fish Biology and Fisheries, 7, 139-172. doi:10.1023/A:1018479526149
83. Mertz, G. and Myers, R.A. (1998) A simplified formulation for fish production. Canadian Journal of Fisheries and Aquatic Sciences, 55, 478-484. doi:10.1139/f97-216
84. Schnute, J.T. and Richards, L.J. (1998) Analytical models for fishery reference points. Canadian Journal of Fisheries and Aquatic Sciences, 55, 515-528. doi:10.1139/f97-212
85. Pope, J.G. (1977) Estimation of fishing mortality, its precision and implications for the management of fisheries. In: Steele, J.H., Ed., Fisheries Mathematics, Academic Press, London, 63-76.
86. Megrey, B.A. (1989) Review and comparison of agestructured stock assessment models from theoretical and applied points of view. American Fisheries Society Symposium, 6, 8-48.
87. Charnov, E.L. (1993) Life history invariants. Some explorations of symmetry in evolutionary ecology. Oxford University Press, Oxford.
88. Russell, E.S. (1931) Some theoretical consideration on the overfishing problem. Journal du Conseil International pour l'Exploration de la Mer, 6, 3-20.
89. Tester, A.I. (1955) Estimation of recruitment and natural mortality rate from age-composition and catch data in British Columbia Herring population. Journal of the Fisheries Research Board of Canada, 12, 649-641. doi:10.1139/f55-035
90. Murphy, G.I. (1960) Estimating abundance from long-line catches. Journal of the Fisheries Research Board of Canada, 17, 33-40. doi:10.1139/f60-003
91. Taylor, C.C. (1962) Growth equations with metabolic parameters. Journal du Conseil International pour l'Exploration de la Mer, 27, 270-286.
92. Jones, R. (1977) Tagging: Theoretical methods and practical difficulties. In: Gulland, J.A., Ed., Fish Population Dynamics: The Implications for Management, Wiley Interscience, New York, 46-66.
93. Baker, T.T., Lafferty, R. and Quinn II, T.J. (1991) A general growth model for mark-recapture data. Fisheries Research, 1, 257-281. doi:10.1016/0165-7836(91)90005-Z
94. Alverson, D.L. and Pereyra, W.T. (1969) Demersal fish explorations in the north-eastern Pacific Ocean—An evaluation of exploratory fishing methods and analytical approaches to stock size and yield forecasts. Journal of the Fisheries Research Board of Canada, 26, 1985-2001. doi:10.1139/f69-188

95. Herrington, W.C. (1948) Limiting factors for fish populations. Some theories and an example. A Symposium on Fish Populations. Proceedings of the Meeting Held at the Royal Ontario Museum of Zoology, Toronto, 10-11 January 1947, 229-283.
96. Schaefer, M.B. (1957) Some considerations of population dynamics and economics in relation to the management of commercial marine fisheries. Journal of the Fisheries Research Board of Canada, 14, 669-681. doi:10.1139/f57-025
97. Paulik, G.J. (1961) Detection of incomplete reporting of tags. Journal of the Fisheries Research Board of Canada, 18, 817-832. doi:10.1139/f57-025
98. Paloheimo, J.E. and Dickie, L.M. (1966) Food and growth of fishes. II. Effects of food and temperature on the relation between metabolism and body weight. Journal of the Fisheries Research Board of Canada, 23, 869- 908. doi:10.1139/f66-077
99. Jones, R. and Johnston, C. (1977). Growth, reproduction and mortality in gadoid fish species. In: Steele, J.H., Ed., Fisheries Mathematics, Academic Press, London, 37-62.
100. Charnov, E.L., Berrigan, D. and Shine, R. (1993) The M/K ratio is the same for fish and reptiles. American Naturalist, 142, 707-711. doi:10.1086/285565
101. Francis, R.I.C.C. (1988) Are growth parameters estimated from tagging and age-length data comparable? Canadian Journal of Fisheries and Aquatic Sciences, 45, 936-942. doi:10.1139/f88-115
102. Fournier, D.A., Sibert, J.R., Majkowski, J. and Hampton, J. (1990) MULTIFAN, a likelihood based method for estimating growth parameters and age composition from multiple length frequency data sets illustrated using data for Southern bluefin tuna (T. maccoyii). Canadian Journal of Fisheries and Aquatic Sciences, 47, 301 317. doi:10.1139/f90-032
103. Yamakawa, T. and Matsuda, H. (1997) Improved bêlehrádek's equation for a comprehensive description of the relationship between environmental factors and metabolic rates. Fisheries Science, 63, 725-773.
104. Cadima, E.L. (2003) Fish stock assessment manual. FAO Fisheries Technical Paper, 393.
105. Graham, M. (1935) Modern theory of exploiting a fishery, and application to North Sea trawling. Journal du Conseil International pour l'Exploration de la Mer, 10, 264-274.
106. Foster, J.J., Ferro, R.S.T. and Reid, A.J. (1977) Analytic approaches in fishing technology. In: Steele, J.H., Ed., Fisheries Mathematics, Academic Press, London, 99-116.
107. Brethes, J.C. and O'Boyle, R.N. (1990) Methodes d'evaluation des stocks halieutiques. Centre International d'Exploitation des Oceans, Halifax, 933.
108. Gayanilo Jr., F.C. and Pauly, D. (1997) FAO-ICLARM fish stock assessment (FiSAT) reference manual. FAO Computerized Information Series (Fisheries), 8, 2.
109. Froese, R. and Pauly, D. (1998) FishBase 98: Concepts, design and data sources. International Center for Living Aquatic Resources Management, Manila.
110. Caddy, J.F. (1983) The cephalopods: Factors relevant to their population dynamics and to the assessment and management of stocks. FAO Fisheries Technical Paper, 231, 416-452.
111. Alverson, D.L. and Carney, M.J. (1975) A graphic review of the growth and decay of population cohorts. Journal du Conseil International pour l'Exploration de la Mer, 36, 133-143.
112. Hoenig, J.M. (1983) Empirical use of longevity data to estimate mortality rates. Fishery Bulletin of the United States of America, 82, 898-902.
113. Beverton, R.J.H. (1992) Patterns of reproductive strategy parameters in some marine teleost fishes. Journal of Fish Biology, 41, 137-160.
114. Hewitt, D.A. and Hoenig, J.M. (2005) Comparison of two approaches for estimating natural mortality based on longevity. Fishery Bulletin of the United States of America, 103, 433-437.
115. Rosenberg, A.A. and Beddington, J.R. (1988) Lengthbased methods of fish stock assessment. In: Gulland, J.A., Ed., Fish Population Dynamics, John Wiley and Sons, Chichester, 83-103.

116. Froese, R. (2006) Cube law, condition factor and weightlength relationships: History, meta-analysis and recommendations. Journal of Applied Ichthyology, 22, 241-253. doi:10.1111/j.1439-0426.2006.00805.x
117. Hoggarth, D.D., Abeyasekera, S., Arthur, R.I., Beddington, J.R., et al. (2006) Stock assessment for fishery management—A framework guide to the stock assessment tools of the Fisheries Management Science Programme (FMSP). FAO Fisheries Technical Paper, 487.
118. Allen, K.R. (1953) A method for computing the optimum size limit for a fishery. Nature London, 172, 210. doi:10.1038/172210a0
119. Allen, K.R. (1967) Some quick methods for estimating the effect on catch of changes in the size limit. Journal du Conseil International pour l'Exploration de la Mer, 31, 111-126.
120. Tanaka, S. (1962) The effect of reduction of fishing effort on Yield. Journal of the Fisheries Research Board of Canada, 19, 521-529. doi:10.1139/f62-035
121. Ebert, T.A. (1982) Longevity, life history and relative body wall size in sea urchins. Ecological Monographs, 52, 353-394. doi:10.2307/2937351
122. Chapman, D.G. (1961) Statistical problems in dynamics of exploited fisheries populations. In: Neyman, J., Ed., Proceedings of the Fourth Berkeley Symposium on Mathematical Statistics and Probability, Vol. 4. University of California Press, Berkeley, 153-168.
123. Ivlev, V.S. (1961) Mathematical analysis of the population dynamics of fish. Israel Program for scientific translation (form the Russian, original published in 1959).
124. Caddy, J.F. (1996) Modelling natural mortality with age in short-lived invertebrate populations: Definition of a strategy of gnomonic time division. Aquatic Living Resources, 9, 197-207. doi:10.1051/alr:1996023
125. Jensen, A.J.C. (1949) The relation between size of mesh and length of fish released. Rapports et Procès-Verbaux des Rèunions Commission Internationale pour l'Exploration Scientifique de la Mer Méditerranée, 125, 65-69.
126. Pope, J.A., Margetts, A.R., Hamley, J.M. and Akyuz, E.F. (1983) Manual of methods for fish stock assessment; part 3: Selectivity of fishing gear. FAO Fisheries Technical Paper, 41, 1-56.
127. Bishop, Y.M.M. (1959) Errors in estimates of mortality obtained from virtual populations. Journal of the Fisheries Research Board of Canada, 16, 73-90. doi:10.1139/f59-009
128. Die, D.J. and Caddy, J.F. (1997) Sustainable yield indicators from biomass: Are there appropriate reference points for use in tropical fisheries? Fisheries Research, 32, 69-79. doi:10.1016/S0165-7836(97)00029-5
129. Schaefer, M.B. (1957) A study of the dynamics of the fishery for yellowfin tuna in the eastern tropical Pacific Ocean. Bulletin of the Inter-American Tropical Tuna Commission, 2, 247-285.
130. Jones, R. (1956) The analyses of trawl haul statistics with particular reference to the estimation of survival rates. Rapports et Procès-Verbaux des Rèunions Commission Internationale pour l'Exploration Scientifique de la Mer Méditerranée, 140, 30-39.
131. Cushing, D.H. and Horwood, J.W. (1977) Development of a model of stock and recruitment. In: Steele, J.-H., Ed., Fisheries Mathematics, Academic Press, London, 21-35.
132. Carrothers, P.J.G. (1980) Estimation of trawl door spread from wing spread. Journal of Northwest Atlantic Fishery Science, 1, 81-90.
133. Doubleday, W.G. and Rivard, D. (1981) Canadian special publication of fisheries and aquatic sciences. Botton Trawl Surveys, 58.
134. Campbell, N.A. and Phillips, B.F. (1972) The von Bertalanffy growth curve and its application to capture-recapture data in fisheries biology. Journal du Conseil International pour l'Exploration de la Mer, 34, 295-299.
135. Katsanevakis, S. (2006) Modelling fish growth: Model selection, multi-model inference and model selection uncertainty. Fisheries Research, 81, 229-235. doi:10.1016/j.fishres.2006.07.002

136. Nikolsky, G.V. (1963) The ecology of fishes. Academic Press Inc., London.
137. Ramm, D.C. and Xiao, Y. (1995) Herding in groundfish and effective path width of trawls. Fisheries Research, 24, 243-259. doi:10.1016/0165-7836(95)00373-I
138. Pauly, D. (1981) The relationships between gill surface area and growth performance in fish; a generalisation of von Bertalanffy's theory of growth. Meeresforschung, 28, 251-282.
139. De Lury, D.B. (1961) On the planning of experiments for the estimation of fish populations. Journal of the Fisheries Research Board of Canada, 8, 281-307.
140. Andersen, K.P. and Ursin, E. (1977) A multispecies extension to the Beverton and Holt theory of fishing, with accounts of phosphorus circulation and primary production. Meddelelser fra Danmarks Fiskeri-Og Havundersogelser, 7, 319-435.
141. Nicholson, M.D. and Pope, J.A. (1977) The estimation of mortality from capture-recapture experiments. In: Steele, J.H., Ed., Fisheries Mathematics, Academic Press, London, 77-85.
142. Sissenwine, M.P., Fogarty, M.I. and Overholtz, W.J. (1988) Some fisheries management implications of recruitment variability. In: Gulland, J.A., Ed., Fish Population Dynamics, John Wiley & Sons, New York, 129-152.
143. Ware, D.M. (1975) Growth, metabolism, and optimal swimming speed of a pelagic fish. Journal of the Fisheries Research Board of Canada, 32, 33-41. doi:10.1139/f75-005
144. Rothschild, B.J. and Suda, A. (1977) Population dynamics of tuna. In: Gulland, J.A., Ed., Fish Population Dynamics, John Wiley and Sons, New York, 309-334.
145. Cushing, D.H. (1981) Fisheries biology: A study in population dynamics. University of Wisconsin Press, Madison.
146. Harden, J.F.R. (1977) Performance and behaviour on migration. In: Steele, J.H., Ed., Fisheries Mathematics, Academic Press, London, 145-170.
147. Barkley, R.A. (1964) The theoretical effectiveness of towed-net samplers as related to sample size and swimming speed of organisms. Journal du Conseil Permanent International pour l'Exploration de la Mer, 24, 146-157.
148. Davis, F.M. (1934) Mesh experiments with trawls 1928- 1933. Fishery Investigation, 14, 1-56.
149. Houghton, R.G. (1977) The fishing power of trawlers in the western English Channel between 1965 and 1968. Journal du Conseil International pour l'Exploration de la Mer, 37, 130-136. doi:10.1093/icesjms/37.2.130
150. Treschev, A.I. (1978) Application of the fished volume method for measuring fishing effort. ICES Cooperative Research Report, 79, 1-55.
151. Anganuzzi, A., Hilborn, R. and Skalski, J.R. (1994) Estimation of size selectivity and movement rates from mark-recovery data. Canadian Journal of Fisheries and Aquatic Sciences, 51, 734-742. doi:10.1139/f94-073
152. Stamps, J.A., Mangel, M. and Phillips, J.A. (1998) A new look at relationships between size at maturity and asymptotic size. American Naturalist, 152, 470-479.
153. Beyer, J.E., Kirchner, C.H. and Holtzhausen, J.A. (1999) A method to determine size-specific natural mortality applied to westcoast steenbras (Lithognathus aureti) in Namibia. Fisheries Research, 41, 133-155. doi:10.1016/S0165-7836(99)00011-9
154. Southward, G.M. and Chapman, D.G. (1965) Utilization of Pacific halibut stocks: Study of Bertalanffy's growth equation. International Pacific Halibut Commission Scientific Report, 39, 1-33.
155. Cortés, E. (2000) Life history patterns and correlations in sharks. Review in Fisheries Science, 8, 299-344.
156. Rothschild, B.J. (1977) Fishing effort. In: Gulland, J.A., Ed., Fish Population Dynamics, John Wiley and Sons, New York, 96-115.
157. Cushing, D.H. (1988) The study of stock and recruitment. In: Gulland, J.A., Ed., Fish Population Dynamics, John Wiley and Sons, Chichester, pp. 105-128.
158. Tomlinson, P.K. and Abramson, N.J. (1961) Fitting a von Bertalanffy growth curve by the least squares. California Fish and Game Fish Bulletin, 116, 1-69.

159. Holden, M.J. (1977) Elasmobranchs. In: Gulland, J.A., Ed., Fish Population Dynamics, John Wiley & Sons, Chichester, 187-215.
160. Francis, R.I.C.C. (1995) The analysis of otolith data—A mathematician's perspective (What precisely is your model?). In: Secor, D.H., Dean, J.M. and Campana, S.E., Eds., Recent Developments in Fish Otolith Research, University of South Carolina Press, Columbia.
161. Lockwood, S.J. (1974) The use of the von Bertalanffy growth equation to describe the seasonal growth of fish. Journal du Conseil International pour l'Exploration de la Mer, 35, 175-179.
162. MacDonald, P.D.M. and Pitcher, T.J. (1979) Age-groups from size-frequency data: A versatile and efficient method of analysing distribution mixtures. Journal of the Fisheries Research Board of Canada, 36, 987-1001. doi:10.1139/f79-137
163. Caddy, J.F. (1991) Death rates and time intervals: Is there an alternative to the constant natural mortality axiom? Reviews in Fish Biology and Fisheries, 11, 109 118. doi:10.1007/BF00157581
164. Roff, D.A. (1984) The evolution of life history parameters in teleosts. Canadian Journal of Fisheries and Aquatic Sciences, 41, 989-1000. doi:10.1139/f84-114
165. Beverton, R.J.H. and Iles, T.C. (1992) Mortality rates of 0-group plaice (Pleuronectes platessa L.), dab (Limanda limanda L.) and turbot (Scophthalmus maximus L.) in European waters: II. Comparison of mortality rates and construction of life-table for 0-group plaice. Netherland Journal of Sea Research, 29, 49-59. doi:10.1016/0077-7579(92)90007-2
166. Gallucci, V.F. and Quinn T.J. II (1979) Reparameterizing, fitting, and testing a simple growth model. Transactions of the American Fisheries Society, 108, 14-25. doi:10.1577/1548-8659(1979)108<14:RFATAS>2.0.CO;2
167. Allen, K.R. (1971) Relation between production and biomass. Journal of the Fisheries Research Board of Canada, 28, 1573-1581.
168. Robson, D.S. (1966) Estimation of the relative fishing power of individual ships. ICNAF Research Bulletin, 3, 5- 14.
169. Arreguín-Sánchez, F. (1996) Catchability: A key parameter for fish stock assessment. Reviews in Fish Biology and Fisheries, 6, 221-242. doi:10.1007/BF00182344
170. Myers, R.A. and Doyle, R.W. (1983) Predicting natural mortality rates and reproduction—Mortality trade-offs from fish life history data. Canadian Journal of Fisheries and Aquatic Sciences, 40, 612-620. doi:10.1139/f83-080
171. McGurk, M.D. (1986) Natural mortality of marine pelagic fish eggs and larvae: The role of spatial patchiness. Marine Ecology Progress Series, 34, 227-242.
172. Alverson, D.L. (1971) Manual of methods for fisheries resource survey and appraisal. Part 1. Survey and charting of fisheries resources. Food and Agriculture Organization of the United Nations, 1-102.
173. Garcia, S.M., Sparre, P. and Csirke, J. (1989) Estimating surplus production and maximum sustainable yield from biomass data when catch and effort time series are not available. Fisheries Research, 8, 13-23. doi:10.1016/0165-7836(89)90037-4
174. Anon. (1948) A symposium on fish populations. Bulletin of the Bingham Oceanographic Collection, New Haven.
175. Wetherall, J.A., Polovina, J.J. and Ralston, S. (1987) Estimating growth and mortality in steady-state fish stocks from length-frequency data. In: Pauly, D. and Morgan, G.R., Eds., Length Based Methods in Fisheries Research, ICLARM, Conference Proceeding, Manila, 53-74.
176. Welch, D.W. and McFarlane, G.A. (1990) Quantifying the growth of female Pacific Hake (Merluccius productus): An example of measuring uncertainty and bias in nonlinear parameter estimation. Canadian Journal of Fisheries and Aquatic Sciences, 47, 672-681.doi:10.1139/f90-074
177. Froese, R. and Binohlan, C. (2000) Empirical relationships to estimate asymptotic length, length at first maturity and length at maximum yield per recruit in fishes, with a simple method to evaluate length frequency data. Journal of Fish Biology, 56, 758-773. doi:10.1111/j.1095-8649.2000.tb00870.x

178. Binohlan, C. and Froese, R. (2009) Empirical equations for estimating maximum length from length at first maturity. Journal of Applied Ichthyology, 25, 611-613. doi:10.1111/j.1439-0426.2009.01317.x

CHAPTER 14

A Novel Lipase as Aquafeed Additive for Warm-Water Aquaculture

Chao Ran, Suxu He, Yalin Yang, Lu Huang, Zhigang Zhou*

Key Laboratory for Feed Biotechnology of the Ministry of Agriculture, Feed Research Institute, Chinese Academy of Agricultural Sciences, Beijing, P. R. China

ABSTRACT

A novel *Acinetobacter* lipase gene *lipG1* was cloned from DNA extracted from intestinal sample of common carp (*Cyprinus carpio*), and expressed in *E. coli* BL21. The encoded protein was 406 amino acids in length. Phylogenetic analysis indicated that LipG1 and its relatives comprised a novel group of true lipases produced by Gram-negative bacteria. LipG1 showed maximal activity at 40°C and pH 8.0 when *p*NP decanoate (C10) was used as the substrate, and remained high activity between 20°C and 35°C. Activity of the lipase was promoted by Ca^{2+} and Mg^{2+}, and inhibited by Zn^{2+} and Cu^{2+}. Moreover, LipG1 is stable with proteases, most commercial detergents and organic solvents. Substrate specificity test indicated that LipG1can hydrolyse *p*NP esters with acyl chain length from C2 to C16, with preference for medium-chain *p*NP esters (C8, C10). Lastly, LipG1was evaluated as an aquafeed additive for juvenile common carp (*Cyprinus carpio*). Results showed that supplementation of LipG1significantly improved the gut and heptaopancreas lipase activity of fish fed with palm oil diet. Consistently, improved feed conversion ratio and growth performance were recorded in the LipG1 feeding group, to levels comparable to the group of fish fed with soybean oil diet. Collectively, LipG1 exhibited good potential as an aquafeed additive enzyme, and deserves further characterization as the representative of a novel group of lipases.

INTRODUCTION

China is the largest aquaculture producer in the world, with a total production of 53.94 million tons of fish in 2012, accounting for nearly 60% of the world total [1]. Consequently, there has been a consistent increase in the demand for aquafeed over time. However, the price of aquafeed is relatively high, which represents one of the main bottlenecks for the development of aquaculture. In aquafeed, dietary lipids provide essential fatty acids and are important source of energy for fish. The optimum dietary lipid levels for tilapia and common carp are as high as 8% and12%, respectively [2]. Fish oil and soybean oil have been used as the lipid source in aquafeed for these fish species. However, due to the high price of both fish oil and soybean oil, as well as the finite nature of fish oil, alternative vegetable oils are prevalent. Among these, palm oil (PO) is the vegetable oil with the highest production volume and lowest price worldwide, and therefore is a very attractive lipid source for reduction of the aquafeed price.

Fatty acids (FA) digestibility in fish decreases with increasing chain length and increases with increasing degree of unsaturation [3, 4, 5]. Also, the digestibility of FA is influenced by dietary lipid source and the level of dietary saturates [6], i.e., high level of saturated fatty acids in the diet may reduce the digestibility of saturated FA, and to a lesser extent, of the monoenes and PUFA [3, 4]. Although research into the use of palm oil in fish diets has shown encouraging results [7–10], supplementation of palm oil in aquafeed may reduce the digestibility of lipid and fatty acids, especially when used at high dietary level and low temperature, due to its high level of saturated fatty acids [3, 4, 8]. The reduction in lipid and fatty acids digestibility could lead to compromised growth and feed utilization efficiency of fish, given a lower level of total lipid in the aquafeed and longer feeding period [4, 8]. The reduction of fatty acids digestibility of high PO diet, especially saturates, was partially due to the poor triacylglycerol (TAG) digestion, as indicated by the increased TAG content in fecal lipids of fish fed high PO diet [3]. Therefore, a potential way to solve the problem is supplementation of exogenous lipase in the diet to improve the lipolysis of TAGs of palm oil.

The studies on the supplementation of exogenous lipase for improvement of lipid digestion and growth performance have rarely been conducted in fish. Samuelsen et al. reported that lipase addition had no effect on growth performance and related parameters of rainbow trout (*Oncorhynchus mykiss*) [11]. Moreover, properties of many available lipases were not suitable as aquafeed additive.

Microbial lipases have been widely utilized due to their availability, substrate specificity, and diversity in catalytic activities [12, 13]. In particular, many novel lipases have been isolated or cloned from *Acinetobacter* spp., showing various beneficial biochemical properties such as stability in organic solvents, broad substrate specificity, cold-adapted ability and stereo-selectivity [14–19]. In this study, a novel *Acinetobacter* lipase gene was cloned from DNA extracted from intestinal sample of common carp, *Cyprinus carpio*, and subsequently expressed in *E. coli*. Biochemical properties of the lipase were characterized, and the application of this lipase as aquafeed additive was evaluated in common carp fed diet with high level of palm oil.

MATERIALS AND METHODS

Ethics statement

All experimental and animal care procedures were approved by Feed Research Institute of Chinese Academy of Agricultural Sciences Animal Care Committee, under the auspices of the China Council for Animal Care (Assurance # 2012-HSX01). MS-222 was used as the anaesthetic. Field study was conducted on private aquaculture pond under the owner's permission.

Strains and media

E. coli DH5α was used for cloning of the PCR product. *E. coli* BL21 was used for expression of the recombinant enzyme. *E. coli* strains were cultured in LB medium (10 g tryptone, 5 g yeast extract, 10 g NaCl per liter) at 37°C.

Cloning of the lipase gene and sequence analysis

By comparing the amino acid sequences of *Acinetobacter* lipases, two highly conserved regions were identified. The degenerate primers were designed based on the sequences of the two regions by CODEHAP primer [20]. The partial *Acinetobacter* lipase gene sequences between the two conserved motifs were amplified from DNA extracted from intestinal sample of common carp by touchdown PCR. The amplified PCR product was cloned into *E. coli* DH5α using pGEM-T Easy (Promega, Madison, USA). Fifty transformants were randomly picked. Plasmids were extracted with the

TIANprep Mini Plasmid Kit (TIANGEN, Beijing, China), and the cloned partial lipase gene sequences were sequenced at Beijing Calculation Centre. One sequence that showed relatively high abundance among the sequenced clones was selected for further study. To obtain the upstream and downstream of the partial sequence of the lipase gene, genome walking PCR was performed with a genome walking kit according to manufacturer's instructions (Takara, Japan). The primers were designed based on the partial sequence (Table 1). The elongated fragment was sequenced at Beijing Calculation Centre. Sequence assembly was performed using Vector NTI Suite 7.0 (InforMax, Gaithersburg, MD, USA). ORF was found using the alternative initiation codon finder in OR finder (http://www.ncbi.nlm.nih.gov/projects/gorf/). The amino acid sequence was analyzed by SignalP 3.0 server [21] to identify the signal peptide. Alignments of the DNA and protein sequences were carried out using BLASTn and BLASTp, respectively (http://www.ncbi.nlm.nih.gov/BLAST/). Phylogenetic tree was generated using the neighbor-joining method by MEGA 5.0 [22].

Table 1. Primers used for gene cloning and expression.

	Sequences	bp
Acinet alkaline F	5'-GGCACATGGCACAACAGGNGTNGSNGA-3'	27
Acinet alkaline R	5'- GAGCTGCATGTCCGCCYTGNSWNTG-3'	25
AG1 usp1	5'- GGTCATCCAGCGATTCGACACTT-3'	23
AG1 usp2	5'- CTGCCACTACGGCGTCAGTAATTG-3'	24
AG1 usp3	5'- TTAATTGACTGCGACTTGGTGCAC-3'	24
AG1 dsp1	5'- TTGTGCACCAAGTCGCAGTCAAT-3'	23
AG1 dsp2	5'- TTAACTGCGGGCTATGTGGTGGG-3'	23
AG1 dsp3	5'- AGTGTCGAATCGCTGGATGACC-3'	22
AG1 dsp2	5'- TTAACTGCGGGCTATGTGGTGGG-3'	23
AG1 dsp3	5'- AGTGTCGAATCGCTGGATGACC-3'	22
lipG1 F	5'-TAGGAATCCCATGATGATGATCATGATTCTTATC-3'	34
lipG1 R	5'-ATTTGCGGCCGCTTGAATTGGCATTAATGTTTGTAC-3'	36
338-GC-f	5'-CGCCCGCCGCGCGCGGCGGGCGGGGCGGGGGCACGGGGGGACTCCTACGGGAGGCAGCAG-3'	60
519 r	5'-ATTACCGCGGCTGCTGG-3'	17

Expression of the recombinant lipase in *E. coli*

The following primers were used for construction of the lipase expression plasmid: lipG1F (with *Eco*RI site) and lipG1 R (with *Not*I site). The PCR products were inserted into pET-28a (+) digested with *Eco*RI and *Not* I. The recombinant plasmid, pET28a—lipG1, was sequenced to confirm the insertion, followed by transformation into *E. coli* BL21 (DE3) cells. Transformed cells were picked from a single colony and grown overnight at 37°C in shake flask containing 3 ml LB broth supplemented with 100 µg/ml kanamycin, followed by inoculation at a dilution of 1:100 in 100 ml fresh LB medium containing kanamycin and aerobic incubation at 37°C. When OD_{600} reached 0.6, isopropyl -β-D-1-thiogalactopyranoside (IPTG) was added to the growth medium at a final concentration of 0.5 mM. After further incubation overnight at 16°C, the cells were harvested by centrifugation at 8,000×g for 10 min at 4°C.

Purification and identification of the recombinant LipG1

The lipase was purified with protocol described in detail previously [19]. The bacterial pellets were resuspended in 15 ml lysis buffer (50 mM Tris-Cl pH 8.0), disrupted by sonication, followed by centrifugation at 10,000 × g for 30 min at 4°C. The concentrated supernatant (crude enzyme) was applied to the Ni-NTA affinity chromatography column (GE Healthcare Life Sciences, Piscataway, USA) which was equilibrated with washing buffer (50mMTris-ClpH8.0, 500 mM NaCl, and 10% glycerol). The enzyme protein was eluted using an imidazole gradient (0, 20, 40, 80, 100, and 200 mM) in washing buffer. After the elution, the imidazoles in the purified enzyme were removed by dialysis in 50 Mm Tris-Cl (pH 8.0). The purified enzyme was analyzed by SDS-PAGE. The single protein band after purification was confirmed by the peptide mass fingerprinting.

Lipase activity assay

Lipase activity was determined using the chromometer method described by Zhang et al. [19]. Unless otherwise described, lipase activity was measured at 40°C with 500 μM *p*NP decanoate in 50 mM Tris-HCl buffer (pH 9.0) containing 1% ethanol. The reaction can begin by the addition of 100 μl of purified enzyme at a concentration of 0.25 mg/ml. The reaction mixture was incubated at 40°C for 10 min, after which the reaction was terminated through addition of 1ml ethanol. Blank reactions were performed in reaction mixtures without the enzyme. The mixtures were centrifuged, and the absorbance of the supernatants was measured at 405 nm with spectrophotometer (Beckman, USA). One unit (U) of enzyme activity was defined as the amount of enzyme that releases 1 μ mol of *p*NP per minute.

Enzyme characterization of recombinant LipG1

The *p*NP decanoate was used to assay the following characteristics of LipG1. The optimal pH was determined at 40°C using buffers with pH from 3.0 to 12.0, i.e., 2 0 mM disodium hydrogen phosphate-citric acid buffer (pH 3.0–8.0), 50 mM Tris—HCl buffer (pH 8.0–10.0), and 20 mM glycine-NaOH (pH 10.0–12.0). To test its pH stability, 100 μl of the lipase (2.5 mg/ml) was preincubated for 1 h at 40°C in 900 μl buffers with pH values ranging from 3.0 to 12.0, and then lipase activity was measured under standard conditions. The enzyme's optimal temperature was determined by measuring its activity at temperatures ranging from 0°C to 60°C, with pH set as 8.0. The thermostability was monitored by preincubation of the enzyme in Tris-HCl buffer (pH 8.0) for 30, 60, 90, and 120 min at 20, 30, 40, and 50°C. Lipase activity was then measured under standard conditions. The effects of different metal ions, chemical reagents, detergents, and organic solvents on enzymatic activity were assessed in 50 mM Tri-HCl buffer (pH 8.0) at 40°C. The reactions contained 1 or 10 mM of NaCl, KCl, $BaCl_2$, $CaCl_2$, $CoCl_2$, $CuCl_2$, $NiCl_2$, $MgCl_2$, $MnCl_2$, $ZnCl_2$, $CdCl_2$; 0.1 or1% Tween-20, Tween-40, Tween-80, Triton X-100, SDS, and cetyltrime-thylammonium bromide (CTAB); 10% or30% methanol, ethanol, isopropanol, capryl alcohol, n-heptane, glycerol and dimethyl sulphoxide (DMSO).

In order to examine the resistance to different proteases, the purified recombinant enzyme (100 μg ml^{-1}) was incubated with 10 μg ml^{-1} trypsin (from bovine, pH 7.6, 25°C, 14,700 U mg^{-1}; Sigma), 250 μg ml^{-1} α-chymotrypsin (type II from bovine, pH 7.8, 25°C, ≥40 U mg^{-1}; Sigma), 500 μg ml $^{-1}$subtilisin A (type VIII from *Bacillus licheniformis*, pH 7.5, 37°C, 10 U mg^{-1}; Sigma), 330 μg ml $^{-1}$proteinase K (type VIII from *B. licheniformis*, pH 7.5, 37°C, 30 U mg $^{-1}$; Amresco, Solon, USA), respectively. After incubation for 1 or 2 h, the residual activity was measured under standard assay conditions. For the control sample, the recombinant enzyme was incubated under the same conditions without protease.

The substrate selectivity of LipG1 for *p*NP esters was assayed using *p*NP acetate (C2), *p*NP butyrate (C4), *p*NP octanoate (C8), *p*NP decanoate (C10), *p*NP laurate (C12), *p*NP myristate (C14), and *p*NP palmitate(C16) as substrates in 50 mM Tris-HCl (pH 8.0) at 40°C. The *p*NP ester substrates except *p*NP palmitate were dissolved in ethanol at a final concentration of 50 mM as the stock solution, and *p*NP palmitate was dissolved in isopropanol. The working concentration range of the *p*NP esters was from 0.01 to 2 mM. Initial velocity versus substrate concentration data were fitted to the Lineweaver—Burk transformation of the Michaelis—Menten equation.

Experimental diets and feeding experiment

Five isonitrogenous and isocaloric experimental diets were formulated containing approximately 34% crude protein, 9% crude lipid (S1 Table).Diet with palm oil as the lipid source was used as negative control (CKn), while soybean oil diet was positive control group (CKp). Palm oil diet were supplemented with 3 U g^{-1} LipG1, 6 U g^{-1} LipG1 and 6 U g^{-1}commercial lipase (Leaveking, Shengzhen, China), giving treatment groups T1, T2 and

M, respectively. The feed pellets were dried under forced air at room temperature for 24 h and then kept at 4°C.

Common carp (*Cyprinus carpio*) were obtained from Tianjing freshwater fish husbandry factory (Tianjing, China). Before starting the experiment, the juvenile common carp were reared in an experimental thermo regulated rearing system and fed with the control diet for 2 weeks to acclimatize to the experimental conditions. Fish of similar sizes (5.66 ± 0.21 g) were randomly distributed into 15 circular glass tanks (30 l) and each glass tank was stocked with 12 fish. Each diet was then randomly assigned to triplicate tanks. The fish were hand-fed to apparent satiation three times daily (at 8:30, 12:30 and 16:30) for 28 days. During the experimental period, the water temperature, dissolved O_2, pH and ammonia content were maintained at 28.0 ± 1.3°C, 7.60 ± 0.28 mg l^{-1}, 7.5 ± 0.29 and 0.10 ± 0.02 mg l^{-1}, respectively.

Sampling and measurements

At the end of the feeding experiment, all of the fish were anesthetized with MS-222. Body weight was measured at batch. The hepatopancreas and gut were rapidly excised from three fish of each tank. The samples of fish from the same tank were pooled, frozen in liquid nitrogen and stored at −80°C. Weight gain (WG), feed conversion ratio (FCR) and survival rate (SR) were calculated.

Lipid-metabolism-related enzymes activities

The hepatopancreas and gut samples were homogenized in 4 volumes of ice-cold buffer (20 mM Tris-HCl, 0.25 M sucrose, 2 mM EDTA, pH 7.4) and centrifuged at 18,000 × g for 10 min at 4°C. Lipase and protease activities were measured in the homogenate with commercial assay kits (Jiancheng Biotech. Co., Nanjing, China).

Adhesive intestinal bacterial community

The DNA was extracted from pooled gut samples in each treatment using the method according to He et al. [23]. The PCR-DGGE method was according to Zhou et al. [24]. The V3 region of the 16S rRNA gene was amplified with the primers 338-GC-f and 519r. DGGE was performed using a D-Code universal Mutation System (BioRad). Purified PCR products were loaded onto denaturing gradients ranging from 40 to 60%, and then electrophoresis was conducted with a constant voltage of 65 V for 15 h in 1×TAE buffer at 60°C. The gels were stained in an ethidium bromide solution (0.5μg/mL in TAE buffer) and destained in distilled water for 20 min, respectively, and viewed by UV transillumination. The excised bands were reamplified and purified, then sequenced.

Statistics

The effects of different diets in the feeding experiment were analyzed with one-way ANOVA, and Duncan's multiple range test was used to compare the means between any two groups. Differences with a *P* value lower than 0.05 were considered as significant. All the statistical analysis was conducted on SPSS 17.0.

NCBI accession numbers

The nucleotide sequence of the chitinase gene (*lipG1*) was deposited in GenBank under the accession number KM925083.

RESULTS

Sequence analysis of LipG1

Bioinfomatics analysis showed that the *lipG1* gene (KM925083) contained an open reading frame with 1,227 nucleotides, encoding a protein of 406 amino acids (Fig A in S2 File). A 20 amino acid residue signal peptide was found at the N-terminal of the lipase with a cleavage position between Cys20 and His21 as predicted by Signal P (Fig A in S2 File). BLASTp analysis indicated that LipG1 was 99% identical to hypothetical protein from *Acinetobacter tandoii* (WP_016167430.1), and was 70%, 68%, 68%, 62% and 61% identical to lipases from *A. baumannii* 348935 (WP_034704719.1), *Acinetobacter* sp. MII (KGH48978.1), *A. lwoffii* SH145(WP_004279025), *A. gyllenbergii* (WP_0328 60322.1) and *A. junii* (WP_005402474.1), respectively. Alignment of *Lip*G1 and its relatives with representatives of subfamily I.1 and I.2 of bacterial true lipases revealed the typical lipase semi-conserved pentapeptide, GXSXG, where the catalytic residue Ser is located, as well as the catalytic residue Asp, at homologous positions in all sequences (Fig 1). The last catalytic residue constituting the lipase consensus catalytic triad, His, was not found at homologous position in LipG1. Nevertheless, a conserved His residue was located in LipG1 and its relatives at the position close to the catalytic His for I.1 and I.2 lipases (Fig 1). Therefore, the consensus lipase catalytic triad in LipG1putatively comprises of Ser 190 (in the motif

GHSQG), Asp 354 and His 386, which requires further experimental confirmation. An HG sequence, which putatively constitutes an oxyanion hole in the three-dimensional protein structure, was also located at homologous position for all sequences (Fig 1).

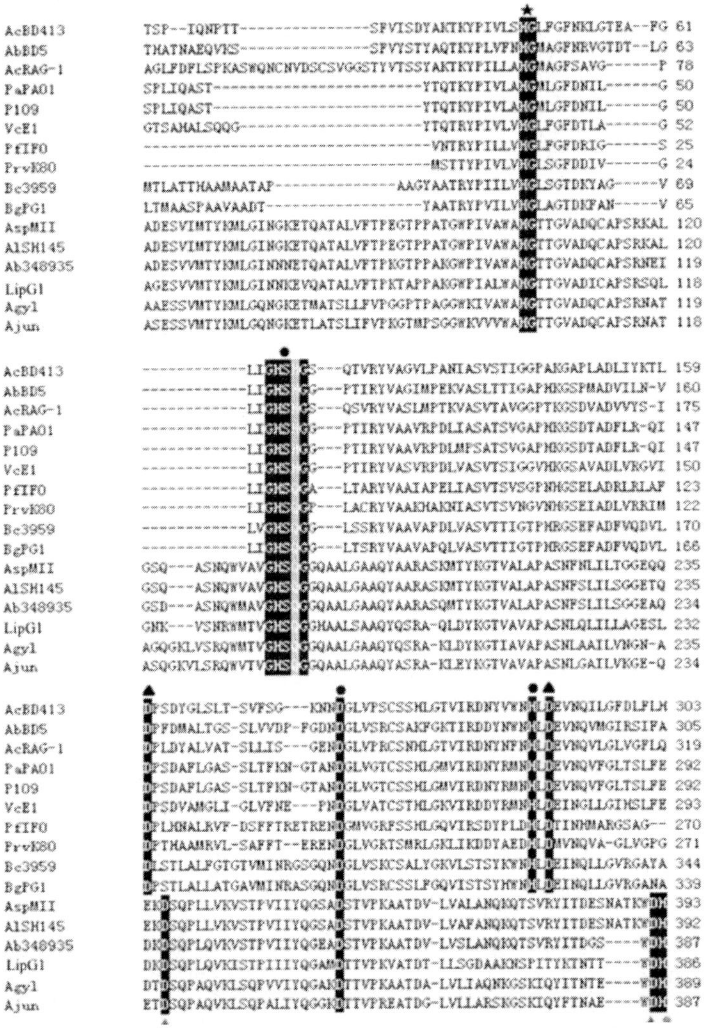

Figure 1. Alignment of amino acid sequences of representative lipases from subfamilies I.1 and I.2 of bacterial lipases and the group of lipases represented by LipG1. Abbreviations and sequence accession numbers: AcBD413 (*Acinetobacter calcoaceticus* BD413), CAA56780.1; AbBD5 (*A. baumannii* BD5), ABW70205.1; AcRAG-1 (*A. calcoaceticus* RAG-1), AAD29441.1; PaPAO1 (*Ps. aeruginosa* PAO1), CAA44997.1; P109 (*Pseudomonas* sp. 109), P26877.1; VcE1 (*V. cholerae* E1), P15493.2; PfIFO

(*P. fragi* IFO-12049), P08658.2; PrvK80 (*Proteus vulgaris* K80), AAB01071.1; Bc3959 (*Burkholderia cepacia* DSM3959), P22088.2; BgPG1 (*Burkholderia glumae* PG1), Q05489.1; AspMII (*Acinetobacter* sp. MII), KGH48978; AlSH145 (*A. lwoffii* SH145), WP_004279025; Ab348935 (*A. baumannii* 348935), WP_034704719.1; LipG1, this study; Agyl (*A. gyllenbergii*), WP_032860322.1; Ajun (*A. junii*), WP_005402474.1. Symbols: ● Catalytic triad residues, ▲ Asp residues involved in calcium binding, ★ HG dipeptide of the oxyanion loop, ○ Putative triad residue, △Putative calcium binding Asp residues.

Enzyme expression, purification, and mass spectrometry analysis

The recombinant LipG1 was expressed in *E. coli* BL21 in the presence of 0.01mM IPTG at 20°C. After Ni-NTA affinity chromatography, the enzymatic activity of recombinant LipG1 was 8908 U mg^{-1} toward C8 *p*NP. The purified enzyme presented as a single band of about 40kDa on SDS-PAGE (Fig B in S2 File), which matched the expected molecular weight. The band was excised from the SDS-PAGE, digested with trypsin, and analyzed by LC—ESI—MS/MS for peptide finger printing. The amino acid sequences obtained from the mass peaks were compared with LipG1. The peptides WMTVGHSQGGHAALSAA, TFTALITAGLRNPNPSLQYSQVFK and TQSNFMTLSPVKKFLDTDSQPLQVKISTPIIIYQ completely matched the sequences of corresponding LipG1 fragments, confirming that the purified protein was recombinant LipG1 (Fig C in S2 File).

Effects of pH and temperature on LipG1 activity

LipG1showed maximum activity at pH 8.0, and retained at least 70% of its maximum activity between pH 7.0 and 8.5 (Fig 2a). LipG1 was stable over a wide pH range, remaining at least 60% of its maximum activity after incubation at pH from 3.0 to 12.0 (Fig 2c). The enzyme activity of LipG1 peaked at 40°C, and retained over 40% of its maximum activity at temperatures from 20°C to 35°C (Fig 2b). The recombinant lipase showed high stability at 20°C and 30°C. However, the activity decreased quickly when incubated at 50°C (Fig 2d).

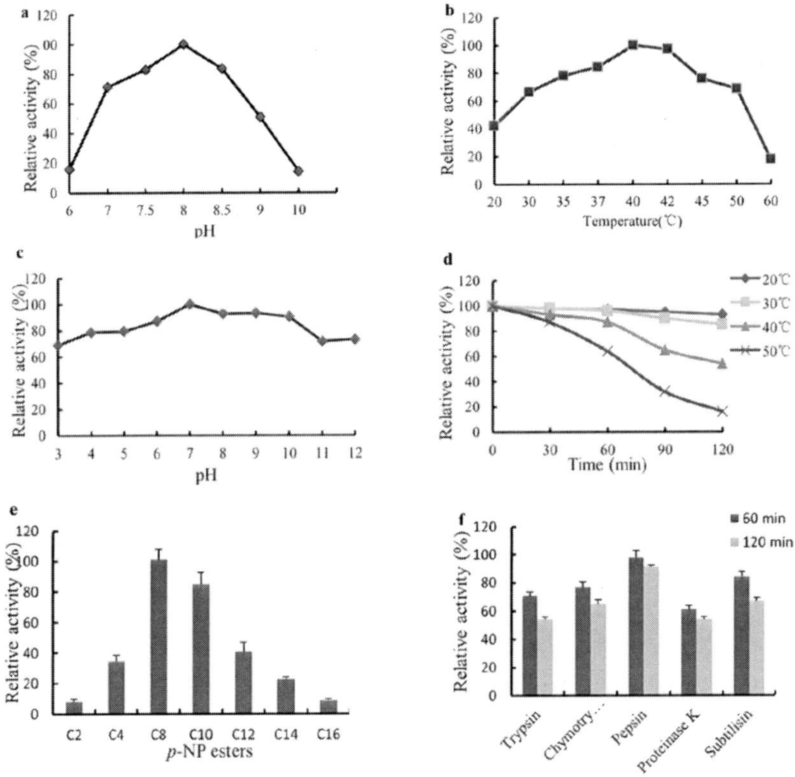

Figure 2. Characterization of purified recombinant LipG1. a. The effect of pH on lipase activity. The activity assay was performed at 37°C in buffers of pH 3.0–12.0 for 10 min. b. The effect of temperature on lipase activity measured in 0.1 M PBS buffer (pH 8.0) for 10 min. c. pH stability of LipG1. After pre-incubating the enzyme at 37°C for 1 h at various pHs, the residual activity was measured in 0.1 M PBS buffer (pH 8.0), 40°C. d. Thermostability of purified LipG1. The enzyme was preincubated at 20, 30, 40, or 50°C in 0.1 M PBS buffer (pH 8.0). Aliquots were removed at specific time points for measurement of residual activity in the same buffer at 40°C. e. Substrate specificity of LipG1 determined with various chain length fatty acid esters. f. Effects of proteases on the activity of LipG1. Each value in the panel represents the mean ± SD (n = 3).

Effects of metal ions, organic solvents, detergents and proteases on LipG1 activity

The effects of different metal ions, organic solvents, and detergents on the activity of LipG1 are shown in Table 2, Tables A and B in S1 File, respectively. At the concentration of 1mM, lipase activity was strongly inhibited byCo^{2+}, Cr^{3+}, Fe^{3+},Pb^{+}, Ag^{+} and EDTA, and enhanced by Ca^{2+}, Mg^{2+}. At the higher concentration of 10 mM, Cu^{2+}, EDTA,Co^{2+}, Zn^{2+} and Mn^{2+}drastically inhibited its activity; Na^{+}, K^{+}, Cr^{3+}, Ni^{2+}, Pb^{+} and Ag^{+} strongly inhibited the activity. Nevertheless, a stimulating effect was still observed for Ca^{2+} and Mg^{2+} (Table 2). The enzyme was stable in various organic solvents at the concentration of 10%, with extended stability at the concentration of 30% for methanol, ethanol, n-heptane, glycerol and DMSO (>75% activity remained). However, isopropanol and capryl alcohol drastically reduced its activity at the concentration of 30% (Table A in S1 File). At the concentration of 0.1%, Tween 80 and CTAB didn't significantly inhibit the activity of LipG1, and the non-ionic detergent Triton X-100 slightly enhanced its activity. Tween 20, Tween 40, SDS reduced its activity by 37.5%, 28.3%, 57.7%, respectively. At the concentration of 1%, all the detergents except Triton X-100 drastically inhibited the lipase activity, with over 50% activity loss observed (Table B in S1 File).

Table 2. Effects of various metal ions on LipG1 activity.

Metal ion	Relative activity (%)	
	1 mM	10 mM
Na^{+}	101.6±2.3	88.7±1.7
K^{+}	107.4±1.9	81.6±2.3
Ca^{2+}	115.5±±4.4	213.9±3.5
Zn^{2+}	97.5±±2.1	37.7±2.1
Li^{+}	104.2±2.5	97.5±2.5
Co^{2+}	50.6±1.2	30.3±0.9
Cr^{3+}	79.6±2.2	70.4±1.7
Ni^{2+}	81.1±3.1	49.2±1.2
Cu^{2+}	84.9±±2.2	0
Mg^{2+}	119.4±2.7	171.0±1.4
Fe^{3+}	66.9±1.8	18.2±0.6
Mn^{2+}	93.7±2.0	15.1±0.9
Pb^{+}	76.6±1.5	56.8±1.3
Ag^{+}	60.6±1.2	43.4±1.5
EDTA	75.5±1.3	11.1±0.4

We also investigated the resistance of recombinant LipG1against proteases (Fig 2f). The enzyme retained over 60% activity after treatment with all proteases at 37°C for 1 h, with over 80% activity retained for subtilisin and pepsin. Also, longer incubation time of 120 min didn't lead to considerable further activity loss for all the tested proteases.

Substrate specificity of LipG1

The substrate specificity of LipG1 was examined using *p*NP esters with different chain lengths. Results showed that the lipase showed activity on a wide range of substrates. LipG1 preferred medium chain fatty acid substrates, and the highest activity was registered for *p*NP octanoate (Fig 2e, S2 Table).

Effect of LipG1 on growth and feed utilization of common carp

Dietary supplementation of LipG1 at 6 U/g significantly increased the final body weight (FBW) and weight gain (WG) of common carp compared with negative control after feeding for 4 weeks. LipG1 at 3 U/g and commercial lipase at 6 U/g marginally increased the growth of carp compared with the negative control. Overall, all the lipase supplementation groups showed similar growth performance ($P > 0.05$), which is comparable to the group with soybean oil (Table 3). Also, significantly lower FCR (feed conversion ratio) was observed in 6 U/g LipG1 group compared with the negative control ($P < 0.05$), and marginally decreased FCR was recorded in 3 U/g LipG1 and the commercial lipase groups, with no significant difference among the lipase addition groups, which were comparable to the soybean oil group(Table 3).

Table 3. Effects of dietary lipase on weight gain, feed conversion ratio and survival rate of common carp (*Cyprinus carpio*) (n = 3).

	CKn	T1	T2	M	CKp
IBW (g)	2.09 ± 0.04^{a}	2.14 ± 0.02^{a}	2.16 ± 0.01^{a}	2.13 ± 0.01^{a}	2.14 ± 0.02^{a}
FBW (g)	4.12 ± 0.12^{a}	4.4 ± 0.05^{ab}	4.58 ± 0.12^{b}	4.44 ± 0.08^{ab}	4.45 ± 0.06^{ab}
WG (%)	96.7 ± 2.17^{a}	109.18 ± 3.90^{ab}	112.32 ± 4.67^{b}	108.4 ± 3.58^{ab}	108.25 ± 1.37^{ab}
FCR	2.03 ± 0.05^{a}	1.80 ± 0.06^{ab}	1.75 ± 0.07^{b}	1.81 ± 0.06^{ab}	1.81 ± 0.02^{ab}
SR (%)	100	100	100	100	100

Values in the same row with different superscript letters (a, b, ab) have significant difference ($P<0.05$). FBW, final body weight (g); Weight gain (%), WG (%) = 100 × (Final body weight–Initial body weight)/Initial body weight; Feed conversion ratio, FCR = feed fed (g)/weight gained (g).

Effect of LipG1 on digestive enzyme activities of common carp

Heptaopancreas trypsin activity was not significantly different among groups, while the gut trypsin activity was similar among the palm oil groups regardless of lipase supplementation, and the activity in soybean oil group was significantly higher (Table 4). Dietary exogenous lipases (T1, T2 and M) significantly increased gut lipase activity compared with the negative control ($P < 0.05$), with an activity comparable to the positive control. The heptaopancreas lipase activity was significantly increased in groups with lipase supplementation at 6 U/g (T2 and M) (Table 4). For both gut and heptaopancreas lipase activity, the maximal value was observed in the 6 U/g LipG1 group.

Table 4. Trypsin and lipase activities in the gut and heptaopancreas of common carp (*Cyprinus carpio*) fed with different diets (n = 3).

Treatments	CKn	T1	T2	M	CKp
Trypsin activity (U/mg)					
Heptaopancreas	127.7 ± 2.4^{a}	$125.3 \pm ,3.8^{a}$	128.7 ± 4.3^{a}	125.3 ± 2.3^{a}	134.3 ± 1.5^{a}
Gut	77.0 ± 1.5^{a}	78.0 ± 2.3^{a}	79.7 ± 3.0^{a}	79.3 ± 2.4^{a}	86.7 ± 1.5^{b}
Lipase activity (U/g)					
Heptaopancreas	62.0 ± 0.6^{a}	65.7 ± 1.2^{ab}	68.7 ± 2.9^{b}	68.7 ± 2.0^{b}	67.3 ± 2.0^{ab}
Gut	41.3 ± 2.0^{a}	49.7 ± 1.8^{b}	58.3 ± 1.5^{c}	54.0 ± 1.7^{bc}	54.0 ± 1.7^{bc}

Values in the same row with common superscript (a, b, ab, c, bc) are not significantly different ($P > 0.05$).

Effect of LipG1 on common carp gut microbiota

The 16S rRNA gene V3 region PCR-DGGE fingerprints of the adhesive intestinal microbiota of different groups are shown in S1 Fig Most of the OTUs of the intestinal microbiota of common carp were assigned to the *Proteobacteria* (17 OTU) and *Firmicutes* (4 OTU) (S3 Table). The gut microbiota was affected by dietary lipid sources, as the microbiota of palm oil diet associated groups were clearly distinguishable from the microbiota of the soybean oil group (Fig 3). Lipase supplementation didn't exert significant change to the intestinal microbiota, and the intestinal microbial profiles of all the lipase supplement groups were similar (Fig 3).

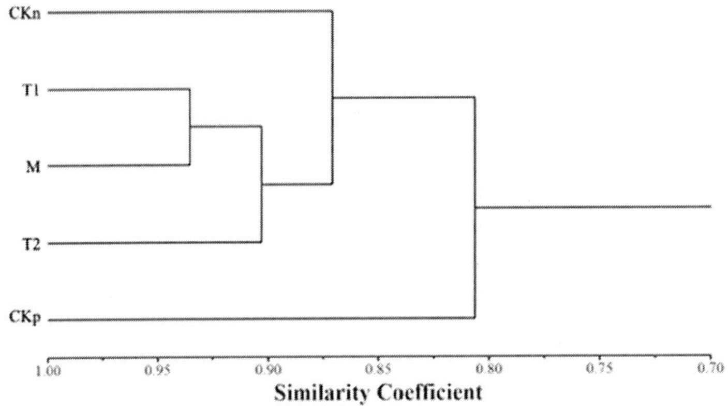

Figure 3. Cluster analysis of the adhesive gut bacterial communities of common carp (*Cyprinus carpio*).

DISCUSSION

Bacterial true lipases comprise the family I of bacterial lipolyitc enzymes, and were further classified into six subfamilies [25]. True lipases from Gram-negative bacteria were assigned into the first three subfamilies. Subfamily I.1 were lipases from *Pseudomonas aeruginosa, Vibrio cholerae, Acinetobacter calcoaceticus, etc*. Subfamily I.2 mainly comprises lipases produced by *Burkholderia*, while I.3 contains enzymes from two distinct species: *Ps. fluorescens* and *Serratia marcescens* [25]. Many lipase genes have been cloned from *Acinetobacter* [14, 15, 17, 18, 19]. The *Acinetobacter* lipases were assigned to subfamily I.1 of bacterial true lipases [25], and were proposed as a separate '*Acinetobacter*' clade based on similarity of amino acid sequences [15, 26]. However, amino acid sequence alignment showed that LipG1 isn't homologous to any of the above mentioned *Acinetobacter* lipases and doesn't belong to the typical '*Acinotebacter*' lipase clade. Based on the phylogenetic tree predicted from multiple sequence alignment of the lipases (Fig 4), and considering the deviation in the position of the catalytic residue His (Fig 1), we propose that LipG1 and its relatives comprise another subfamily of true lipases from Gram-negative bacteria. The lipases homologous with LipG1 in the BLASTp list all originated from genomic sequence analysis rather than experimental characterization. Also, around half of the hits with over 60% identity in the BLASTp list were designated as hypothetical proteins from *Acinetobacter* strains. Therefore, LipG1 represents a novel group of lipases that deserves further characterization.

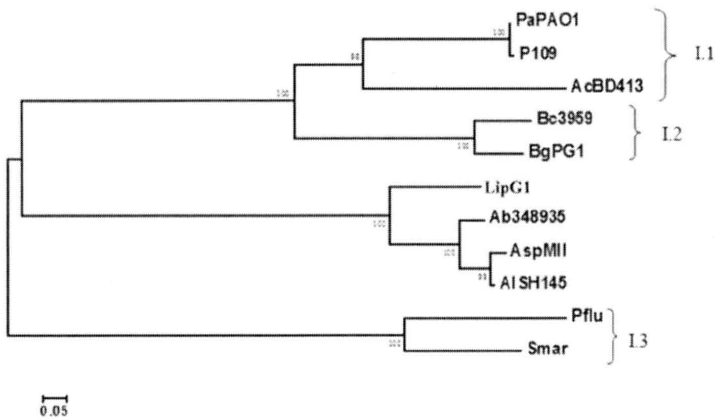

Figure 4. Phylogenetic tree of true lipases produced by Gram-negative bacteria, including LipG1 and its relatives. Abbreviations and sequence accession numbers: PaPAO1 (*P. aeruginosa* PAO1), CAA44997.1; P109 (*Pseudomonas* sp. 109), P26877.1; AcBD413 (*Acinetobacter calcoaceticus* BD413), CAA56780.1; Bc3959 (*Burkholderia cepacia* DSM3959), P22088.2; BgPG1 (*Burkholderia glumae* PG1), Q05489.1; LipG1, this study; Ab348935 (*A. baumannii* 348935), WP_034704719.1; AspMII (*Acinetobacter* sp. MII), KGH48978; AlSH145 (*A. lwoffii* SH145), WP_004279025; Pflu (*P. fluorescens* SIK W1), BAA02012.1; *Smar* (*Serratia marcescens*), BAA02519.1.

LipG1 is most active at 40°C, and retained over 40% of its optimal activity at temperatures from 20°C to 35°C, the temperature range for warm-water aquaculture. Activity of LipG1 peaked at pH 8.0, and showed strong activity in a narrow alkaline pH range, which is similar with other *Acinetobacter* lipases [14, 16, 17]. Remarkably, LipG1 exhibited stability over a wide range of pH from 3.0–12.0. In contrast, incubation of many lipases at acidic pH lower than 5.6 resulted in inactivation [16, 26]. The stability of LipG1 at acidic pH is a useful characteristic as feed additive, as the enzyme has to pass through the acidic gastric environment of the host. In line with other lipases from *Acinetobacter*, the activity of LipG1 was activated by Ca^{2+} and Mg^{2+}, while inhibited by Zn^{2+} and Cu^{2+} [16, 18, 19, 27]. Particularly, the positive effect of Ca^{2+}on enzyme stabilization and activity was common in *Acinetobacter* lipases, most probably due to a function of the Ca^{2+}-binding pocket, leading to correct active-site configuration [26]. The Ca^{2+}-binding Asp residues were conserved in subfamily I.1 and I.2 lipases [15, 25]. Alignment didn't show putative Ca^{2+}-binding Asp residues at the conserved positions in LipG1 (Fig 1). However, in both regions, an

Asp residue was located in the close position, which probably constitutes the Ca^{2+}-binding pocket (Fig 1). Another notable characteristic of LipG1 is the strong resistance to proteases, which is rarely reported for *Acinetobacter* lipases [28]. Collectively, the ion, protease resistance and strong pH stability of LipG1 supported it as a candidate aquafeed additive enzyme. Moreover, the stability of lipase in the presence of surfactants and organic solvent suggests its potential application in industry.

The overall lipid digestibility may be reduced by high saturates diets [3], such as the diet with palm oil as the sole lipid source. The lowered lipid digestibility may compromise the growth performance of fish, as indicated in the data of our study, where the weight gain and feed efficiency were both marginally reduced in the group with palm oil as the lipid source, compared with the soybean oil group. The feeding experiment data in this study showed that supplementation of exogenous lipase efficiently improved the heptaopancreas and gut lipase activity of common carp. The increased lipolytic activity led to improved feed efficiency and growth performance of the palm oil group of fish, which were comparable to the group with soybean oil as the lipid source. Moreover, lipase supplementation didn't exert significant change to the intestinal microbiota, indicating that it doesn't have negative effects on the intestinal health status as reflected by the microbiota. Although several *Acinetobacter* lipase genes were cloned from the intestinal DNA sample of common carp, *Acinetobacter* sp. was recovered only in T1 group in the DGGE fingerprint, suggesting the variation of its occurrence in the microbiota of common carp. Also, the phenomenon may be attributed to the limitation of DGGE in describing the microbiota of ecological samples, especially for low abundance species. Study on the effects of exogenous lipase on lipid digestion and growth is rare in aquatic animals. Koven reported that the supplementation of lipase increased the incorporation of oleic acid in tissues of gilthead seabream larvae [29]. In adult rainbow trout, no effect on growth performance and related parameters were registered with lipase addition [11]. In both research, traditional lipid source was used, and the issue of high dietary palm oil supplementation was not addressed. To our knowledge, this is the first study investigating the effect of supplementation of exogenous lipase on the lipid digestibility and performance of fish fed diet with high level of palm oil. The working temperature range, stability at broad pH range, and resistance to metal ions and proteases of LipG1 make it a suitable exogenous lipase as aquafeed additive. Study is underway to further improve the lipolytic efficiency of LipG1 on palm oil.

In this study, the lipase LipG1 exhibited efficient lipolytic activity at environmental factors generally encountered in the warm water aquaculture, and resistance against acidic pH, metal ions and proteases. Amino acid sequence analysis showed that LipG1 and its relatives represent a novel group of uncharacterized lipases. Supplementation of LipG1significantly improved the gut lipase activity, feed conversion ratio, and growth performance of common carp fed diet with high amount of palm oil, to a level comparable to the group of fish fed with soybean-oil-based diet. Collectively, these results paved the way for further application of LipG1 as aquafeed additive enzyme.

ACKNOWLEDGMENTS

The authors would like to thank Yang Deng for the fish work.

FUNDING: This work was supported by the National Science and Technology Support Program Project of China (2012BAD25B02 and 2013BAD10B01) and Beijing earmarked fund for Modern Agro-industry Technology Research System (SCGWZJ 20121104-4). Z.Z. was the author who received the funding. The URL of funder's websites is not available. The funders had no role in study design, data collection and analysis, decision to publish, or preparation of the manuscript.

REFERENCES

1. FAO. Fishery statistical collections. 2014. Rome Retrieved from: http://www.fao.org/fishery/statistics/global-aquaculture-production/en
2. Tocher DR. Fatty acid requirements in ontogeny of marine and freshwater fish. Aquacult Res. 2010; 41: 717–732
3. Ng WK, Campbell PJ, Dick JR, Bell JG. Interactive effects of dietary palm oil concentration and water temperature on lipid digestibility in rainbow trout, *Oncorhynchus mykiss*. Lipids. 2003; 38: 1031–1038. pmid:14669967
4. Ng WK, Sigholt T, Bell JG. The influence of environmental temperature on the apparent nutrient and fatty acid digestibility in Atlantic salmon (*Salmosalar* L.) fed finishing diets containing different blends of fish oil, rapeseed oil and palm oil. Aquacult Res. 2004; 35:1228–1237
5. Johnsen RI, Grahl-Nielsen O, Roem A. Relative absorption of fatty acids by Atlantic salmon *Salmosalar* from different diets, as evaluated by multivariate statistics. Aquacult Nutr. 2000; 6: 255–261
6. Caballero MJ, Obach A, Rosenlund G, Montero D, Gisvold M, Izquierdo MS. Impact of different dietary lipid sources on growth, lipid digestibility, tissue fatty acid composition and histology of rainbow trout, *Oncorhynchus mykiss*. Aquaculture. 2002; 214: 253–271
7. Ng WK, Tee MC, Boey PL. Evaluation of crude palm oil and refined palm olein as dietary lipids in pelleted feeds for a tropical bagrid catfish *Mystusne murus* (Cuvier and Valenciennes). Aquacult Res. 2000; 31: 337–347
8. Torstensen BE, Lie O, Froyland L. Lipid metabolism and tissue composition in Atlantic salmon (*Salmosalar* L.)–Effects of capelin oil, palm oil, and oleic acid-enriched sunflower oil as dietary lipid sources. Lipids. 2000; 35: 653–664. pmid:10901428

9. Varghese S, Oommen OV. Long-term feeding of dietary oils alters lipid metabolism, lipid peroxidation, and antioxidant enzyme activities in a teleost (*Anabastestudineus* Bloch). Lipids. 2000; 35: 757–762. pmid:10941876
10. Bell JG, Henderson RJ, Tocher DR, McGhee F, Dick JR, Porter A, et al. Substituting fish oil with crude palm oil in the diet of Atlantic salmon (*Salmosalar*) affects muscle fatty acid composition and hepatic fatty acid metabolism. J Nutr. 2002; 132: 222–230. pmid:11823582
11. Samuelsen T, Isaksen M, McLean E. Influence of dietary recombinant microbial lipase on performance and quality characteristics of rainbow trout, *Oncorhynchus mykiss*. Aquaculture. 2001; 194:161–171
12. Jaeger KE, Dijkstra BW, Reetz MT. Bacterial biocatalysts:molecular biology, three-dimensional structures, and biotechnological applications of lipases. Annu Rev Microbiol. 1999; 53:315–351 pmid:10547694
13. Kademi A, Fakhreddine L, Baratti J. Purification and characterization of a thermostable esterase from the moderate thermophile *Bacillus circulans*. Appl Microbiol Biotechnol. 2000; 54:173–179 pmid:10968629
14. Kok RG, van Thor JJ, Nugteren-Roodzant IM, Brouwer MBW, Egmond MR, Nudel CB, et al. Characterization of the extracellular lipase, LipA, of *Acinetobacter calcoaceticus* BD413 and sequence analysis of the cloned structural gene. Mol Microbiol. 1995; 15: 803–818. pmid:7596283
15. Sullivan ER, Leahy JG, Colwell RR. Cloning and sequence analysis of the lipase and lipase chaperone-encoding genes from *Acinetobacter calcoaceticus* RAG-1, and redefinition of a Proteobacterial lipase family and an analogous lipase chaperone family. Gene. 1999; 230: 277–285. pmid:10216267
16. Snellman EA, Sullivan ER, Colwell RR. Purification and properties of the extracellular lipase, LipA, of *Acinetobacter* sp. RAG-1. Eur J Biochem. 2002; 269: 5771–5779. pmid:12444965
17. Han SJ, Back JH, Yoon MY, Shin PK, Cheong CS, Sung MH, et al. Expression and characterization of a novel enantioselective lipase from *Acinetobacter* species SY-01.Biochimie. 2003; 85:501–510. pmid:12763309
18. Park IH, Kim SH, Lee YS, Lee SC, Zhou Y, Kim CM, et al. Gene cloning, purification, and characterization of a cold-adapted lipase produced by *Acinetobacter baumannii* BD5. J Microbiol Biotechnol. 2009; 19:128–135. pmid:19307760
19. Zheng X, Chu X, Zhang W, Wu N, Fan Y. A novel cold-adapted lipase from Acinetobacter sp. XMZ-26: gene cloning and characterisation. Appl Microbiol Biotechnol. 2011; 90:971–980. pmid:21336927
20. Timothy M. CODEHOP-mediated PCR—A powerful technique for the identification and characterization of viral genomes. Virol J. 2005; 2: 1–24. pmid:15769292
21. Nielsen H, Engelbrecht J, Brunak S, von Heijne G. A neural network method for identification of prokaryotic and eukaryotic signal peptides and prediction of their cleavage sites. Int J Neural Syst. 1997; 8:581–599 pmid:10065837
22. Tamura K, Peterson D, Peterson N, Stecher G, Nei M, Kumar S. MEGA5: molecular evolutionary genetics analysis using maximum likelihood, evolutionary distance, and maximum parsimony methods. Mol Biol Evol. 2011; 28(10): 2731–2739. pmid:21546353
23. He S, Zhou Z, Liu Y, Cao Y, Meng K, Shi P, et al. Effects of dietary flavomycin and florfenicol as antibiotic growth promoters on the intestinal autochthonous microbiota of hybrid tilapia (*Oreochromis niloticus*). Arch Microbiol. 2010; 192:985–994 pmid:20844867
24. Zhou Z, He S, Liu Y, Shi P, Huang G, Yao B. The effects of dietary yeast culture or short-chain fructo-oligosaccharides on the intestinal autochthonous bacterial communities in juvenile hybrid tilapia *Oreochromis niloticus*. J World Aquacult Soc. 2009; 40:450–459.
25. Arpigny JL, Jaeger KE. Bacterial lipolytic enzymes: classification and properties. Biochem J. 1999; 343: 177–183. pmid:10493927

26. Snellman EA, Colwell RR. Acinetobacter lipases: molecular biology, biochemical properties and biotechnological potential. J Ind Microbiol Biotechnol. 2004; 31(9): 391–400. pmid:15378387
27. Ahmed EH, Raghavendra T, Madamwar D. An alkaline lipase from organic solvent tolerant Acinetobacter sp. EH28: Application for ethyl caprylate synthesis. Bioresour Technol. 2010; 101(10): 3628–3634. pmid:20096565
28. Wang H K, Shao J, Wei YJ, Zhang J, Qi W. A Novel low-temperature alkaline lipase from *Acinetobacter johnsonii* LP28 suitable for detergent formulation. Food Technol Biotechnol. 2011; 49(1): 96–102
29. Koven WM, Kolkovski S, Tandler A, Kissil G, Sklan D. The effect of dietary lecithin and lipase, as a function of age, on *n*–9 fatty acid incorporation in the tissue lipids of *Sparusaurata* larvae. Fish Physiol Biochem. 1993; 10: 357–364. pmid:24214374

CHAPTER 15

Maintaining Economic Value of Ecosystem Services Whilst Reducing Environmental Cost: A Way to Achieve Freshwater Restoration in China

Mingli Lin[1,2], Zhongjie Li[1], Jiashou Liu[1*], Rodolphe E. Gozlan[3], Sovan Lek[4], Tanglin Zhang[1], Shaowen Ye[1], Wei Li[1], Jing Yuan[1]

[1]*State Key Laboratory of Freshwater Ecology and Biotechnology, Institute of Hydrobiology, Chinese Academy of Sciences, Wuhan, China,*

[2]*Sanya Institute of Deep-sea Science and Engineering, Chinese Academy of Sciences, Sanya, China,*

[3]*UMR207 IRD, CNRS 7208-MNHN,-UPMC, Muséum National d'Histoire Naturelle, Paris, France,*

[4]*UMR 5174 EDB, CNRS-University Paul Sabatier, 118 route de Narbonne, Toulouse, France*

ABSTRACT

Freshwater fisheries are central to food security in China and this remains one of the most important priorities for the growing human population. Thus, combining ecosystem restoration with economics is pivotal in setting successful conservation in China. Here, we have developed a practical management model that combines fishery improvement with conservation. For six years, a ban on fertilizer and a reduction of planktivorous fish stocking along with the introduction of both mandarin fish *Siniperca chuatsi* and Chinese mitten crab *Eriocheir sinensis* was apparent in Wuhu Lake, a highly eutrophic lake located in the middle reaches of the Yangtze River. Annual fish yield decreased slightly after the change in management, whereas fisheries income increased 2.6 times. Mandarin fish and Chinese mitten crab accounted for only 16% of total fisheries production but for 48% of total fisheries income. During this six year period, water clarity increased significantly from 61 cm to 111 cm. Total nitrogen, total phosphorus and chlorophyll decreased significantly from 1.14 to 0.84 mg/L, 0.077 to 0.045 mg/L, and 21.45 to 11.59 μg/L respectively, and macrophyte coverage increased by about 30%. Our results showed that the ecological status of shallow lakes could be rapidly reversed from eutrophic to oligotrophic using simple biomanipulation, whilst maintaining fisheries economic value. It also offers a better approach to shallow fisheries lake management in Asia where traditionally the stocking of Chinese carp and use of fertilizers is still popular.

INTRODUCTION

China has now become the most significant contributor to world aquaculture with an annual production of 41 million tonnes (i.e. 62 percent of global aquaculture production in 2012) [1]. However this rapid economic development has also led to a range of environmental changes directly impacting fisheries [2]. For example, lake aquaculture (i.e. 998 000 ha) represents the main part of inland fisheries in China [3] and with traditional techniques that include application of mixed fertilizers (ammonium, phosphate and nitrate based) along with intensive stocking of carp (e.g. grass carp *Ctenopharyngodon idellus*, planktivorous silver carp *Hypophthalmichthys molitrix* and bighead carp *Aristichthys nobilis*) [4] has caused a significant decrease in macrophytes [5,6], increased nutrient load and reduced water quality [7]. Due to the rapid recent eutrophication of Yangtze River floodplain lakes [8], a set of sustainable management strategies that combine economic sustainability of fisheries and the conservation of freshwater ecosystems is necessary [9].

It is known that the specific and functional diversity of aquatic ecosystems, in particular lotic ones, is regulated by a set of "bottom-up" and "top-down" processes [10,11]. Thus, the current use of fertilizers in these shallow lakes lead to a bottom up eutrophication of these ecosystem via an increase of nutriment load alongside a bottom up impact on macrophytes and plankton via the choice of fish species stocked [10,11]. For example, carp species typically forage on benthos and they put many nutrients back into the water column. This, along with increased nutrient load via the production of faeces, will sustain planktonic algal blooms and reduce the community of aquatic macrophytes, limiting their access to nutrients in the substrate and reducing light level in the water column. Thus, changing community structure through the introduction of top predators with less impact on nutrient load coupled with a reduction in fertilizer and carp species could lead to a whole food web trophic cascade and result in improved water quality and increased macrophyte abundance [12]. However, a knowledge gap remains regarding the long-term economic and ecological sustainability of using this type of bio-chemical manipulation as a tool for lake restoration [13–16]. There are few examples of lake restoration via bio-manipulation processes [14,15,17], mostly in North American and European Lakes but that combine both ecosystem restoration with fisheries economics. The added challenge within an East Asian context is to maintain the economic value of the fisheries in addition to the lake restoration, as this aspect is central to the local population endorsing any new management options. In effect, previous research in shallow lakes of the Yangtze floodplain only focused on estimating the stocking capacity of piscivorous fish, the biomass and production of small fish as well as bioenergetics of mandarin [18–24].

Here using an un-replicated large-scale long-term experiment on an existing eutrophic shallow lake, we tested the effect of nutrient control and change in community composition (i.e. fish, crustaceans) on the ecological status of the lake and the economics of the associated fisheries. Specifically, we tested if 1) using simple rules arising from our current theoretical understanding of "bottom-up" and "top-down" processes, we could rapidly improve the water quality and macrophyte abundance of shallow lakes and 2) we could achieve this improved ecological status without impacting the economic value of the fisheries based on the use of replacement species of high market value. [18–20]

MATERIALS AND METHODS

Ethics statement

The studied lake belongs to the Chinese government and is rented to the private Wuhu Lake Fishery Management Company for management (188# Dazui Village, Liuzhi Street, Huangpi District, Wuhu City, Hubei Province, China). This is the company, which has given us the permission of water quality sampling and given us the fishery management data. We confirm that no fish were caught for the purposes of research.

Study area

The study was carried out between 2006 and 2011 in Wuhu Lake located in the middle reaches of the Yangtze River, Hubei Province, China (Fig 1). This is a very typical large (i.e. 2133 ha) but shallow lake (mean depth 2.6 m) disconnected by a sluice gate from the Yangtze River in 1952. It originally hosted a wide diversity of submersed macrophytes (mainly *Ceratophyllum demersum, Vallisneria spiralis, Trapa bispinosa* and *Potamogeton wrightii*), the baseline biomass was 3800g/m^2 [25], but almost disappeared when traditional fisheries started post 1995. By contrast the density of phytoplankton increased from 89.3×10^{4}ind/L in 1995 to 1726×10^{4}ind/L in 2005 [25]. Since 1995, the lake was managed to culture domesticated carp species including *C. idellus, H. molitrix,* and *A. nobilis* at stocking densities of 20–55 kg/ha, with the additional production of mandarin fish *Siniperca chuatsi.* To support the fish culture, plankton biomass was enhanced by the addition of non-organic fertilizers (417 kg/ha ammonium bicarbonate and superphosphate annually during 2002 to 2006) and organic fertilizers (1327 kg/ha brewer's grains and animal faeces annually during 2005–2007).

Biomanipulation

At the beginning of each year an agreed biomass per species (fish and crab) is stocked and fertiliser added. At the end of the year, all species are harvested and sold. Each year is thus a replicated experiment with pre-set species composition and biomass. Here, two replications (2006–07) consist of 'classic' stocking of carp and fertiliser (pre biomanipulation), followed by four replications post biomanipulation (2008–11). In 2008, we started a combined top-down (1) and bottom up (2) bio-manipulation of the lake by 1) stopping the stocking of *C. idellus,* decreasing the biomass of *H. molitrix, A. nobilis* from 8.9 ± 0.9 × 10^4 kg to 5.6 ± 1.5 × 10^4 kg and introducing the Chinese mitten crab *Eriocheir sinensis* a species of high economic value (Table 1); 2) stopping the fertilisation of the lake (i.e. no fertilisers or brewer's grain) and replanting *Vallisneria spiralis* to recover submerged macrophytes to provide refuge and food for the Chinese mitten crabs. The stocking biomass of the coin-sized mitten crab (i.e. ~ 5 g/ind.) was 1.58 ± 0.05 ×10^4 kg and approximately 20 ± 5.3 × 10^4 mandarin fingerlings (~ 3 cm total length). Both stocking densities followed recommendations from [23,26]. The traditional fishery model and our ecosystem-based model are presented as Fig 2.

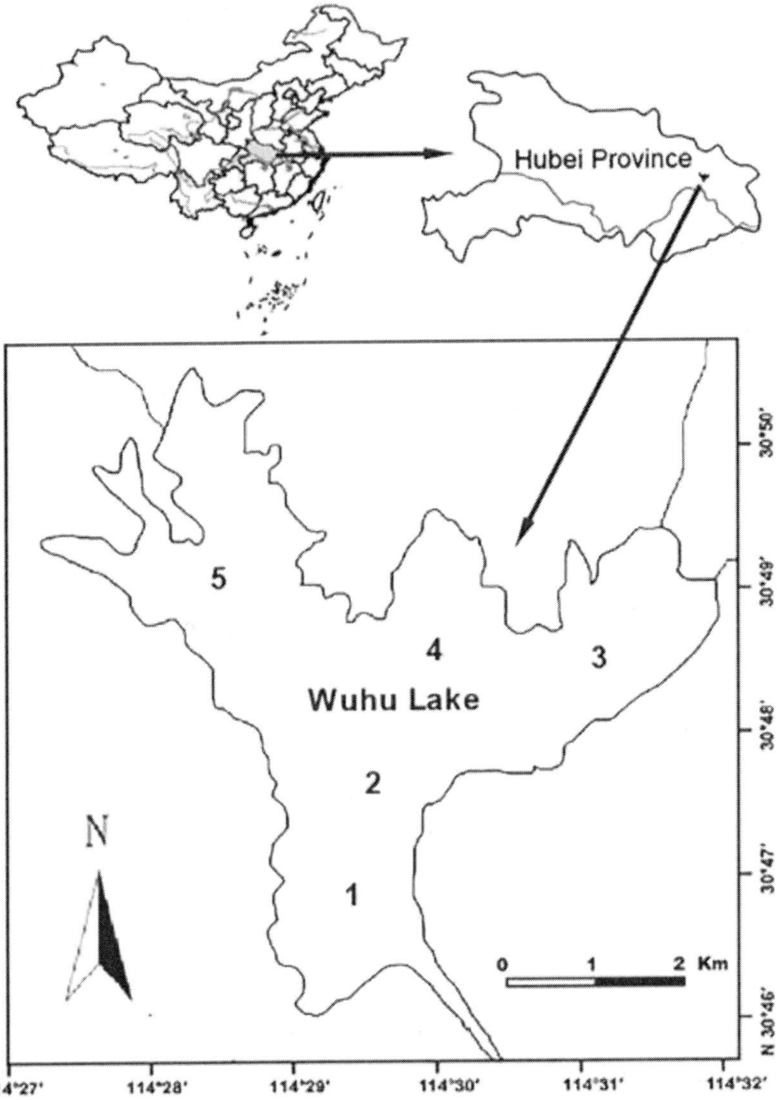

Figure 1. Location of Wuhu Lake located in the middle reaches of the Yangtze River, Hubei Province, China. Sample sites are displayed as number on the map.

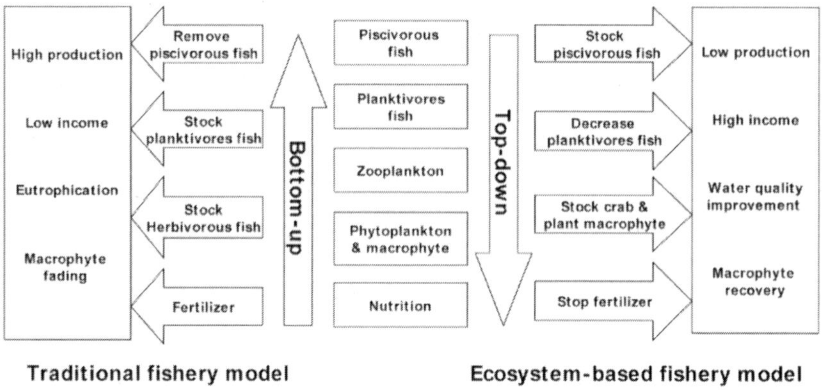

Figure 2. Traditional fishery model and ecosystem-based fishery model in Yangtze lake.

Table 1. Stocking number and biomass of carp, Mandarin fish and Chinese mitten crab pre (2006–2007) and post bio-manipulation (2008–2011) of Lake Wuhu.

Year	Silver carp and bighead carp 10^4kg	Mandarin fish 10^4ind	Chinese mitten crab 10^4kg
2006	8	4.8	0
2007	9.8	4	0
2008	3.6	25	1.7
2009	9.8	4	1.5
2010	3.6	25	1.6
2011	5.2	26	1.5

Sample collection and analysis

Wuhu Lake Fishery Management Company is a partnership company, and the sale data of different fish species were calculated precisely as they related to the profit share. Therefore, catch data of the crab and fish from 2006 to 2011 were directly obtained from sale records. Water quality was measured during May and October during 2006 to 2011 from five constant sampling sites in the lake (Fig 1). Water transparency (SD) was measured by Secchi Disc. In order to measure the content of total nitrogen (TN), total nitrogen (TN) and chlorophyll a (Chl a), 5 L water were sampled in each site and then transported to the laboratory. TN, TP and Chl a content was quantified immediately by Alkaline potassium persulfate digestion-UV spectrophotometric, Ammonium molybdate spectrophotometric and the homochromy spectrophotometric methods respectively [27]. All measuring procedures strictly followed the protocol of Chinese national standards [28] and finished within 24 h after sampling to avoid error cause by water quality changing.

In order to determine the coverage rate of submersed macrophytes, the whole lake was divided evenly into several polylines and then 48 sites were randomly selected above these lines. All these sites were recorded by GPS and remained fixed for the length of the study. Submersed macrophytes were sampled by scythes (1/5 m^2) 2–3 times at each site just above the sediment. After scything, plants were gathered with a 425 μm handnet and put into plastic bags. In the laboratory, the plants were rinsed, rid of superfluous water and wet weight taken. The rate of macrophyte coverage was calculated as an occurence ratio across all 48 sampling sites.

Mean and standard error (s.e.) were calculated and the environment data testing for normality used a Shapiro-Wilk test. Independent Sample T-test (parametric) and Mann-Whitney-Wilcoxon Test (non-parametric) tested the significance of the differences in water quality before and after the start of the bio-manipulation. All analyses were performed with R [29].

Fishery economic value

In this study, two factors determined the economic value of the fishery. One was the yield and the other one was the market price ($\$.kg^{-1}$) for each individual fish species. As the market price changed each year, the economic value did not necessarily follow the yield data.

RESULTS

Environmental change

Following the change in the fishery management in 2008, there was a significant increase in water clarity and overall quality ($P < 0.05$; Table 2, Fig 3). Macrophyte coverage also increased with almost no submersed macrophytes observed in 2006 to 31% submersed macrophyte (mainly *V. spiralis*) coverage by summer 2010. Annual macrophyte cover was only estimated pre and post biomanipulation treatment to avoid disturbance of the macrophyte recolonisation process.

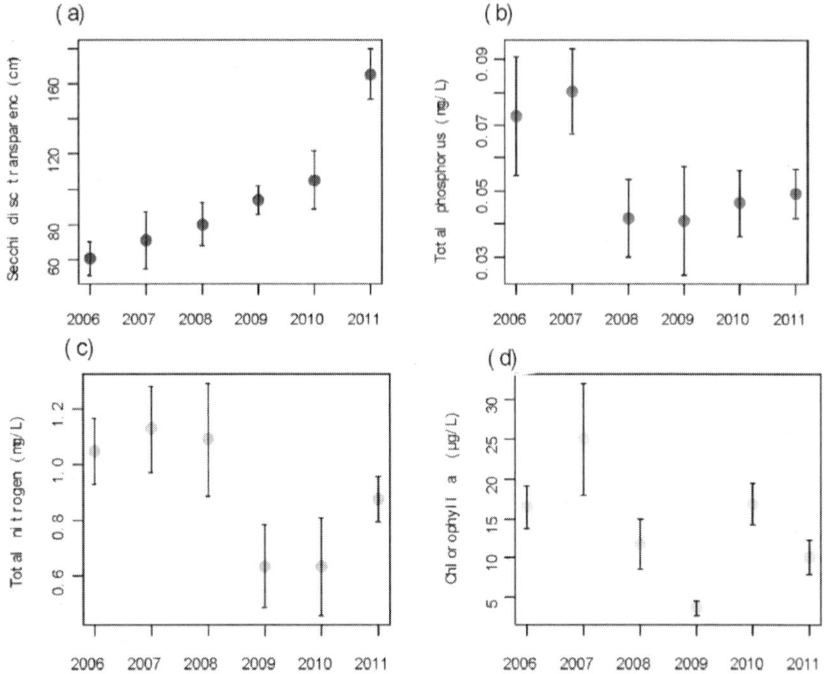

Figure 3. Water quality parameters of Wuhu Lake (China) before and after the start of bio-manipulation in 2008.

Table 2. Mean water parameters and standard error pre (2006–2007) and post bio-manipulation (2008–2011) of Lake Wuhu.

Water parameters	Pre bio-manipulation		Post bio-manipulation		*p* value	Test	df (W)
	Mean	SE	Mean	SE			
Water clarity (cm)	65.8	9	111.2	9.6	0.002	t	26
Total phosphorus (mg/L)	0.077	0.01	0.045	0.005	0.038	MWW	(400)
Total nitrogen (mg/L)	1.14	0.11	0.84	0.07	0.004	MWW	(361)
Chlorophyl a (µg/L)	21.45	4.5	11.59	1.46	0.025	MWW	(369)

Statistical test (t = t-test; MWW = Mann-Whitney-Wilcoxon test), significance (*P*) and degree of freedom (df) are included.

Fishery catch

Within four years following the start of the management change, the overall annual fish yield decreased by 3% from pre biomanpulation mean ± s.e. = 785000 ± 5000 kg to post biomanpulation 763000 ± 30000 kg; the Chinese carp, including *H. molitrix*, *A. nobilis C. idellus* and *Cyprinus carpio* decreased from mean ± s.e. = 630400 ± 67000 kg to 477625 ± 161640 kg. Following the start of the management change, the total of mandarin fish and Chinese mitten crab established a sustainable annual crop at mean ± s.e. = 20400 ± 3706 kg and 93600 ± 22124 kg respectively (Fig 4) and the overall yield of mandarin fish and Chinese mitten crab accounted for about 15% of the total fisheries production.

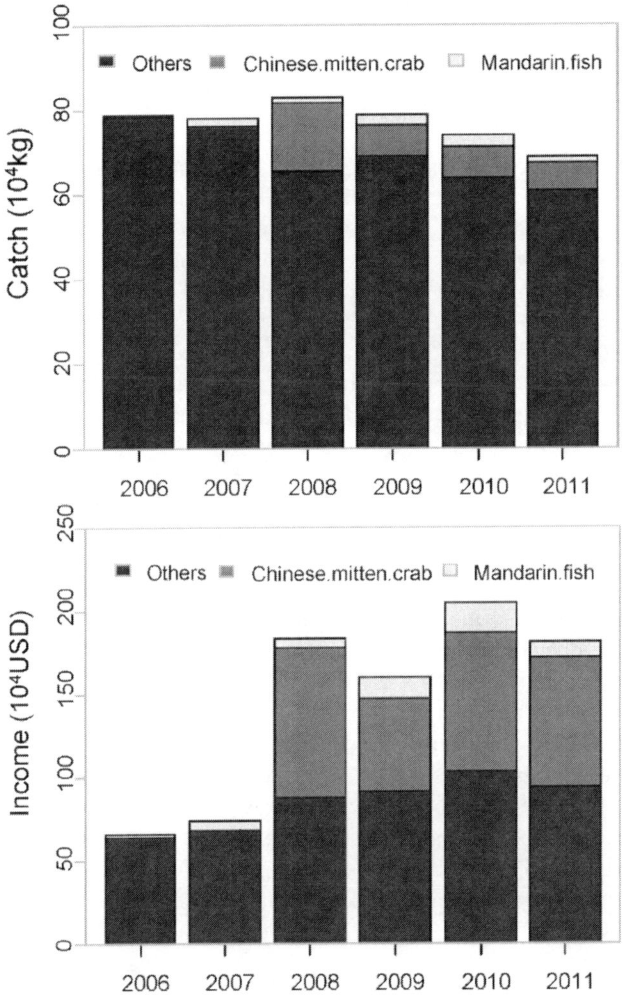

Figure 4. Comparison of fisheries catch and income between pre bio-manipulation and post bio-manipulation in Wuhu Lake (2008 was the start of the change in the management practice). Price of Chinese mitten crab has doubled between 2008 and 2010.

Economic value

The fisheries income increased significantly following the bio-manipulation (Fig 4), with an annual average income of about $1.83 million USD, 2.6 times that of 2006–2007 (i.e. ~$0.70 million). During this period, mandarin fish and Chinese mitten crab alone accounted for about mean ± se = 48% ± 2 of the total income generated.

DISCUSSION

Within four years, coinciding with the change in management, the transparency of Wuhu Lake water has more than doubled, the macrophyte biomass has increased by thirty percent and the economic value of the fishery has tripled. The main aspect underpinned by these results is that in a Chinese context, improving ecological status of lakes does not have to be achieved to the detriment of ecosystem service sustainability. On the contrary, it is the consideration of the economic dimension that will allow the long-term sustainability of biodiversity and ecological restoration. Lake Wuhu is very typical of thousands of lakes on the floodplain of the Yangtze River that were converted into fisheries in 1980s [30]. Animal protein and food security has been and remains one of the most important priorities for the growing population of China and innovation in environmental management is central to achieving that subtle balance between conservation and economics.

In aquatic ecosystems, the flux of energy via predator-prey interaction is generally directed from small-sized, abundant organisms to rare, larger bodied species [31] with a high turnover at the lower levels of a foodweb, whereas larger animals fix nutrients for longer time periods, making them unavailable. However in these complex foodweb structures the presence of keystone species [32], such as carp species [33], exerts a disproportionate influence on the pattern of species diversity in a community, in particular if they are stocked at high densities, removing macrophytes and lowering water clarity [5, 6]. Until now, stocking *H. molitrix* and *A. nobilis* along with lake fertilization has been the dominant management model of shallow lakes fisheries in the middle and lower reaches of the Yangtze River, allowing stable yield and low-risk profit but leading to nutrient enrichment of the lakes and associated problems of eutrophication [8]. This study goes further than previous research on these shallow lake fisheries of the Yangtze [18–24] to show that a careful selection of high economic value species can achieve high economic value while improving the water quality in a traditional fishery lake. As the application of fertilizers along with intensive stocking of carp are still popular in China, this study is pivotal in demonstrating to local stakeholders as well as policy makers that in a freshwater fisheries context the economic and environmental aspects are not necessarily antagonistic.

This bio-manipulation indicates that with a slight change in the composition of top predators and stopping using fertilizers, sustained by a reduction of carp production that limits the bio-engineering effect of fine particle re-suspension in the water column, macrophytes can establish large biomass nitrogen levels decrease. Nutrient control, especially TP load, restricts phytoplankton growth and thus helps water clarity [14,15,17,34], whilst apex predators preying on zooplanktivorous species allowed zooplankton growth and therefore increased phytoplankton grazing also leading to an increase in water clarity [16,35]. Similar effects were observed in the Great Lakes after the accidental introduction of another ecosystem engineer the zebra mussel *Dreissena polymorpha* [36], and in Lake Mendota after piscivore stocking [37]. Based on previous research [12], we speculated that mandarin fish triggered a so-called "trophic cascade" whilst the removal of grass carp and end of the fertilization process triggered a bottom-up effect [38]. The two effects are not contradictory and both possibly leading to synergetic effects such as the reduction of eutrophication. A traditional measure for a trophic cascade is a change in plant biomass [39], which can be taken as a measure of productivity. Here, despite not having a direct measure of the annual re-growth of the macrophyte community, the overall biomass of the macrophytes in Lake Wuhu pre and post bio-manipuation increased markedly.

However, the stocking balance needs to be carefully weighed as over stocking of Chinese mitten crab could lower the abundance of macrophytes and negatively impact the fisheries yield and thus income [26]. In light of this, we suggest that the stocking density used in this bio-manipulation experiment corresponds to the upper stocking limit. Although we will never restore the integrity of Wuhu Lake's ecosystem as it will always be managed due to socio-economic forces, it is still important to integrate the dimension of resilience in the future management of these lakes [40]. Resource based systems like fisheries should be kept in a stable state that guarantees optimal exploitation, also known as imposed resiliency [41]. However, stable systems do not have to be resilient and a higher biodiversity, although artificially driven, will be a key driver of resilience in these shallow lakes in future years.

REFERENCES

1. FAO. The State of the World's Fisheries and Aquaculture. Food and Agriculture Organization of the United Nations. 2014; 20pp

2. Zhao Y, Gozlan RE, Zhang C. Current state of freshwater fisheries in China. In: Craig J, editors. Freshwater Fisheries. Oxford: Wiley-Blackwell publisher. 2015; 1–19pp
3. FBMA (Fisheries Bureau of the Ministry of Agriculture). China Fishery Statistical Yearbook. Beijing: Agricultural Press. 2010; 158p
4. Liu J, He B. Cultivation of Freshwater Fishes in China. Beijing: Science Press. 1992 pp. 381–424
5. Chen H. Effects of fish culture in ecosystem of Lake Donghu. Acta Hydrobiologica Sinica. 1989; 13: 359–368
6. Sun G, Sheng L, Feng J, Lang Y, Li Z. Relationship between fishery and eutrophication in Chinese lakes. Journal of Northeast Normal University 1999; 19: 74–78
7. Xia WK, Wu XP, Liu JS, Li ZJ, Zhang TL, Ye S, et al. Fishery resource and preliminary exploitation scheme of Wuhu Lake. Reservior Fishery. 2007; 27: 38–40
8. Jin X, Xu Q, Huang C. Current status and future tendency of lake eutrophication in China. Science in China (Series C): Life Sciences. 2005; 48: 948–954
9. Li X, Shu S. Toward a coordinated development of fishery and aquatic environment on the middle and lower basins of yangtze river. Resources and Environment in the Yangtze Valley. 1997; 6: 7–11
10. McQueen DJ, Johannes MRS, Post JR, Stewart TJ, Lean DRS. Bottom-up and top-down impacts on freshwater pelagic community structure. Ecological Monographs. 1989; 59: 289–309
11. Hunter MD, Price PW. Playing chutes and ladders: heterogeneity and the relative roles of bottom-up and top-down forces in natural communities. Ecology. 1992; 73: 724–732
12. Carpenter SR, Kitchell JF, Hodgson JR. Cascading trophic interactions and lake productivity. BioScience. 1985; 35: 634–639
13. Lampert W, Sommer U. Limnoökologie. Stutgart: Thieme. 1993; 440pp
14. Mehner T, Kasprzak P, Wysujack K, Laude U, Koschel R Restoration of a stratified lake (Feldberger Hausee, Germany) by a combination of nutrient load reduction and long-term biomanipulation. International Review of Hydrobiology. 2001; 86: 253–265
15. Jeppesen E, Meerhoff M, Jacobsen BA, Hansen RS, Søndergaard M. Restoration of shallow lakes by nutrient control and biomanipulation-the successful strategy varies with lake size and climate. Hydrobiologia. 2007; 581: 269–285
16. Carpenter SR, Cole JJ, Kitchell JF, Pace ML. Trophic cascades in lakes: lessons and prospects. In: Terborg HJ, Estes JA, editors. Trophic Cascades. Washington DC: Island Press. 2009; pp. 1–8
17. Beklioglu M, Ince O, Tuzun I. Restoration of the eutrophic Lake Eymir, Turkey, by biomanipulation after a major external nutrient control. Hydrobiologia. 2003; 489: 93–105
18. Ye SW, Li ZJ, Lek-Ang S, Feng G, Lek S, Cao WX. Community structure of small fishes in a shallow macrophytic lake (Niushan Lake) along the middle reach of the Yangtze River, China. Aquatic Living Resources. 2006; 19: 349–359
19. Xie SG, Cui YB, Zhang TL, Fang RL, Li Zj. The Spatial Pattern of the Small Fish Community in the Biandantang Lake—A Small Shallow Lake Along the Middle Reach of the Yangtze River, China. Environmental Biology of Fishes. 2000; 57: 179–190
20. Xie SG, Cui YB, Li ZJ. Small fish communities in two regions of the Liangzi Lake, China, with or without submersed macrophytes. Journal of Applied Ichthyology. 2001; 17: 89–92. pmid:21171455
21. Liu JS, Cui YB, Liu JK. Resting metabolism and heat increment of feeding in mandarin fish (*Siniperca chuatsi*) and Chinese snakehead (*Channa argus*). Comparative Biochemistry and Physiology Part A: Molecular & Integrative Physiology. 2000; 127: 131–138
22. Liu J, Cui Y, Liu J. Food consumption and growth of two piscivorous fishes, the mandarin fish and the Chinese snakehead. Journal of Fish Biology. 2005; 53: 1071–1083

23. Cui YB, Li ZJ. Fishery Resources and Conservation of Environment in Lakes of the Changjiang River Basin. Beijing: Science Press. 2005; 313–340pp
24. Jin G, Li ZJ, Xie P. Ecological theories for sustainable fishery in Lakes and an example. Journal of Lake Science. 2003; 15: 69–75
25. Wang SM, Dou HS. Lakes of China. Beijing: Science Press. 208–209pp
26. Wang HZ, Wang HJ, Liang XM, Cui YD. Stocking models of Chinese mitten crab (*Eriocheir sinensis*) in Yangtze lakes. Aquaculture. 2006; 255: 456–465
27. Lorenzen CJ Determination of chlorophyll and pheo-pigments: spectrophotometric equations. Limnology and Oceanography. 1967; 12: 343–346
28. Huang X, Chen W, Cai Q. Standard Methods for Observation and Analysis in Chinese Ecosystem Research Network-Survey, Observation and Analysis of Lake Ecology. Beijing: Science Press. 1999; 1–247pp
29. R Development Core Team. R: A Language and Environment for Statistical Computing. Vienna: R Foundation for Statistical Computing. 2011; Available: http://www.Rproject.org. pmid:21238596
30. Jiang J, Huang Q, Sun Z. Analysis of ecological environment of lake-wetland in Yangtze River basin. Ecology and Environment. 2006; 15: 424–429
31. Ings TC, Montoya JM, Bascompte J, Bluthgen N, Brown L, et al. Review: Ecological networks—beyond food webs. Journal of Animal Ecology. 2009; 78: 253–269. pmid:19120606
32. Paine RT. The pisaster-tegula interaction: prey patches, predator food preference, and intertidal community structure. Ecology. 1969; 50: 950–961.
33. Parkos JJ, Santucci V, Hahl D. Effects of adult common carp (*Cyprinus carpio*) on multiple trophic levels in shallow mesocosms. Canadian Journal of Fisheries and Aquatic Sciences. 2003; 60:182–192
34. Bornette G, Puijalon S. Response of aquatic plants to abiotic factors: a review. Aquatic Sciences. 2011; 73: 1–14. pmid:20716861
35. Carpenter SR, Kitchell JF, Hodgson JR, Cochran PA, Elser JJ, Elser MM, et al. Regulation of lake primary productivity by food web structure. Ecology. 1987; 68: 1863–1876
36. Gozlan RE. Introduction of non-native freshwater fish: is it all bad? Fish and Fisheries. 2008; 9: 106–115
37. Lathrop RC, Johnson BM, Johnson TB, Vogelsang MT, Carpenter SR, Hrabik TR, et al. Stocking piscivores to improve fishing and water clarity: a synthesis of the Lake Mendota biomanipulation project. Freshwater Biology. 2002; 47: 2410–2424
38. Lammens EHRR. The central role of fish in lake restoration and management. Hydrobiologia. 1999; 136: 191–198
39. Polis GA, Sears ALW, Huxel GR, Srong DR, Maron J. When is a trophic cascade a trophic cascade? Trends in Ecology and Evolution. 2000. 15: 473–475. pmid:11050351
40. Holling CS. Resilience and stability of ecological systems. Annual Review of Ecology and Systematics. 1973; 4: 1–23
41. de Leo GA, Levin S. The multifaceted aspects of ecosystem integrity. Conservation Ecology. 1997; 1: 3

CHAPTER 16

Comparing Trawl and Creel Fishing for Norway Lobster (Nephrops norvegicus): Biological and Economic Considerations

Ana Maria Leoca´dio[1¤], David Whitmarsh[2†], Margarida Castro[3*]

[1]*Faculdade de Cie^ncias e Tecnologia, Universidade do Algarve, Campus Gambelas, Faro, Portugal,*

[2]*Department of Economics, Portsmouth Business School, University of Portsmouth, Portsmouth, United Kingdom,*

[3]*CCMAR, Centre of Marine Sciences, Universidade do Algarve, Campus Gambelas, Faro, Portugal*

ABSTRACT

This study compares the fishing activity and landings of the trawl and creel fisheries for Norway lobster (*Nephrops norvegicus* (L.)) off the Portuguese coast, and evaluates the financial viability of two vessels typical of each fleet. Crustacean trawlers are part of an industrial fleet that, besides *Nephrops*, targets deep water shrimps. Creels are used by a multi-gear, multi-target artisanal fleet, fishing only in areas unavailable to trawlers and, when catching *Nephrops*, set specifically to target this species. Trawlers have in recent years contributed with 85% of the landings in weight, but only 74% in value (2005–2009 average). Despite smaller landings, the *Nephrops* creel fishery provides individuals of larger size and in better condition, thereby obtaining higher unit prices. Economic viability was also higher for the creel vessel, with trawling being only viable if major costs (such as labor and fuel) are covered by the revenue from other target species (e.g., the rose shrimp). At present, *Nephrops* populations on the South and SW coast are subject to intense fishing and to a recovery plan. The possibility of reallocation of some of the fishing effort directed at *Nephrops* from trawlers to creels is discussed in terms of the conservation of the resource and economic return.

INTRODUCTION

The Norway lobster (Nephrops norvegicus (L.)) is the most valuable invertebrate resource caught in EU waters with a total first sale annual value of around 431 million Euros, corresponding to 74 thousand tonnes for the 2007–2008 average (EUROSTAT, http://epp.eurostat. ec.europa.eu /portal/page/portal/fisheries/data/database). In 2008 the UK and Ireland accounted for over 70% of the landings; Portugal ranked 13th with estimated landings of 247 tones (EUROSTAT). Throughout Europe, this species is caught mainly by trawling. Despite the relatively minor importance of Portuguese landings in the context of global European catches, first sale values are considerably higher in Portugal than elsewhere in Europe. While the average price per kg in 2008 was 7.51 € for Europe, it was 23.95 € in Portugal. This is in part due to the quality of the product; Norway lobsters are sold either fresh, refrigerated with ice (from the trawling fleet), or alive (from the artisanal fleet).

Portuguese Norway lobster landings have decreased abruptly from annual totals of around 1500 tons in 1987 to 329 tons in 1989. Subsequently, reported landings have not recovered to previous levels, staying below 400 tons since the 1990s (Figure 1A) (DGPA - General Directorate for Fisheries and Aquaculture, statistical information, http://www. dgpa.mina gricultura.pt/portal/page?_pageid=33,46256&_dad=portal&_schema=PORT AL&g_d=11050&cboui=11050). Management recommendations are produced by ICES (International Council for the Exploration of the Sea) and the assessment carried out in 2001 [1] considered that this drop in landings, since the 1980s, was the result of a severely declining stock, due to over-exploitation. A recovery plan was approved in December 2005 (Council Regulation (CE) N° 2166/2005) imposing limits on the caches and effort and defining closed areas. A minimum 70 mm diagonal mesh is enforced for codends of trawlers targeting Norway lobster and there are no requirements for selective devices in the trawling nets. A month closure (January) is imposed the operation of this fleet is also restricted under a general regulation that prohibits trawling in areas within 6 miles from the shore. The minimum landing size for Nephrops is 20 mm carapace length.

The limitations to creeling are the number of traps per vessel (varying from 500 to 1000 according to the length of the vessel), the mesh size, that when targeting Nephrops is 30 to 50 mm square mesh or a 40 mm distance between bars and forbiddance of landing berried females. The most common traps used for Nephrops are built with an iron frame wrapped with 40 mm rigid plastic netting.

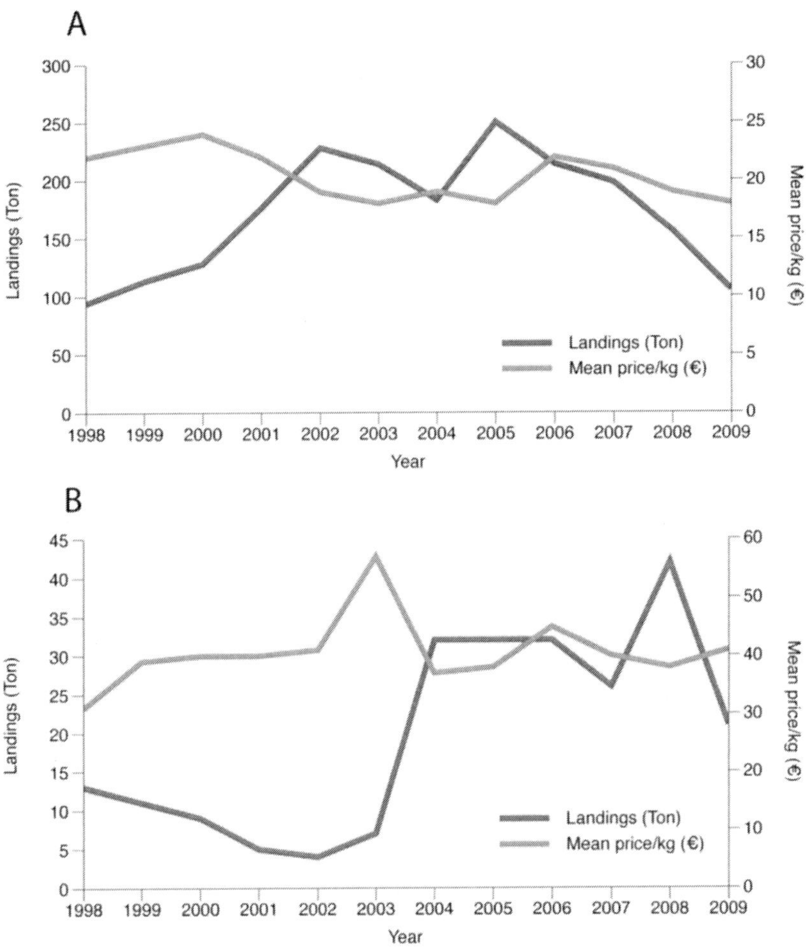

Figure 1. Evolution of the landings in weight and value. Annual landings in tons and mean price per Kg, for the trawl (A) and creel (B) fleets landing *Nephrops*, 1998–2007. Trawl landings exceed creel landings in weight. However, creel landings always obtained higher prices per Kg. Prices were converted from nominal to real terms using the Harmonized Index of Consumer Prices (HICP at EUROSTAT [15]).

In Portugal most of Nephrops landings came from the crustacean trawling fleet, composed at present of approximately 40 vessels that along the last two decades have transferred part of the fishing effort towards the rose shrimp (Parapenaeus longirostris), today the most important species, and the red shrimp (Aristeus antenatus). This change in effort accounts, in part, for the reduction in Nephrops landings, but unfortunately no specific studies exist that allow the quantification of effort changes.

Creels were introduced roughly 30 years ago and have since accounted for a small proportion of total landings, but from 2003 onwards (Figure 1B), the importance of this fishery increased, reaching and maintaining levels of around 15% of the landings in weight and over 26% in value (average 2004–2009 DGPA database). At present, there are roughly 10 artisanal vessels that use creels (DGPA database), baited usually with mackerel or bogue (100–150 g per creel and ~0.30€ per kg). Creeling in restricted to areas where trawlers cannot operate. These are areas with large boulders on muddy bottoms and areas within 6 miles from the coast. In these areas, very large Nephrops support a highly valuable creel fishery that markets live individuals. All the creeling fishing grounds are located on the southwest coast (Figure 2). In the South coast, the lack of trawl free sites has discouraged the practice of creeling. In Europe, creels have been implemented locally to fish Nephrops in some inshore areas around the west coast of Scotland [2], Sweden [3], north-eastern Adriatic and Ionian Sea [4] and since 1980, after the complete prohibition of trawling, in inshore areas, off the Faeroe Islands [5].

In face of the possibility that present levels of Nephrops trawling may be excessive, resulting in recommendations to stop or decrease catches [6]–[14], the use of creels as an alternative should be seriously considered. This is the rationale for the present work, which makes a comparative assessment of the biological and economic characteristics of both gear types, by thoroughly investigating the fishing pattern, composition of the catches, values of costs and returns for one trawler and a vessel using creels, representing well their fleets.

MATERIALS AND METHODS

Ethics statement

No specific permits were required for the described field studies. Data resulted from the observation of fisheries catches, onboard fishing vessels, at port landings or at fish auctions. In all cases the sampled individuals were obtained through regular fishing operations, our sampling only involved the measurements of carapace length and sex identification.

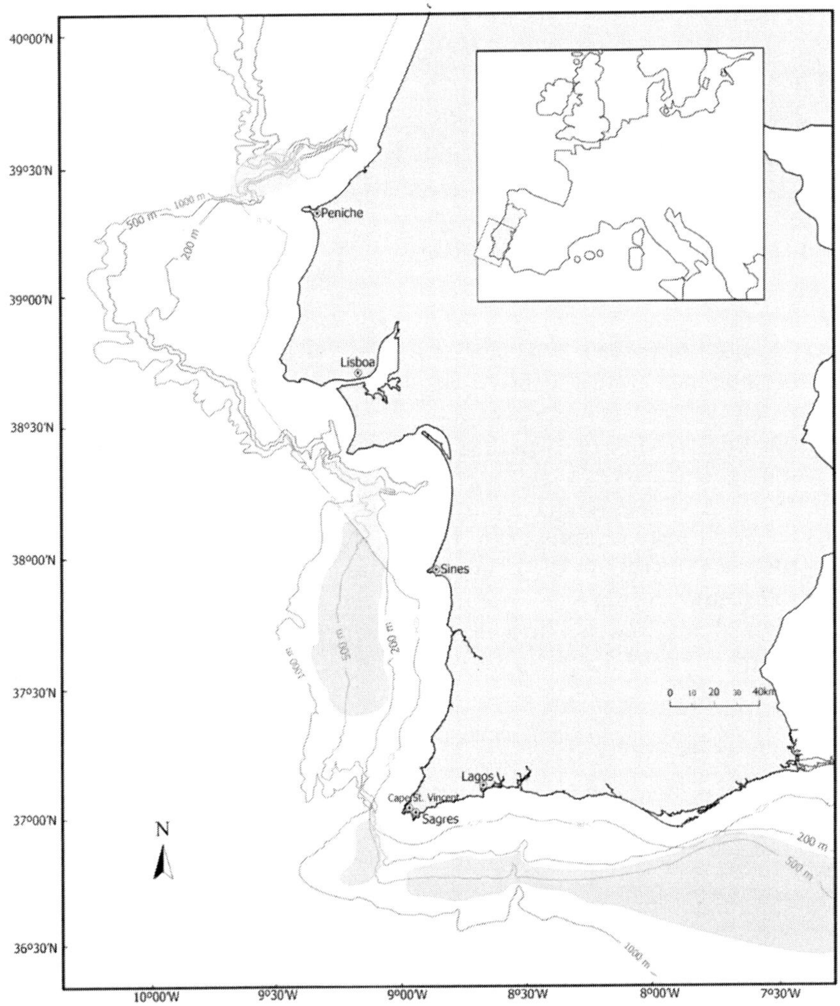

Figure 2. Map showing the location of the fishing grounds. Traditional fishing grounds are shown: fishing grounds explored with trawling in blue and creeling in yellow. Sampling for this work was done NE of Peniche (creels) and off Lagos on the south coast (trawl). Sources of information for mapping the fishing grounds: [32]–34.

The characteristics of the trawl and creel *Nephrops* fisheries considered were: fishing effort targeting *Nephrops*, catch information (structure, condition and value) and financial information. In addition to global values for both fleets, a typical vessel from each one (with average catches similar to those of the fleet they represent) was chosen, and as much information as possible was gathered for each one. For these two vessels, detailed information regarding biological and economic aspects was obtained, including on board and at the port sampling of the landings, and extensive interviews with crew members and managers of the fishing companies that own the vessels.

Fishing activity and landings

General characteristics of the trawl and creel *Nephrops* fishing are presented in Table 1. Information on the Portuguese commercial fisheries was obtained from the DGPA database, and includes individual data by vessel: type of fishing license, total number of trips per year, landings per trip, gross revenue per trip, species landed per trip and fishing port used for landing. This information is given on a trip-by-trip basis, between 1998 and 2007. The sale values of landings (Euros), presented in nominal terms, were converted to real terms using the Harmonized Index of Consumer Prices (HICP at EUROSTAT [15]). The HICPs, are economic indicators constructed to measure the changes over time in the prices of consumer goods and services acquired by households, and provide the official measure of consumer price inflation in the euro-zone [15].

Table 1. Main differences between trawl and creel *Nephrops* fishery.

	Trawl fishery	Creel fishery
Vessels size (meters)	>25	<20
Fuel consumption	High	Low
Number of vessels	~30	~10
Area	Offshore (6 miles limit)	Areas inaccessible to trawling
Duration of fishing trip (days)	1 to 3	1 (usually once a week)
Target species	Mixed crustacean fishery	Selective for *Nephrops*
Selectivity	Low	High
Discards (of total catch)	>60%(1)	Almost non-existent(2)
Environmental impacts	High(3)	Low(4)

Source: DGPA database,
(1)[29–31],
(2)[4,28,35,36],
(3)[37–41] and,
(4)[25,27,28].

On board sampling of the trawler's catches occurred during summer 2007, on the south coast of Portugal at depths ranging from 350 to 650 m, using commercial nets with 70 mm cod-end mesh size. Catch structure for the vessel using creels was obtained at several landing ports on the southwest coast, also during the summer of 2007. Catches were obtained at depths ranging from 600 to 650 m using 1000 creels with 40 mm mesh size. Biological sampling included the measurement of standard length with calipers, (carapace length measured in mm). During biological sampling, the condition of *Nephrops* caught was measured by a scale indicating vitality [16], defined from 0 to 2 (0=no signs of life; 1=some movement of the appendages; 2=strong signs of life, assuming aggressive posture).

Nephrops is sold at the fish auction of Vila Real de Santo António port, sorted into 4 size categories of different value, category 1 being the largest and most expensive and 4 the smallest and cheapest. Landings already sorted by size category were sampled at the fish auction in June and July of 2007. The probability of a length class falling into a given size category was estimated and extrapolated to the catches sampled in this work.

Financial viability

The basic criterion used to assess the financial sustainability of the trawl and creel fishery was based on a measure of financial viability of capital investment, the net present value (NPV). NPV is defined as the present value of net cash flows. It is a standard method for using the time value of money to appraise long-term projects [17].

For both trawls and creels only a part of the fishing effort is allocated to *Nephrops*. For the trawler it was estimated that 1/3 of the fishing effort was allocated to *Nephrops* (based on proportion of trawling time targeting *Nephrops*). For the creel vessel the effort allocated for *Nephrops* is estimated to be 1/5 of the total fishing effort (proportion of days fishing this species).

The financial viability of the two vessels, for the *Nephrops* fishery only, was compared using the net present value (NPV), representing the financial worth of investment of fishing (comparisons of costs and revenues). The NPV was calculated using the expression:

$$NPV = \sum_{i=0}^{i=n} \frac{R_i - C_i}{(1+r)^i}$$

Where R_i is the revenue earned in year i, C_i is the costs incurred in year i, r is the discount rate (representing the opportunity of capital) and n is the investment time horizon in years.

The time horizon considered was of 10 years. The 5% discount rate used was based on recommendations from the European Commission [18] to determine NPV. The input data used to calculate the NPV was obtained for the sampled vessels for the year 2007, by interviewing the owners and captains and complemented with information extracted from the DGPA landings database. Data includes running costs (proportional to vessel utilization rate), fixed costs (those incurred regardless of the amount of fishing activity) and labor costs based on a share of net revenue (gross revenue minus running costs) for the artisanal fishery, and a fixed value for the trawl fishery. Data regarding price per kg, fishing days per year and catch rate (kg per fishing day) were also included.

The decision regarding which costs and revenues should be included was entirely based on the demands of *Nephrops* fishery: fuel (fishing and travelling) and lubricants for trawlers and bait and fuel (traveling) for the creel fishing. Other costs were allocated to *Nephrops* fishing proportionally to the fraction of the fishing effort dedicated to catch *Nephrops* (1/3 for the trawler and 1/5 for the creel vessel).

Sensitivity analysis was conducted to assess the financial viability of the *Nephrops* fishery in both vessels by changing the variables that affect NPV one at a time [17]. The results of the sensitivity analysis are summarized in a switching value, the level of change in a given variable required for the NPV to become zero, under the assumption that all other variables remain at present values.

RESULTS

Characterization of the landings

The technical characteristics of the two vessels under study were obtained from the DGPA and are presented in Table 2. The landings of individual vessels were analyzed in detail for the years 2005–2007 in order to understand how representative of their fleets were the two vessels chosen.

Of the 35 trawlers that landed Nephrops, the fleet targeting this species was considered to be composed of vessels that simultaneously satisfied two criteria: the proportion of crustaceans in the landings was over 15% and the total amount of Nephrops was over one ton per year. A total of 17 vessels satisfied both criteria (accounting for 73% of Nephrops trawling landings), with average annual landings from 3.4 to 16.7 tons in weight and 54 and 323 thousand Euros in value. For the same period, the trawler selected for this work had average annual landings of 8.4 tons with a value

Table 2. Technical characteristics of the vessels selected.

Characteristics	Trawl vessel	Creel vessel
Crew (number)	6–7	6
Total length (meters)	24.80	17.08
Engine (HP)	600	134.23
Gross tonnage	215	26.71
Construction year	2000	1993

Source: DGPA database.

of 202 thousand Euros. The artisanal fleet that landed Nephrops between 2005 and 2007 was composed of 29 vessels. It was considered that the fleet targeting Nephrops included all the vessels with landings in the three years, a total of 5 units. Two more were added that, despite landing Nephrops for only two of the years considered, had average landings of over one ton per year and represented 58% of the landings in those years. For the combination of 2005 to 2007, these 7 units were responsible for over 90% (both in weight and value) of the Nephrops catches landed by the artisanal fleet. The range of annual landings for these vessels was 1.0 to 13.2 tons and 31 to 367 thousand. The artisanal vessel selected had average landings of 2.9 tons in weight and 248 thousand Euros in value. Overall landings (for all vessels engaged in the Nephrops fishing) and mean annual prices per kg, for the last 12 years with available data, are given in Figure 2. Trawling landings, after an increase until 2002, remained above average until 2005 and have since decreased to the low values observed at the beginning of the period considered here. In 2009 the Nephrops landings of the crustacean fleet were around 100 tons. For the creel fishery, landings rose sharply until 2003 and have remained high until 2006, and suffered a drop from 42 to half that value afterwards. This reduction did not affect individual vessels yields and was due to the removal of 2 of the most important vessels from this fleet segment (DGPA database).

The length distributions and size categories of the catches for the sampled vessels are presented in Figure 3 and values for first sale in Table 3. The structure of the catches combines information from the size distributions obtained on board and from the length categories, obtained at the VRSA fish auction. Trawl catches were composed of small individuals (ranging from 20 to 58 mm carapace length, with a mean size of 37 mm) while creel catches were composed of large individuals (ranging from 45 to 84 mm with a mean size of about 58 mm). Ninety percent of the trawl catches were traded in size category 4 (lowest value, Table 3), and category 1 (highest value) was absent while 13% of the creel catches were sold in category 1, 71% were in category 2 and 16% in category 3. Regarding the conditions upon arrival on deck (Table 4) in trawl catches 16% of the individuals had no sign of life (39% with some movements and 45% very active) while creel caught individuals all arrived on deck showing clear signs of life (30% with some movements and 70% very active).

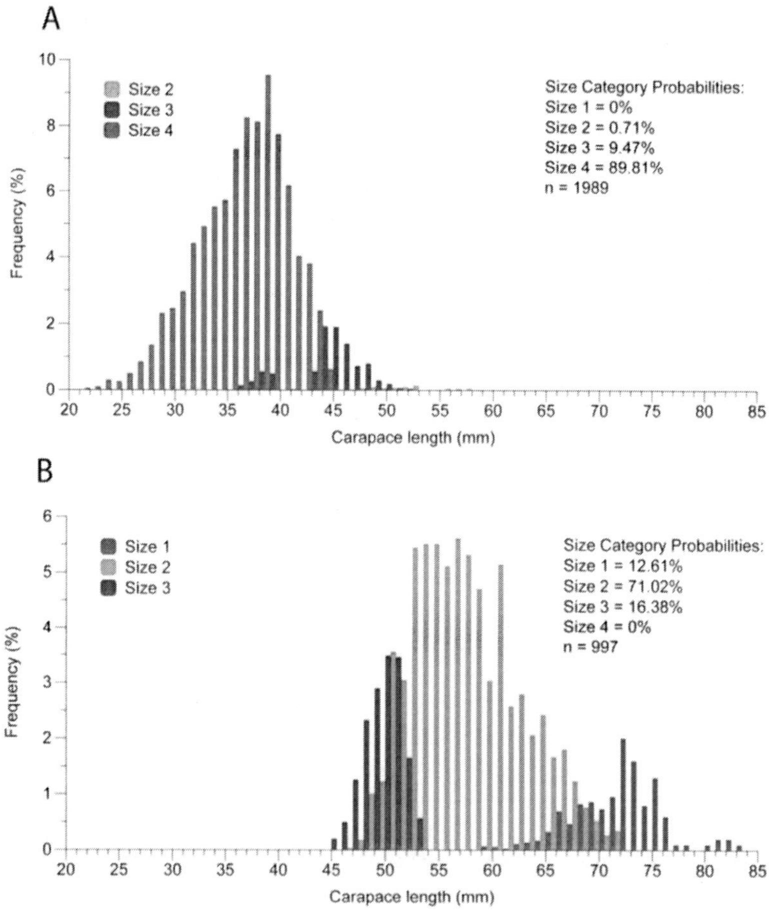

Figure 3. Size structure of the landings. Size frequency distribution (in percentage) of trawl (A) and creel (B) catches, including size probabilities for each size category. Trawl catches are composed of small individuals while creel catches are composed of large individuals. Ninety percent of the trawl catches are traded in size category 4 (lowest value) and category 1 (highest value) is absent. On the other hand, 13% of the creel catches are sold in size category 1, while 71% are in size category 2 and 16% in size category 3. The size distributions for each size category were sampled at the Vila Real de Santo António fish auction and result from the onboard separation of the catch.

Table 3. Approximate price and corresponding size ranges for *Nephrops*.

Size categories	CL (mm)	Approximate range (€/Kg)
1	>60	47–130(*)
2	47–72	23–67
3	36–53	7–24
4	<44	2–10

Shows range of values of first sale (Euros/kg/size category) and carapace length (mm).
Source: data collected at the auction market of Vila Real de Santo António (VRSA) in June and July of 2007.
(*)can reach up to 200€/kg at particular times.

Table 4. Condition of *Nephrops* at arrival on deck.

Condition scale	0	1	2
Trawl	16 (2.7)	39 (3.6)	45 (3.6)
Creels	0	30 (6.3)	70 (6.3)

Description of the condition (in percentage) on arrival on deck for trawl (net mesh size = 70 mm, n = 186) and creel catches (net mesh size = 40 mm, n = 53). Condition scale: 0 = no movements, 1 = some movements, 2 = strong life signals (aggressive posture). Standard error values are in parenthesis.

Economic valuation

The annual average selling price of Norway lobster (per kg) landed in 2007 was very different for the studied vessels: 15€ for the trawler (standard deviation 5.8) and 80€ for the creel vessel (standard deviation 20.8). The catch per unit effort (kg/day of fishing) was 26 kg for the trawler and 70 kg for the creel vessel. The trawler landed Nephrops 198 days against 45 days for the artisanal vessel. The annual running cost was 76 545€ for the trawler (sum of fuel and lubricants) and 18 000€ for the artisanal vessel (sum of fuel and bait), with total annual fixed costs of 35 457€ and 12 000€ respectively (Table 5). In 2007, the average fuel consumption was 2134 liters per day for the trawler and just 238 liters per day for the creel vessel.

Table 5. Baseline values used in the Net Present Value (NPV) analysis and switching values from sensitivity analysis.

Variable	Units	Present	Switching value
Creels			
Price	€ per kg	80.19	53.53
Vessel utilization	Fishing days per year	45	30
Catch rate	kg per day fishing	70.5	47.1
Running costs	€ per day	400	2 279
Fixed costs	€ per year	12 000	96 569
Labor costs	N/A		
Crew share	proportion net revenue	0.50	0.94
Financial support (subsidies)	N/A		
Taxes and Social security	Proportion gross revenue	0.17	0.83
NPV (with discount rate at 5%)	€	737 585	
Trawl			
Price	€ per kg	14.84	32.56
Vessel utilization	Fishing days per year	198	434
Catch rate	kg per day fishing	26.1	57.4
Running costs	€ per day	387	−57
Fixed costs	€ per year	35 457	−56 255
Labor costs	€ per year	69 714	−21 998
Crew share	N/A		
Financial support (subsidies)	€ per year	13 217	104 930
Taxes and Social security	N/A		
NPV (with discount rate at 5%)	€	−799 893	

NPV is the present value of net cash flows and the switching values summarize the sensitivity analysis, representing the value required for the NPV to become zero, under the assumption that all other variables remain unchanged.
Source: landings data from DGPA database, economic information provided by the companies owning the vessels selected.

The total benefit of fishing for Norway lobsters with creels was estimated at 737 585 € considering the NPV over 10 years at 5% discount rate. With the same criterion, Nephrops trawling was an unprofitable fishery, with losses of the order of 799 893 €. However, when the NPV for the trawler was estimated for the total catch, where the rose shrimp represents the most important fraction, there were considerable financial gains, with a benefit estimated at 1 041 201 €. The sensitivity of these results to the input variables was tested based on the switching value (Table 5). Results reveal that profitability for creels was not marginal and that this fishery would still be profitable with a drop by 30% of any one of the revenue-determining variables (price, catch rate, or vessel utilization). For the trawler, by contrast, any one of these variables would need at least to double in order for the Nephrops component of its activity to generate a positive NPV.

Table 6 shows values indicating the importance of the Nephrops in weight (10.4% for the creels and 8.5% for the trawl) and value (51.6% for the creels and 12.7% of the trawl) and the profitability of the activity using as indicators net revenue per kg Nephrops caught. Fishing for Nephrops with the trawler was not a profitable operation, with losses of 20.29 euros per kg of Nephrops caught. These losses decreases to 6.82 if labor costs not considered (totally allocated to the shrimp part of the operation).

Table 6. Importance of *Nephrops* in the fishery and efficiency of the gears.

Indicators	Creels	Trawl
Proportion *Nephrops* in weight	10.4%	8.5%
Proportion *Nephrops* in value	51.6%	12.7%
Net revenue/Kg *Nephrops* landed (Euros)	26.66	−17.72
Net revenue/Kg *Nephrops* landed (no labor) (Euros)		−4.25

Importance of *Nephrops* in weight and value and revenue per kg of *Nephrops* caught for 2007. For the trawler net revenue was calculated with and without labor costs.

DISCUSSION

When comparing the two gears used to catch *Nephrops* off the Portuguese coast, through a detailed analysis of two vessels representing trawling and creel fishing, marked differences can be observed in exploitation pattern and size and condition of the landings. In both cases *Nephrops* is only targeted for part of the time (roughly 1/5 of the effort for the creel vessel and 1/3 for the trawler). For the trawler, *Nephrops* fishing was subsidiary with respect to the main target species, the rose shrimp while for the artisanal vessel, fishing *Nephrops* with creels was an essential valuable complement.

The unit value of landings for *Nephrops* caught by creels is superior to that for trawls, due to the larger size and better condition of the individuals (sold alive), as was verified in other studies comparing the same gears [19], [20]. Trawls tend to capture a wide range of sizes, including small individuals down to the minimum landing size (20 mm CL), while no large animals (above 60 mm CL) were caught, probably due to the exhaustive exploitation of the south coast stock. Creels captured few animals below 45 mm CL.

The differences in size structure may result from two sources: level of exploitation of the stocks and behaviour when interacting with the gear. Higher fishing mortality and smaller size of first capture in trawling are likely to result in the disappearance of large sizes. Trawling catches are influenced by biological rhythms of burrow occupation and emergence, since *Nephrops* are caught when out of their burrows [21]–[23]. Creel catches are not influenced by burrow emergence patterns in the same way as trawl catches, but are still dependent on individual animal behavior likely related to feeding activity when attracted to bait in creels [24].

Observations with video [25] showed that larger individuals were more likely to enter further into the creel eye and be trapped. Similar results were observed in an aquarium study were the largest *Nephrops* consistently displaced smaller individuals and were the only ones to enter the creel [26]. It is also possible that dominance and aggressive behavior of larger animals inhibits smaller individuals from entering the creels or force them to escape through the mesh (rigid plastic square mesh with 4 cm side).

The presence of large sizes in creel catches may indicate that these populations are subjected to lower mortality rates than those from areas were trawlers operate.

The impact of creeling on benthic communities is minimal compared to trawling [25], [27], [28], an aspect relevant for conservation of biodiversity. Creels are very selective for the target species, catching almost exclusively *Nephrops* [25] while trawling produces large quantities of by-catch composed mainly of fish [29]–[31]. The discarding of targeted crustacean species is usually due to damage of the individuals in the net and/or during the sorting process. Nine percent of *Nephrops* are discards in the trawl fishery [31].

The financial analysis revealed a non-profitable trawl *Nephrops* fishery. On the other hand, the fishing the Norway lobster with creels was economically viable. Results from this study suggest that if the trawling fleet depended more on fishing *Nephrops*, the profitability of fishing activity would be compromised. On the other hand, a fishing day targeting *Nephrops* by an artisanal vessel should be very profitable. Here, costs are low and revenues are high; for the sampled vessels the fuel consumption is low (9 times less when compared with the trawler) and mainly justified by the journey to and from the fishing area, as fuel used during the hauling of creels is almost insignificant.

Similar results, in terms of comparative returns by trawls and creels, were found in Sweden [3], [28] where advantages in replacing bottom trawling with creels were reported due to less seafloor impact, less fish by-catch and undersized *Nephrops* discarded and a decrease in fuel consumption. In the Adriatic Sea, creels were found to be more ecologically sustainable due to lower discard levels, but were also shown to have lower economic viability due to the high scavenger activity in the area, that reduces the lifespan of the bait [4].

The maintenance of a trawl fishery for *Nephrops*, of apparently little economic interest, can be understood with a global analysis of this multi-target fishery. Nowadays, the rose shrimp has become the main target species, due to remarkably high first sale prices (up to 200€ per Kg at certain times of the year). This species provides most of the economic return of the crustacean trawling fleet (around 70% from 1996 to 2007, DGPA database).

Given the economic interest of fishing *Nephrops* with creels, one might expect the increase of this activity. At present, the biggest limitation for the expansion of the creel fishery in Portugal is gear conflict. Trawlers operate outside the 6 miles limit and *Nephrops* is uncommon within the 6 miles limit. Creel fishing is therefore restricted to sheltered areas where trawlers cannot operate, due to irregular and rocky bottoms. There are not many of these areas and territorialism or lack of information among fishermen might also hold back the expansion of creel use. This situation limits the proportion of fishing effort allocated by the artisanal fleet to *Nephrops* creel fishing.

Considering the current condition of the Portuguese *Nephrops* populations in the South and Southwest coasts (under a recovery plan) and the negative impacts that trawling has on the environment, creels may provide a more sustainable and profitable alternative to trawls, contributing to the recovery of the resource. This could be achieved without detrimentally affecting the rose shrimp fishery that occurs on the slope at shallower depths than *Nephrops*[32] and almost exclusively on the South coast. A transition regime could include, for example, the exclusion of trawlers from the SW coast, leaving the deep water crustaceans (including *Nephrops*) to be exploited by creels, while maintaining the present regime on the South coast (Figure 2). This would imply a reduction in the number of trawlers overall, to avoid relocation of trawl fishing effort to the South coast. Examples of the coexistence of both sectors can be found in Scotland, with different levels of conflict [25] and informal assignment of fishing grounds to both gears.

This is a preliminary study and consequently there are still large gaps in the present knowledge of how *Nephrops* creel fisheries exploit the stock and how this varies in different fishing areas. In this study it was not possible to compare the exploitation pattern of both gears directly, as trawling and creeling take place in different areas and exploit different *Nephrops* populations. Future studies should also include demand and supply analyses, to evaluate the needs of the market and changes that might occur from switching from trawls to creels such as decrease in total catch and increase in the fraction of high value lobsters. Sensitivity analysis could usefully be supplemented by formal risk analysis in order to generate probability distributions of the NPV outcomes.

ACKNOWLEDGMENTS

The authors thank the captains and owners who mostly kindly provided data regarding costs and revenues of both fleets, Sofia Erzini and Rita Castilho for drawing figures and Karim Erzini for reviewing the text. Data provided by General Directorate for Fisheries and Aquaculture (Direcção Geral das Pescas e Aquicultura – DGPA) and Docapesca – Portos e Lotas, S. A. is gratefully acknowledged.

REFERENCES

1. ICES (2001) Report of the Working Group on *Nephrops* Stocks. 516 p.ICES2001Report of the Working Group on *Nephrops* Stocks.ICES CM 2001/ACFM:16. p. 516 p.
2. Adey J, Smith I, Atkinson R, Tuck I, Taylor A (2008) Ghost fishing' of target and non-target species by Norway lobster Nephrops norvegicus creels. Mar Ecol Prog Ser 366: 119–127.J. AdeyI. SmithR. AtkinsonI. TuckA. Taylor2008Ghost fishing' of target and non-target species by Norway lobster Nephrops norvegicus creels.Mar Ecol Prog Ser366119127
3. Eggert H, Ulmestrand M (2000) A bioeconomic analyses of the Swedish Fishery for Norway Lobster (*Nephrops norvegicus*). Mar Resour Econ 14: 225–244.H. EggertM. Ulmestrand2000A bioeconomic analyses of the Swedish Fishery for Norway Lobster (*Nephrops norvegicus*).Mar Resour Econ14225244
4. Morello E, Antolini B, Gramitto M, Atkinson R, Froglia C (2009) The fishery for *Nephrops norvegicus* (Linnaeus, 1758) in the central Adriatic Sea (Italy): Preliminary observations comparing bottom trawl and baited creels. Fish Res 95: 325–331.E. MorelloB. AntoliniM. GramittoR. AtkinsonC. Froglia2009The fishery for *Nephrops norvegicus* (Linnaeus, 1758) in the central Adriatic Sea (Italy): Preliminary observations comparing bottom trawl and baited creels.Fish Res95325331
5. ICES (1990) Report of the Working Group on *Nephrops* Stocks. 322 p.ICES1990Report of the Working Group on *Nephrops* Stocks.ICES. ICES CM 1990/Assess. 16. p. 322 p.
6. ICES (2003) Report of the Working Group on *Nephrops* Stocks. 630 p.ICES2003Report of the Working Group on *Nephrops* Stocks.ICES CM 2003/ACFM:18. p. 630 p.
7. ICES (2004) Report of the Working Group on *Nephrops* Stocks. 436 p.ICES2004Report of the Working Group on *Nephrops* Stocks.ICES CM 2004/ACFM:19. p. 436 p.
8. ICES (2005) Report of the ICES Advisory Committee on Fishery Management. 1418 p.ICES2005Report of the ICES Advisory Committee on Fishery Management.ICES Advice 2005, Book 1–11. p. 1418 p.
9. ICES (2006) Report of the ICES Advisory Committee on Fishery Management. 113 p.ICES2006Report of the ICES Advisory Committee on Fishery Management.ICES Advice 2006. p. 113 p.
10. ICES (2007) Report of the ICES Advisory Committee on Fishery Management. 94 p.ICES2007Report of the ICES Advisory Committee on Fishery Management.ICES Advice 2007. p. 94 p.
11. ICES (2008) Report of the ICES Advisory Committee. 122 p.ICES2008Report of the ICES Advisory Committee.ICES Advice 2008. p. 122 p.

12. ICES (2009) Report of the ICES Advisory Committee. 77 p.ICES2009Report of the ICES Advisory Committee.ICES Advice 2009. p. 77 p.
13. ICES (2010) Ecoregion Bay of Biscay and Western Iberian Seas; Stock: *Nephrops* in Division IXa. pp. 57–72.ICES2010Ecoregion Bay of Biscay and Western Iberian Seas; Stock: *Nephrops* in Division IXa.5772In: Report of the ICES Advisory Committee. ICES Advice 2010. p. p.
14. ICES (2011) Ecoregion Bay of Biscay and Western Iberian Waters; Stock: *Nephrops* in Division IXa, In: Report of the ICES Advisory Committee. pp. 45–49.ICES2011Ecoregion Bay of Biscay and Western Iberian Waters; Stock: *Nephrops* in Division IXa, In: Report of the ICES Advisory Committee.4549ICES Advice 2011. p. p.
15. Office for Official Publications of the European Communities (2004) Harmonized Indices of Consumer Prices (HICPs), a short guide for users. Luxembourg: European Commission. 15 p.Office for Official Publications of the European Communities2004Harmonized Indices of Consumer Prices (HICPs), a short guide for users. Luxembourg: European Commission.Office for Official Publications of the European Communities. p. 15 p.
16. Castro M, Araújo A, Monteiro P, Madeira A, Silvert WFR (2003) The efficacy of releasing caught *Nephrops* as a management measure. Fisheries Research 65: 475–484.M. CastroA. AraújoP. MonteiroA. MadeiraWFR Silvert2003The efficacy of releasing caught *Nephrops* as a management measure.Fisheries Research65475484
17. Lumby S (1984) Investment Appraisal. United Kingdom: Van Nostrand Reinhold. S. Lumby1984Investment AppraisalUnited KingdomVan Nostrand Reinhold
18. European Commission (2006) Guidance on the methodology for carrying out cost-benefit analysis. 23 p.European Commission2006Guidance on the methodology for carrying out cost-benefit analysis.The New Programming Period 2007–2013. Directorate-General Regional Policy. Working Document No. 4. p. 23 p.
19. Eriksson S (2006) Differences in the condition of Norway lobsters (*Nephrops norvegicus* (L.)) from trawled and creeled fishing areas. Marine Biology Research 2: 52–58.S. Eriksson2006Differences in the condition of Norway lobsters (*Nephrops norvegicus* (L.)) from trawled and creeled fishing areas.Marine Biology Research25258
20. Ridgway I, Taylor A, Atkinson R, Chang E, Neil D (2006) Impact of the capture method and trawl duration on the health status of the Norway lobster, *Nephrops norvegicus*. J Exp Mar Biol Ecol 339: 135–147.I. RidgwayA. TaylorR. AtkinsonE. ChangD. Neil2006Impact of the capture method and trawl duration on the health status of the Norway lobster, *Nephrops norvegicus*.J Exp Mar Biol Ecol339135147
21. Aguzzi J, Sardà F (2008) A history of recent advancements on *Nephrops norvegicus* behavioral and physiological rhythms. Rev Fish Biol Fish 18: 235–248.J. AguzziF. Sardà2008A history of recent advancements on *Nephrops norvegicus* behavioral and physiological rhythms.Rev Fish Biol Fish18235248
22. Newland P, Chapman C (1989) The swimming and orientation behavior of the Norway lobster, *Nephrops norvegicus* (L.), in relation to trawling. Fish Res 8: 63–80.P. NewlandC. Chapman1989The swimming and orientation behavior of the Norway lobster, *Nephrops norvegicus* (L.), in relation to trawling.Fish Res86380
23. Newland P, Neil D, Chapman C (1982) Escape swimming in the Norway lobster. J Crust Biol 12: 342–353.P. NewlandD. NeilC. Chapman1982Escape swimming in the Norway lobster.J Crust Biol12342353
24. Bell M, Redant F, Tuck I (2006) *Nephrops* species. In: Philips B, editor. Lobsters: Biology, Management, Aquaculture and Fisheries. Blackwell Publishing Ltd. pp. 412–461.M. BellF. RedantI. Tuck2006*Nephrops* species.B. PhilipsLobsters: Biology, Management, Aquaculture and FisheriesBlackwell Publishing Ltd412461editor. pp. pp.
25. Adey J (2007) 474 p.J. Adey2007Aspects of sustainability of creel fishing for Norway lobster, *Nephrops norvegicus* (L.), on the west coast of Scotland [PhD]: University of Glasgow. p. 474 p.

26. Livingstone M (2001) A study into the effectiveness of escape panels for Norway lobster, *Nephrops norvegicus* (L), with application to the Torridon fishery. Edinburgh: Scottish Agricultural College. M. Livingstone2001A study into the effectiveness of escape panels for Norway lobster, *Nephrops norvegicus* (L), with application to the Torridon fisheryEdinburghScottish Agricultural College
27. Kinnear J, Barkel P, Mojsiewicz W, Chapman C, Holbrow A, et al. (1996) Effects of *Nephrops* creels on environment. 11 p.J. KinnearP. BarkelW. MojsiewiczC. ChapmanA. Holbrow1996Effects of *Nephrops* creels on environment.Fisheries Research Services, Scottish Office Agriculture, Environment and Fisheries Department, Marine Laboratory. Report No. 2/96. p. 11 p.
28. Ziegler F, Valentinsson D (2008) Environmental life cycle assessment of Norway lobster (*Nephrops norvegicus*) caught along the Swedish west coast by creels and conventional trawls - LCA methodology with case study. Int J Life Cycle Ass 13: 487–497.F. ZieglerD. Valentinsson2008Environmental life cycle assessment of Norway lobster (*Nephrops norvegicus*) caught along the Swedish west coast by creels and conventional trawls - LCA methodology with case study.Int J Life Cycle Ass13487497
29. Borges T, Erzini K, Bentes L, Costa M, Gonçalves J, et al. (2001) By-catch and discarding practices in five Algarve (southern Portugal) métiers. J Appl Ichthyol 17: 104–114.T. BorgesK. ErziniL. BentesM. CostaJ. Gonçalves2001By-catch and discarding practices in five Algarve (southern Portugal) métiers.J Appl Ichthyol17104114
30. Erzini K, Costa M, Bentes L, Borges T (2002) A comparative study of the species composition of discards from five fisheries from Algarve (southern Portugal). Fish Manag Ecol 9: 31–40.K. ErziniM. CostaL. BentesT. Borges2002A comparative study of the species composition of discards from five fisheries from Algarve (southern Portugal).Fish Manag Ecol93140
31. Monteiro P, Araújo A, Erzini K, Castro M (2001) Discards of the Algarve (southern Portugal) crustacean trawl fishery. Hydrobiologia 449: 267–277.P. MonteiroA. AraújoK. ErziniM. Castro2001Discards of the Algarve (southern Portugal) crustacean trawl fishery.Hydrobiologia449267277
32. Viriato A, Figueiredo M (1991) Topografia submarina dos fundos de crustáceos da vertente Algarvia. Lisbon, Portugal: Relatórios Científicos e Técnicos do INIP, n° 43. 31 p.A. ViriatoM. Figueiredo1991Topografia submarina dos fundos de crustáceos da vertente AlgarviaLisbon, PortugalRelatórios Científicos e Técnicos do INIP, n° 43p. 31 p.
33. Figueiredo M (1989) Distribuição batimétrica do lagostim e espécies associadas de interesse commercial, ao longo da costa continental portuguesa. Lisbon, Portugal: INIP. Relatórios Científicos e Técnicos do INIP, n°12. 53 p.M. Figueiredo1989Distribuição batimétrica do lagostim e espécies associadas de interesse commercial, ao longo da costa continental portuguesaLisbon, PortugalINIP. Relatórios Científicos e Técnicos do INIP, n°12p. 53 p.
34. Viriato A, Figueiredo M, Figueiredo I, Correia J (1996) Atlas de apoio à pesca de arrasto na vertente continental portuguesa. Lisbon, Portugal: 51 p.A. ViriatoM. FigueiredoI. FigueiredoJ. Correia1996Atlas de apoio à pesca de arrasto na vertente continental portuguesaLisbon, Portugalp. 51 p.
35. Bergmann M, Wieczorek S, Moore P, Atkinson R (2002) Discard composition of the *Nephrops* fishery in the Clyde Sea area, Scotland. Fish Res 169–183.M. BergmannS. WieczorekP. MooreR. Atkinson2002Discard composition of the *Nephrops* fishery in the Clyde Sea area, Scotland.Fish Res169183
36. Evans S, Hunter J, Wahju E, Wahju R (1994) Composition and fate of the catch and bycatch in the Farne Deep (North Sea) *Nephrops* fishery. ICES J Mar Sci 51: 155–168.S. EvansJ. HunterE. WahjuR. Wahju1994Composition and fate of the catch and bycatch in the Farne Deep (North Sea) *Nephrops* fishery.ICES J Mar Sci51155168
37. Dounas C, Davies I, Triantafyllou G, Koulouri P, Petihakis G, et al. (2007) Large-scale impacts of bottom trawling on shelf primary productivity. Cont Shelf Res 27: 2198–

2210.C. DounasI. DaviesG. TriantafyllouP. KoulouriG. Petihakis2007Large-scale impacts of bottom trawling on shelf primary productivity.Cont Shelf Res2721982210

38. Jones J (1992) Environmental impact of trawling on the seabed: a review. N Z J Mar Freshwater Res 26: 59–67.J. Jones1992Environmental impact of trawling on the seabed: a review.N Z J Mar Freshwater Res265967

39. Nilsson H, Rosenberg R (2003) Effects on marine sedimentary habitats of experimental trawling analyzed by sediment profile imagery. J Exp Mar Biol Ecol 285–286: 453–463.H. NilssonR. Rosenberg2003Effects on marine sedimentary habitats of experimental trawling analyzed by sediment profile imagery.J Exp Mar Biol Ecol285–286453463

40. Olsgard F, Schaanning M, Widdicombe S, Kendall M, Austen M (2008) Effects of bottom trawling on ecosystem functioning. J Exp Mar Biol Ecol 366: 123–133.F. OlsgardM. SchaanningS. WiddicombeM. KendallM. Austen2008Effects of bottom trawling on ecosystem functioning.J Exp Mar Biol Ecol366123133

41. Queirós A, Hiddink J, Kaiser M, Hinz H (2006) Effects of chronic bottom trawling disturbance on benthic biomass, production and size spectra in different habitats. J Exp Mar Biol Ecol 335: 91–103.A. QueirósJ. HiddinkM. KaiserH. Hinz2006Effects of chronic bottom trawling disturbance on benthic biomass, production and size spectra in different habitats.J Exp Mar Biol Ecol33591103

CHAPTER 17

A Spatial Method to Calculate Small-Scale Fisheries Effort in Data Poor Scenarios

Andrew Frederick Johnson[1*], Marcia Moreno-BaÂez[1,2], Alfredo Giron-Nava[1], Julia Corominas[3], Brad Erisman[4], Exequiel Ezcurra[5], Octavio Aburto-Oropeza[1]

[1] *Marine Biology Research Division, Scripps Institution of Oceanography, La Jolla, CA, United States ofAmerica,*

[2] *Centro para la Biodiversidad Marina y la ConservacioÂn, La Paz, Mexico,*

[3] *Universitat PoliteÂcnicade Catalunya, Barcelona, Spain,*

[4] *Marine Science Institute, University of Texas, Austin, Texas, United States of America,*

[5] *UC MEXUS, University of California Riverside, Riverside, California, United States of America*

ABSTRACT

To gauge the collateral impacts of fishing we must know where fishing boats operate and how much they fish. Although small-scale fisheries land approximately the same amount of fish for human consumption as industrial fleets globally, methods of estimating their fishing effort are comparatively poor. We present an accessible, spatial method of calculating the effort of small-scale fisheries based on two simple measures that are available, or at least easily estimated, in even the most data-poor fisheries: the number of boats and the local coastal human population. We illustrate the method using a small-scale fisheries case study from the Gulf of California, Mexico, and show that our measure of Predicted Fishing Effort (PFE), measured as the number of boats operating in a given area per day adjusted by the number of people in local coastal populations, can accurately predict fisheries landings in the Gulf. Comparing our values of

PFE to commercial fishery landings throughout the Gulf also indicates that the current number of small-scale fishing boats in the Gulf is approximately double what is required to land theoretical maximum fish biomass. Our method is fishery-type independent and can be used to quantitatively evaluate the efficacy of growth in small-scale fisheries. This new method provides an important first step towards estimating the fishing effort of small-scale fleets globally.

INTRODUCTION

Global marine fisheries are under pressure from increasing demands for protein, driven by rapidly growing human populations [1,2]. Many commercial stocks remain either fully or over-exploited and show continued declines [3], whilst rules and regulations governing them continue to tighten, driving new fisheries development [4]. Fisheries biomass is often extracted from areas only known by fishers themselves, and is recorded at the point of landing or where the first sale of the catch is made. For this reason, fishing locations often remain elusive, off the radar and difficult to police [5]. Understanding where fish are caught has never been more important. The amount of fishing (the effort) and the area over which fishing potentially takes place (the range) are critical measures for successful fisheries management and making realistic predictions about the wider ecological consequences of fishing [6], although for some specific fishery scenarios this is not always the case [7]. These measures provide important details for management, particularly spatially-explicit approaches such as protected areas or fisheries closures. They also allow more accurate estimates of the economics of fishing regarding time at sea and fuel usage [8,9] and aid in understanding some of the social factors involved in driving fisheries [10].

Landed catch biomass data are now widely recorded for commercial fisheries and are a central component in the design of many fisheries management strategies. Whether or not they are a useful measure of stock status, however, remains undecided [11]. Effort data for fisheries remain comparatively sparse and are often concentrated around the industrial fleet, relying on electronic logbook-ID-type tracking systems such as VMS (Vessel Monitoring System) and AIS (Automatic Identification System) [12]. Methods of calculating fishing effort ultimately depend on the intended applications of such information [13] and different resolutions of data will yield different descriptions of effort [14]. Although this highlights the importance of understanding why such measures are needed, data availability will often be the primary factor determining which measures and calculations can be made [6]. Considering the economic and the multinational interests often involved, fisheries privacy and data release issues are commonly more of an impediment than the existence of data. Small-scale fisheries often lack data on the number of vessels, gear types used and landing records. This makes the calculation of fishing effort difficult and may bring into question the accuracy of any such results gained for such fleets [13].

The current gap in knowledge for small-scale fishing effort needs addressing if the collateral impacts of such fisheries are to be quantitatively estimated. The importance of understanding these impacts is reiterated when we consider that landings for human consumption from small-scale fisheries are approximately equal to that of the industrial fleet [15]. As McCluskey and Lewison note, *"Small scale does not necessarily mean small impact"* [13]. The approaches that estimate fishing effort for small-scale fisheries are currently labor- and data-intensive, require close cooperation with the fishing sector and involve considerable sampling error and subjectivity, as they do for many data-poor fishery scenarios [16].

The term "effort" encompasses a broad range of meanings and complexities, from nominal measures such as durations of extraction and amounts or sizes of extractive gear to effective measures that standardize effort such as the rate of fish capture or the Catch Per Unit of Effort (CPUE). Such detailed information of a system is often unavailable. For this reason, instead of attempting to design a method that estimates effort based on many case-specific assumptions, we present a new method to estimate fishing effort across broad spatial scales, accounting for data limitations commonly encountered for small-scale fisheries. We did not address temporal scales in the example presented here because of a lack of data, which is a common issue in many other data-poor settings. To illustrate our method we use small-scale fisheries from the Gulf of California, Mexico as a case-study.

We use the term Predicted Fishing Effort (PFE) to describe our new measure, as it estimates the amount of fishing activity (based on the number of boats) within an estimated fishing range. We hypothesize that the human population and the numbers of fishing boats together determine the fishing effort in a given area and the subsequent amount of fish landed locally. PFE is therefore measured as the number of boats per day in an area (in this study a grid cell of 500 km^2). If PFE estimates are compared to total commercial fisheries landings, our method can also be used to 1) predict the present amount of fish biomass extraction from areas of previously unknown activity; 2) evaluate the potential for future growth in a fishery; and 3) provide an estimate of the most cost-effective number of fishing boats per area without diminishing returns.

METHODS

Unless the relationship between local human populations and fishing boats are perfectly auto-correlated and track each other linearly and normally (i.e. more humans always mean more boats and vice versa), both measures are necessary to calculate a reliable estimate of PFE. By accounting for both measures, areas with high numbers of boats but low numbers of local populations overall (e.g. isolated fishing camps), or areas where there may be low numbers of boats close to extremely high local populations (e.g. major cities with relatively few per-capita fishing vessels) are accounted for.

Data origin

A schematic flow-through diagram illustrating each step in the process of designing and calculating PFE is presented in Fig 1. Human population data were obtained from Mexico's National Institute of Statistics and Geography (INEGI) [17]. These data were geo-referenced by community. A 5 km buffer inland from the coastline was created to select the majority of the populous within the coastal communities likely to have direct links to the local fishing industry (Fig 2A). The number and location of every coastal in-water small-scale fishing boat in the Gulf was calculated from three different sources: two coastal over-flight surveys conducted by the World Wildlife Fund in 2006 [18] and an estimate using coastal images of the Gulf taken from Google Earth in 2009 [19] (Fig 2B). Small-scale boats were defined as any traditional Mexican panga fishing vessel. These are out-board engine-driven, open vessels up to approximately 27 feet (8.2 meters) in length, usually operated by a maximum of 3 crew members with little to no mechanical fishing equipment on board. It is noteworthy that no actual field research was undertaken and therefore no ethics clearance or permits were necessary for data collection. All the data used and produced in these analyses are open source and available from http://datamares.ucsd.edu/eng/projects/fisheries/spatial-distribution-of-potential-fishing-effort-in-the-gulf-of-california-mexico/.

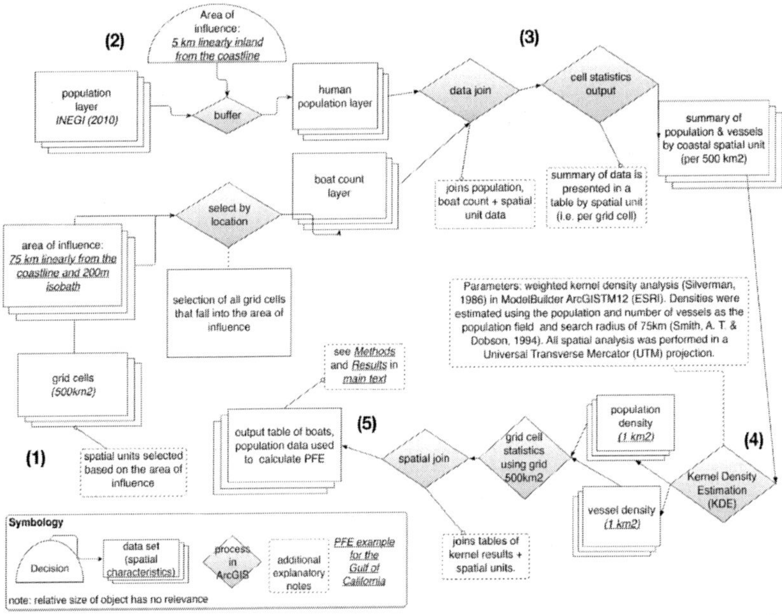

Figure 1. Schematic flow-through diagram describing each stage of the method used to calculate Predicted Fishing Effort (PFE).

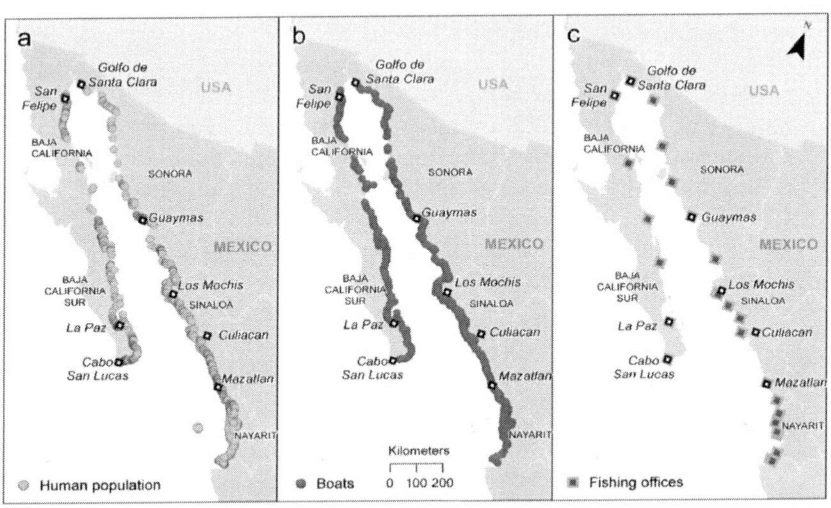

Figure 2. Map of the case-study area, the Gulf of California, used to design the model of Predicted Fishing Effort (PFE).

Spatial analysis and data distribution

There is a substantial difference between the areas fished by the small-scale (coastal) and the industrial (off-shore) fleets within the Gulf of California [20]. The distances travelled by fishers to fishing grounds can therefore vary widely and are determined by factors including target species, time of year, coastline complexity, water depth and weather. The average linear distance travelled by small-scale fishers to their fishing ground in the upper Gulf, was 75 km between 2009 and 2016 [21]. We used this measure to create a buffer delineating the area fished by the small-scale fleet, within which to calculate the PFE. The resultant area that could potentially be fished (herein referred to as the *area of influence*) corresponded well to other published estimates of small-scale vessel activity in the Gulf of California [22]. In ideal scenarios where data are not limited, such areas of influence can be calculated on a case-by-case basis and based on accurate estimates of average areas of activity. Over-conservative buffer areas that are too small will concentrate PFE estimates, whilst buffer areas that are too large will spread PFE estimates too thinly.

To calculate a measure of effort spread throughout the area of influence we created a grid and, within the center of each cell, placed estimated values of the number of boats and human population. We used a conservative grid cell size of 500 km^2 cells based on the estimated potential fishing area per panga vessel in the Gulf [23]. Only grid cells that intersected with the area of influence were included for further analysis. Values of population and boats were assigned to the centroid of each corresponding cell (herein referred to as the *raw data*). Data on human population size and the number of boats covered a total of 133 grid cells. We used kernel density estimates (KDEs) to spread the data for the human population, the number of boats and the mean total annual catch across the full area of influence (565 grid cells of 500 km^2). We computed the KDEs with the Spatial Analyst Extension in Model Builder ESRI ArcGIS 10.2 by using the following function [24]:

$$f(u) = \frac{1}{nh} \sum_{i=1}^{n} K(u)$$

Where n represents the total number of input points, *h* represents the search radius (fishing range) of fishers from the departure point (set as 75 km), *K* represents the Kernel function (described below) and *u* represents the input variable (human population, number of boats or total catch (tonnes)). The output of this function has the same units as the input variable, but normalized per unit area (cell-size parameter = 1 km²). The Kernel function is described as: $K(u) = \frac{3}{4}(1 - u^2) \quad |x - x_i| \leq h$

$$K(u) = 0$$

$$|x - x_i| > h$$

This function describes that for each data point with coordinates x_i from the original dataset, the KDE assigns densities around it according to a quadratic distribution (with a maximum at the original location of the data point) and up to a distance equal to h (75 km) in the interpolated grid with coordinates x for each cell. We transformed the 1 km² resolution data into the 565 grid cells of 500 km² by calculating the mean KDE of each cell (S1 Fig). All the analyses were performed using a Universal Transverse Mercator (UTM) projection.

To build a model of PFE based on human population and the number of boats, we hypothesized that there was a significant relationship between these two variables. To test this hypothesis we ran an Ordinary Least Squares (OLS) regression. This was undertaken for both the raw data (n = 133) and the post-KDE data matrix (n = 565) to ensure KDE did not alter the relationship between the variables. Significant slope and intercept statistics were then used to build the final PFE model in which PFE was a function of human population and the number of boats. In certain cases or locations, this relationship may not be significant, for example, in very remote or transitory fishing scenarios where the local human population may not relate to the number of boats in the area. To understand if a location may deviate from this linear norm of more population equals more boats, simple regression analyses between the two measures can be used. Outlying cases can either be noted and investigated further, or if justified, excluded from further analysis. In cases with many such non-normal cases, different non-linear regression models may be used to see if a significant relationship between human populations and the number of boats exists. Final significant models should to be incorporated into the PFE estimate using the scaling coefficient of the chosen model (see results for example). If no such relationship is found, then our method should not be used to calculate PFE. We do, however, encourage researchers in data-poor scenarios to persist in their calculations of fishing effort and note that the more that is known about how a fishery and how fishers behaves, the better equipped a team is to investigate ways to accurately estimate fishing effort. The suitability of the human population and boat estimates within areas of influence should be thoroughly evaluated a priori.

Designing and validating an estimation of effort

The relationship between boats and human population.

The intensity of fishing depends, in principle, on the demand for fish whether for consumption (local markets) or revenue (non-local markets) purposes. In theory, two variables can consequently be used as indicators of this demand, and hence proxies for PFE: (a) the number of boats in each segment of the coast, and (b) the human population concentrated along the same coastal sector. However, these two variables are in all likelihood related to each other, so they cannot be used as additive predictors of PFE as they may in turn be aliased. We therefore first analyzed how the number of boats scaled-up as the coastal population increases, using allometric equations (or power functions, of the type $boats = k.population^{m}$, or, in logarithmic form, $\log(boats) = \log(k) + m. \log(population)$). We found a significant, positive relationship between the logarithm of the number of fishing boats and the logarithm of coastal human population, both pre- ($r^2 = 0.40$, $F_{1,131} = 89.51$, $p < 0.001$, $m = 0.34 \pm 0.03$) and post- ($r^2 = 0.65$, $F_{1,563} = 1039$, $p < 0.001$) KDE (Fig 3A and 3B), yielding a scaling coefficient $m = 0.42 \pm 0.013$ (we added a value of 1 to both the number of boats and the number of inhabitants to avoid the indetermination of log-zero).

The slope for the least squares regression in Fig 3B represents the log-log relationship for the allometric scaling between the number of boats and the population: $\log(boats) = 0.432 \times \log(population)$. We forced the intercept through the origin given that the absence of people or boats in a location would result in no fishing activity. The equation can be rewritten as $\log(boats) = \log(population^{0.43})$. Further, by taking the exponent of both sides, it can be written in a power function form as $boats = population^{0.43}$. That is, the population-driven number of boats in any sector of the coast can be predicted by the population elevated to an exponent of 0.432.

Estimating PFE.

The number of boats counted (*boats*) and the number of boats estimated from the population ($population^{0.43}$) were linearly related with slope = 1. We therefore combined both variables into a single interaction term to estimate the PFE as the geometric mean of both estimates ($PFE = (boats \times population^{0.43})^{0.5}$). Because both estimates are aliased, combining them into a single variable reduces the degrees of freedom of the predictors, making the resulting model more parsimonious. By using a geometric mean, we make sure that extreme cases (high numbers of boats with low coastal populations or vice versa) are accounted for without skewing PFE values. Finally, because in the previous section we showed that *boats* and $population^{0.43}$ are linearly related and hence additive in log-form, it follows that in exponent form they should be combined as

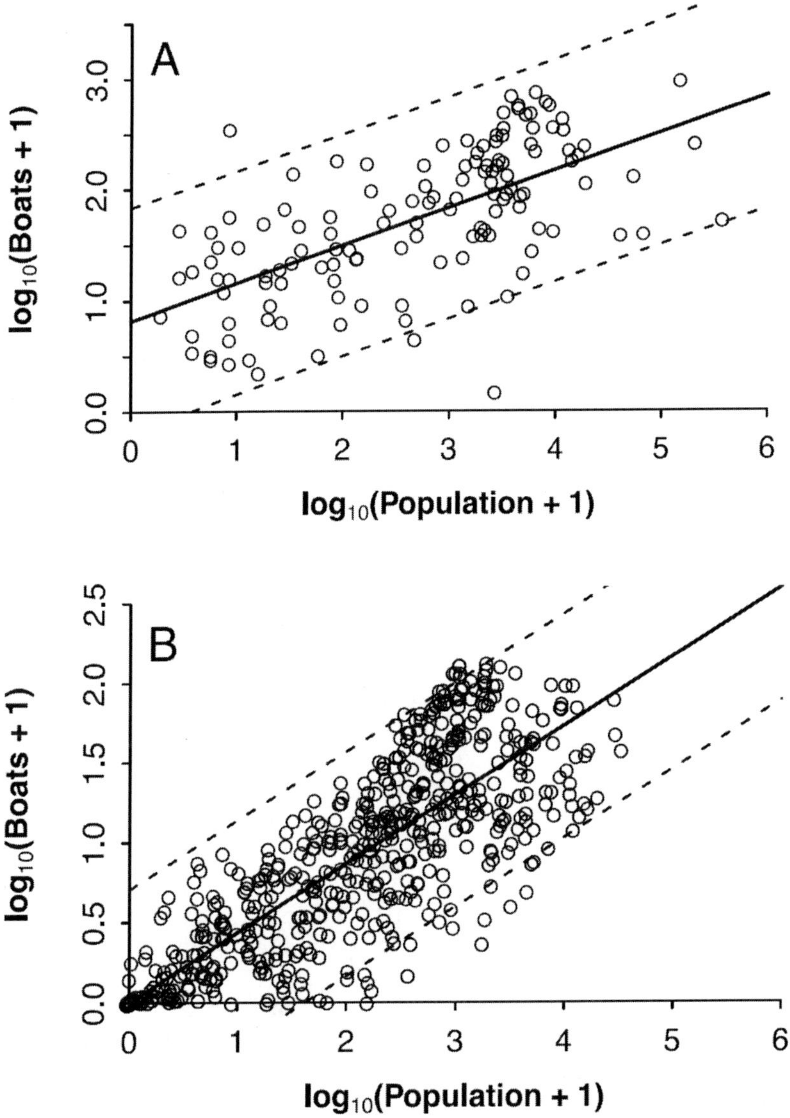

Figure 3. Relationship Between Human Coastal Population And Number Of Small Scale Fishing Boats Pre (A) And Post (B) Kernel Density Estimation. Points represent the raw data, solid lines the fitted model and dashed lines the 95% confidence intervals (see also S1A and S1B Fig for mapped post KDE data).

multipliers. The unit of PFE is therefore the number of boats adjusted by the number of people in local coastal populations. In this case, PFE is calculated using the mean boat count from the three daily snapshots of boat activity in the Gulf. Overall, PFE is measured as the adjusted number of boats per day in a given area (in this case a grid cell of 500km^2).

Validating the estimation of PFE.

To validate our measure of PFE we used vessel-tracking data from the upper Gulf of California region. A local collaboration with fishers using hand-held GPS trackers [25] provided us with real travel-tracks of fishing vessels for an average fishing day in the area. From these tracks, we calculated the total number of times fishing vessels fished (see S2 Fig) inside the same grid cell layer that we used for the PFE calculation. We then compared our estimated values of PFE to these real fishing event totals for each grid cell in the upper Gulf of California. OLS linear regression and a Pearson's correlation of these values were used to determine if a linear relationship between the real and the predicted measures exist and to note the significance of the correlation between the two variables.

Predicting landings from PFE

We collated annual fisheries data (2001–2013) from the Gulf of California from the Mexican National Commission of Fisheries (CONAPESCA). These data included all small-scale fisheries landings from 31 fishing offices across the whole Gulf (Fig 2C), which collectively register commercial landings from five states: Baja California, Baja California Sur, Sonora, Sinaloa, and Nayarit. These fishing offices are located at major ports and important fishing communities, and provide the most detailed fisheries data available. An independent database that included monthly landings for around 300 species of fish and invertebrates from all 31 offices was obtained from CONAPESCA headquarters in Mazatlan, Sinaloa, Mexico. Fishers are either affiliated to cooperatives or work for private companies that operate based on licenses or concessions granted by CONAPESCA, per resource. All information generated by this federal entity (e.g., human resources, infrastructure, the number of licenses by fishing resource) is included in the National Fisheries Register.

KDE was used to estimate spatially explicit values of mean total annual fisheries landings data (tonnes per 500 km^2) between 2001 and 2013 for each cell within the area of influence. By comparing PFE to mean total annual landings of all species in the region, we hypothesized that our model of PFE could be used to accurately predict total fisheries landings in the Gulf of California. We tested the relationship with a non-linear least squares regression. It should be noted that because we only consider mean total annual fisheries landings, the PFE estimate is fishery-type independent. If differences between types of fishery need investigation, the method would need to separate different types of fishing boats in the boat counts, calculate a separate PFE for each with fishery-specific assumptions on the range of the fishing boats and separate catch by fishery when comparing predicted landings to PFE estimates.

Fisheries catch is total weight of fish landed, hence the variable is related to fish and/or human demographic processes. In terms of differential calculus, a per capita rate of change is equal to the rate of change of the logarithm of the variable ($(1/y)\ dy/dx = d\ \ln(y)/dx$). We therefore first explored the relationship between PFE and log(catch), to look at the relative change in catch as PFE increases. Once the log model was defined and tested, we transformed it into its arithmetic form to make predictions about expected catch given a certain PFE in any given sector of the coast. The final model ($catch = K * e^{-b/PFE}$) was fitted to the data using non-linear regression with multiplicative error estimations. To avoid the influence of spatial autocorrelation in our model parameters estimation, we performed 500 random draws of 250 data points each from the PFE and the catch data. For each draw we fitted a logistic curve and estimated K, b and their respective 95% confidence intervals. None of the evaluated fits from the draws showed significant spatial autocorrelation (Mantel test). The final parameters and confidence intervals were calculated as the mean of the 500 estimates generated from the random draws. To approximate normality and homogenize variances, all data was $\log_{10}(x+1)$ transformed after KDE was applied. All models were run using R (2016) and the *lm* and *nlm* packages.

RESULTS

We estimated that at any one time there were approximately 17,839 ± 109.42 (standard error of the mean) fishing pangas operating in the coastal waters of the Gulf (between 2006 and 2009). A total of 1,399,951 people lived in coastal communities within 5 km of the coastline in 2010 [17], the majority of them in the 31 cities with fishing offices reporting to CONAPESCA. In 2008, the Mexican National Institute of Statistics reported that 56,113 people (4% of the total coastal population surrounding the Gulf of California) earned profits directly from the fisheries and aquaculture sector in the Gulf of California. Assuming 2 to 3 fishers work per panga, this represents 64% to 95% of this sector.

The final PFE estimates for the Gulf of California (Fig 4) suggest that fishing effort was highest close to highly populated human settlements on the coast and/or areas where the number of fishing vessels was high. The highest PFE values spread from the Guaymas (in the north) to Mazatlán (in the south) along the Sonoran and Sinaloan coastline (eastern side of the Gulf of California). La Paz and Cabo San Lucas, the largest cities in this region located towards the south of the Baja California Peninsula, had high values of PFE. San Felipe and Santa Clara in the upper Gulf of California also had high PFE values because of the high numbers of fishing boats in these communities compared with the size of their local populations.

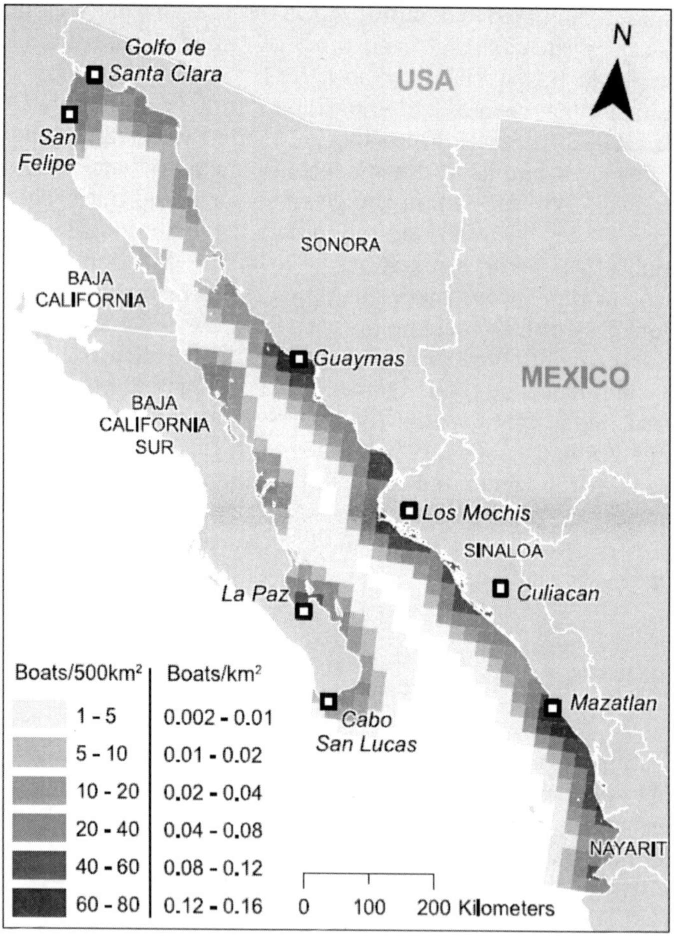

Figure 4. The Predicted Fishing Effort (PFE) presented in 500 km^2 grid cells (n = 565).The unit of PFE is the number of boats adjusted by the number of people in local coastal populations.

Does PFE relate to catch?

The relationship between PFE and log(*catch*) was asymptotic, in which the relative rate of change (the slope of the fitted trend) decreased as PFE increased, of the type log(*catch*) = log(*K*) −*b*/PFE, where *K* is the maximum total catch and *b* is a scaling or slope parameter describing how fast total catch approaches *K* as PFE increases (Fig 5A). Clearly, as PFE increases total catch approaches *K*, while as PFE decreases total catch approaches zero.

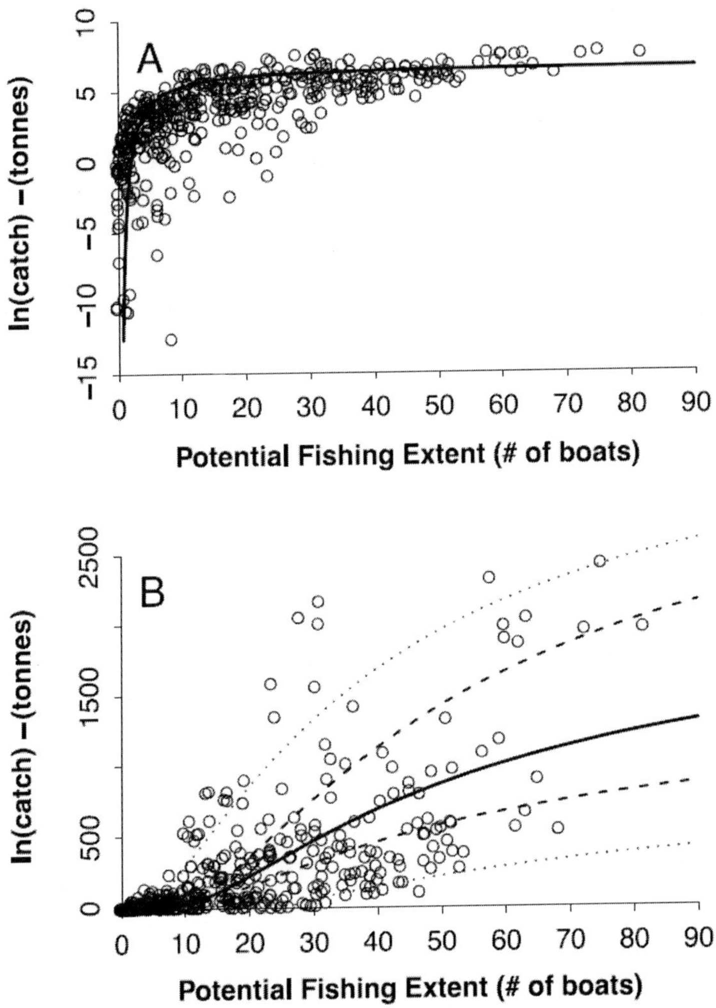

Figure 5. Relationship between Predicted Fishing Effort (# of boats per 500 km²) and mean total annual catch.

In logarithmic form (A) and arithmetic form (B). Points represent raw data (one value per 500 km^2 grid cell), solid lines are the fitted non-linear models, the dashed lines are the 95% prediction intervals of the fit, and the outer dotted lines show one standard deviation for the regression residuals. Note the funnel-shaped errors: as PFE increases so does dispersion in the data.

The fit between PFE and mean total annual catch for the arithmetic model was highly significant ($r^2 = 0.57$; $p < 0.001$) with parameter values K = 2204 tonnes per 500 km^2 per year (95% CI [1451, 3629]) and b = 47.0 (95% CI [31.9, 67.0]; Fig 5B). The fit supports our hypothesis that the use of human population and the number of boats to create a model of PFE can predict fisheries landings in the Gulf of California, whilst accounting for situations that deviate from the linear relationship commonly assumed between local population and the number of boats (small populations with relatively high numbers of boats and large populations with relatively few boats). While PFE has no maximum value (i.e., mathematically it can grow indefinitely), the predicted total catch is bounded between zero and K. The fitted function does not show uniform variances; rather, the variability of the relationship between PFE and total catch increases considerably as PFE grows, exacerbating uncertainty that investing in more boats will increase landings and profits.

The increase in total catch starts to plateau at PFE values of > 23.5 boats per 500km^2, highlighting the importance of considering K as the maximum total catch per grid cell per year. This inflexion point seen in Fig 5B occurs at total catch values of 298 tonnes per 500 km^2 per year. After this point, the rate of increase of total catch per number of boats begins to plateau. This suggests that the number of boats at any moment required to reach total catch levels with no further, significant increase in total catch is approximately 23.5 per 500km^2. Extrapolating this further to the total area of influence (the area we predict is utilized by small-scale fishers in the Gulf = 565 cells = 565*500 = 282,500 km^2), diminishing catch returns begin when the number of pangas reaches 13,277 at any given moment. This value should, however, be taken with caution as it implies that all areas have the same historical level of fishing. It is much more plausible that different areas have historically different levels of fishing and are therefore at different levels of stock status. On the premise that some of the sites have experienced some level of fishing before our data were collected and are not in a pristine, non-fished condition, the extrapolated value likely overestimates the number of boats required to produce a level of catch at the inflection point of the model.

It should be noted that we validated the PFE model by testing our PFE values against fishing event values from the upper Gulf of California, where we had access to real tracking information for the small-scale fleet in the area (Fig 6A). OLS regression and a Pearson's correlation between our estimated PFE values and the real fishing frequency per cell showed a highly significant relationship ($r^2 = 0.43$; $p < 0.001$, $\rho = 0.65$, df = 28) (Fig 6B). Two points that lay outside of the 95% prediction limits of the regression were identified as grid cells in which local fishers stop to clean fish on their way back to port, thus replicating the same movement as at the start of a fishing event.

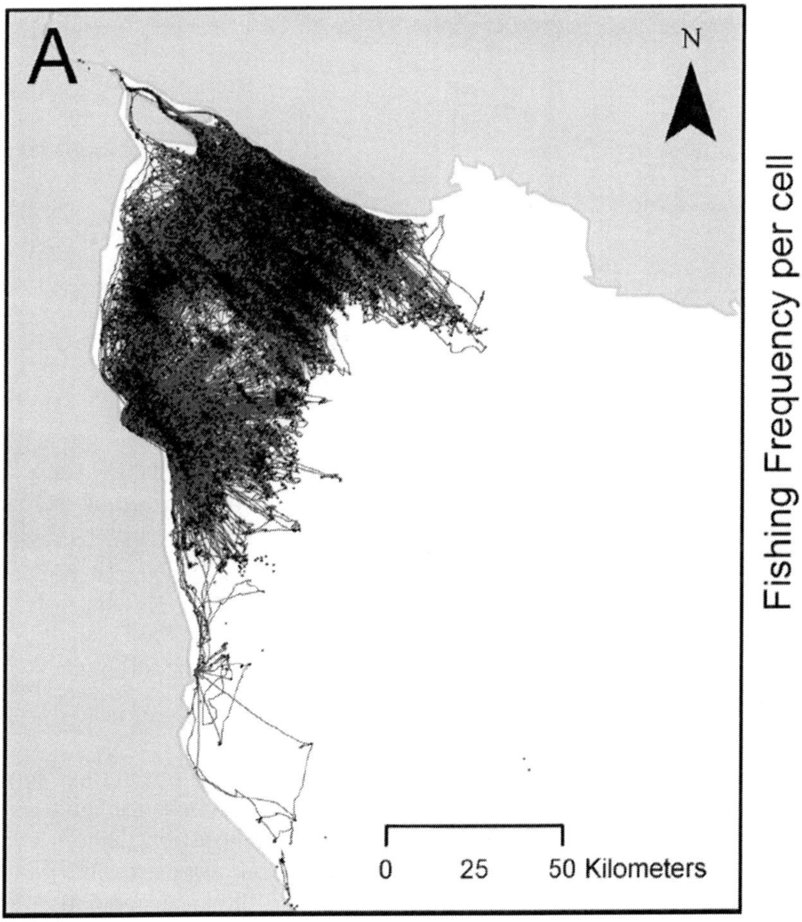

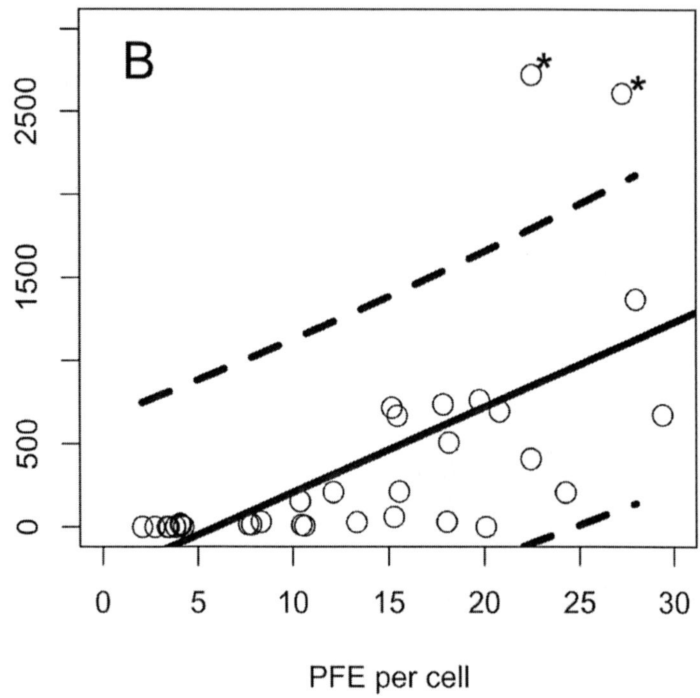

Figure 6. Map of the upper gulf of california showing fishing track lines of small-scale fishing vessels (a) and the relationship between pfe per cell and fishing frequency per cell (b). Asterisks denote outliers, solid line represents the fitted model and broken lines represent 95% confidence intervals.

DISCUSSION

Estimating fishing effort is a necessary requisite for the successful management of marine fisheries. An almost ubiquitous problem with small-scale fisheries is that unless boats are fitted with tracking devices, once they depart on fishing trips, they are for all practical purposes untraceable [13]. By using KDE and the relationship between the local population and the number of boats, we present a new method of estimating the amount of fishing effort within a given area, when supplementary data regarding the fishery are limited or non-existent. In the study area used to build our estimate of fishing effort, PFE was highest in areas of high population density, but also in areas with high numbers of fishing boats relative to local human populations. When incorporating data on locally landed catch, our PFE estimate can also be used to broadly assess the overall efficacy of growth of a small-scale fishery and predict the catch from areas where catch data are unavailable but information on the number of boats and the human population is accessible. Based on our estimates, at any one time in the Gulf of California there were 4,562 more panga vessels operating than the maximum number (13,277) estimated to produce total catches without diminishing returns. We show that the small-scale fishing fleet is at over capacity considering the amount of fish recorded in official catch reports.

Similar methods to the one we present here could also be used to build models that apply to other natural resource extractions that exhibit predictive relationships between local human populations and the number of production units (e.g. boats, diggers, machines etc.) used in the extractive activity, as well as those between the resource extracted and the number of production units measured. Our new method is aimed as a starting point for researchers working in data-poor scenarios tasked with calculating metrics of fishing effort. However, many assumptions and considerations must be acknowledged before the method can be applied in other areas.

Considerations

The design of our model and the use of the PFE to predict catch and evaluate fisheries growth is based on two significant positive relationships: 1) human population versus the number of boats and 2) the PFE versus the total catch. Satisfying these relationships is a valid assumption for fisheries [16] and to our knowledge there is no literature that suggests a lack of support for the hypothesis that increased coastal human populations generally lead to increased numbers of fishing vessels. An increased catch as a result of increased effort is a common, reasonable assumption which is widely supported until a system is overexploited and a stock begins to collapse [26].

Although our model only requires two simple measures, how these are derived may have a significant bearing on the outputs from the method. For example, the accuracy of estimates for the number of boats is likely to vary widely depending on how such boat-count data are collected. In our example, we benefited from aerial survey estimates, which are unlikely to be available in many small-scale fishery scenarios. Observer data for moored boats and boats within site of land are the most probable methods of boat number estimation in many cases. We believe that using additional, reliably sourced data to calculate mean values is a more robust method than using single year data values. For this reason, we calculated a mean of 3 estimates of the number of boats spanning 4 years, as well as more than a decade of catch data to estimate the total mean annual catch per unit area (tonnes / 500km^2 / year). Our final PFE estimates are therefore a better general reflection of fishing effort and its relationship to mean annual total landed catch per area for the Gulf than if we were to rely on single-value estimates. If increased temporal accuracy is required for an effort estimate, it will be necessary to collect data on fishing boats, human population and biomass landed within the time period for which estimates are required. The accuracy and potential error associated with each of these variables, and the amount of fish landed should still, however, always be thoroughly acknowledged.

The method we present assumes a reasonable knowledge of the mean distance travelled by fishers during an average trip and the extent of local human communities likely to be closely involved with local small-scale fishing fleets. Our validation method benefited from high resolution, voluntary tracking data supplied by fishers [21] and the isolated and discrete coastal communities along the Baja California peninsula. Such a scenario is rare, but approximate estimates of distances travelled by fishers can be calculated by time at sea in conjunction with additional knowledge of fishing behavior and vessel range, perhaps gleaned through simple questionnaires [27]. It is important to note that by extrapolating our validation from a specific location (the Upper Gulf) to the whole Gulf, we missed potential differences in the productivity or fishing activity of different areas. An ideal scenario would be to have multiple sets of tracking data from all over the Gulf. This, however was not available and, as with much of our approach, the idea was to make a best attempt considering such data-limited scenarios.

As a model estimate, our PFE should be applied and interpreted with a degree of caution based on the aforementioned assumptions as well as the following limitations. The method does not account for protected areas. Protected areas, however, can easily be demarcated when defining the overall area of influence. Additionally, only using small-scale boats in our calculation of PFE excluded industrial fisheries from our estimates. Conclusions made regarding the collateral impacts of local fisheries should only be drawn for the fleet for which boat estimates were taken and should acknowledge activity from those vessels not included in boat number estimates built into the primary model. Addressing these limitations and assumptions is an area that will likely prove fruitful in further developing the method to increase its accuracy and reliability. The method will also benefit from more road testing in places with different types of fisheries, economic strategies and regulatory environments. Having said this, we believe that our PFE estimates for the Gulf of California accurately captures the effort of the small-scale fleet [28,29].

Fishing effort in the Gulf of California

By building a model to estimate PFE we were able to estimate and visualize where small-scale fishers are most active in the Gulf of California. Comparison of PFE estimates to total landed catch data showed that future investment in the Gulf's fisheries may not be economically or ecologically viable. Assuming consistent landed values of fish, the catch plateau at around 23.5 boats per 500 km^2 per day and indicated there may be little economic sense in adding more boats close to, or past this point in many areas of the Gulf. In areas with PFE values over 23.5, economic and ecological maxima may have already been reached. Increasing fishing in these areas will likely yield minimal additional landings while increasing economic costs and damage to exploited stocks and local ecosystems. This means that more biomass per grid cell is being removed than we built into the model. Fisheries in the areas of the Gulf with high PFE values are therefore likely to underperform, like many other global fisheries. In total, we estimate that 17,839 small-scale boats operated at any given moment in the Gulf between 2006 and 2009, 34% more than would produce maximum total catch without diminished returns. This value increases to 88%, almost double over capacity, when considering more recent estimates (2010) of total boats operating (25,000 boats approx. [30]). We therefore clearly demonstrated overcapacity in the Gulf's small-scale fishing fleet. Our estimates are also highly conservative as they do not account for a growing fleet [30], illegal or unreported fisheries, bycatch or the ecosystem impacts of the commercial fleet, all of which will affect standing stock biomass. Future fisheries growth in the Gulf of California should therefore be addressed very carefully with a long-term outlook that accounts for ecological sustainability, economic efficiency and the effort of the local fishing fleets.

CONCLUSIONS

We have presented a new method that allows small-scale fisheries effort to be estimated based on simple measures attainable, in even the most data-poor scenarios. If the same step-by-step methodology is used, all assumptions backed by robust data and limitations acknowledged, this method will facilitate predicting fishing effort for other data-poor fisheries. Estimates of human population are now commonplace and easily attainable as are estimates of landed catch in FAO-type databases [31], albeit at regional and national scales. We foresee the main data limitation coming from estimates of the number of fishing boats present in a given area. New, high-resolution satellite technologies appear the best way to address this issue in areas where accurate estimates may be difficult to otherwise calculate. When images from these technologies become easily accessible and available in all areas, the challenge will then be to source workforces or automated methods to make counts of the boats. Resultant effort estimates will prove important in highlighting areas that experience high fisheries impacts, and places that may be candidates for protection. Although we do not suggest that our model is exhaustive and fully incorporates all variability in natural and human systems, we believe it is a robust and valid start to elucidating the fishing effort of small-scale fisheries. This new method of effort estimation moves us away from data-intensive and laborious methods of investigating this important measure. Our method predicts the ecological impacts of increasingly concentrated fishing efforts, as well as the economic potential for future expansion. We hope that our PFE method can shed some light on the effects of small-scale fisheries globally. This will benefit management and conservation decisions if presented objectively and in a timely manner, particularly in data-poor scenarios.

ACKNOWLEDGMENTS

We would like to thank CONAPESCA, WWF and T. Plomozo for access to their databases. This work is part of an ongoing effort by the Gulf of California Marine Program (http://gulfprogram.ucsd.edu) to understand the exploitation of resources from the Sea of Cortez and promote the sustainable management of marine fish stocks in the area. We would also like to thank authors contributing to dataMares (http://datamares.ucsd.edu/eng/) whose articles greatly aid the interpretation and dissemination of these results. AGN was funded by CONACYT (CVU 579904) and Fulbright Garcia-Robles (LASPAU ID 20140963) doctoral program fellowships. AFJ was supported by NSF grant DEB-1632648 (2016). We would also like to thank C López-Sagástegui and CR Taft whose comments greatly strengthened this manuscript.

FUNDING: The authors would like to thank the following funders for financial support: CONACYT - the Mexican National Council of Science and Technology (CVU 579904) to AGN; The Fulbright-García Robles Program (LASPAU ID 20140963) doctoral program fellowships to AGN; The Commission for Mexico and the United States (COMEXUS); The Helmsley Charitable Trust (http://helmsleytrust.org/); The International Community Foundation (https://icfdn.org/); The National Science Foundation (https://www.nsf.gov/), grant DEB-1632648 (2016) to AFJ; The Nature Conservancy (http://www.nature.org/); The Packard Foundation (https://www.packard.org/); and The Walton Family Foundation (http://www.waltonfamilyfoundation.org/).

REFERENCES

1. Watson RA, Cheung WWL, Anticamara JA, Sumaila RU, Zeller D, Pauly D. Global marine yield halved as fishing intensity redoubles. Fish and Fisheries. 2012;14: 493–503
2. Rice JC, Garcia SM. Fisheries, food security, climate change, and biodiversity: characteristics of the sector and perspectives on emerging issues. ICES Journal of Marine Science. 2011;68: 1343–1353
3. Ye Y, Cochrane K, Bianchi G, Willmann R, Majkowski J, Tandstad M, et al. Rebuilding global fisheries: the World Summit Goal, costs and benefits. Fish and Fisheries. 2012;14: 174–185
4. Anderson SC, Mills Flemming J, Watson R, Lotze HK. Rapid Global Expansion of Invertebrate Fisheries: Trends, Drivers, and Ecosystem Effects. Bograd SJ, editor. PLoS ONE. 2011;6: e14735–9. pmid:21408090
5. Marsden AD, Sumaila UR. Tracking flows of fisheries products: the case of Pacific halibut in the Canadian economy. Fisheries Research. 2005;73: 259–264
6. Hinz H, Murray LG, Lambert GI, Hiddink JG, Kaiser MJ. Confidentiality over fishing effort data threatens science and management progress. Fish and Fisheries. 2012;14: 110–117
7. Erisman BE, Allen LG, Claisse JT, Pondella DJ II, Miller EF, Murray JH, et al. The illusion of plenty: hyperstability masks collapses in two recreational fisheries that target fish spawning aggregations. Can J Fish Aquat Sci. 2011;68: 1705–1716
8. Cox SP, Walters CJ, Post JR. A Model-Based Evaluation of Active Management of Recreational Fishing Effort. North American Journal of Fisheries Management. 2003;23: 1294–1302
9. Dichmont CM, Pascoe S, Kompas T, Punt AE, Deng R. On implementing maximum economic yield in commercial fisheries. Proceedings of the National Academy of Sciences. 2010;107: 16–21
10. Guillemot N, Léopold M, Cuif M, Chabanet P. Characterization and management of informal fisheries confronted with socio-economic changes in New Caledonia (South Pacific). Fisheries Research. 2009;98: 51–61
11. Hilborn R, Branch TA, Pauly D. Does catch reflect abundance? Nature; 2013
12. Lee J, South AB, Jennings S. Developing reliable, repeatable, and accessible methods to provide high-resolution estimates of fishing-effort distributions from vessel monitoring system (VMS) data. ICES Journal of Marine Science. Oxford University Press; 2010;67: fsq010–1271
13. McCluskey SM, Lewison RL. Quantifying fishing effort: a synthesis of current methods and their applications. Fish and Fisheries. Blackwell Publishing Ltd; 2008;9: 188–200
14. Piet GJ, Quirijns FJ, Robinson L. Potential pressure indicators for fishing, and their data requirements. ICES Journal of Marine Science 2007
15. Jacquet J, Pauly D. Funding Priorities: Big Barriers to Small-Scale Fisheries. Conservation Biology. 2008;22: 832–835. pmid:18637910
16. Stewart KR, Lewison RL, Dunn DC, Bjorkland RH, Kelez S, Halpin PN, et al. Characterizing Fishing Effort and Spatial Extent of Coastal Fisheries. Unsworth RKF, editor. PLoS ONE. 2010;5: e14451–8 pmid:21206903
17. INEGI. Censo de Población y Vivienda 2010. Mexico City; 2011. Available: http://www.inegi.org.mx/est/contenidos/proyectos/ccpv/cpv2010/

18. Rodríguez-Valencia JA, Lópezz-Camacho M, Crespo D, Cisneros-Mata MA. Tamaño y distribución espacial de las flotas pesqueras ribereñas del Golfo de California en el año 2006. [Internet]. 23 Jun 2008 [cited 6 Sep 2016] pp. 1–23. Available: http://awsassets.panda.org downloads rep_tamanio_distribucion_ flotas_ pesqueras_080710.pdf
19. Plomozo Lugo T. Propuesta para la regionalización de la pesca ribereña en el Golfo de California. Aburto-Oropeza O, editor. 2010. pp. 1–80
20. Lluch-Cota SE, Aragón-Noriega EA, Arreguín-Sánchez F, Aurioles-Gamboa D, Jesús Bautista-Romero J, Brusca RC, et al. The Gulf of California: Review of ecosystem status and sustainability challenges. Progress in Oceanography. 2007;73: 1–26
21. Moreno-Báez M, López-Sagástegui C, Mascareñas-Osorio I, Erisman BE, Aburto-Oropeza O. Trackers Program. In: dataMares [Internet]. 1 Jan 2014 [cited 1 Jan 2015]. Available: http://datamares.ucsd.edu/eng/projects/trackers/
22. Moreno-Báez M. Mapping the Human Dimensions of Small-Scale Fisheries in the Northern Gulf of California. 2010
23. Corominas J. El SIG como herrramiientta de valoración del estado de las pesquerías del Golfo de California, México. 2012
24. Silverman BW. Density estimation in action. Density Estimation for Statistics and Data Analysis. Springer US; 1986. pp. 120–158
25. Erisman BE, Paredes GA, Plomozo-Lugo T, Cota-Nieto JJ, Hastings PA, Aburto-Oropeza O. Spatial structure of commercial marine fisheries in Northwest Mexico. ICES Journal of Marine Science. 2011;68: 564–571
26. Worm B, Hilborn R, Baum JK, Branch TA, Collie JS, Costello C, et al. Rebuilding Global Fisheries. Science. American Association for the Advancement of Science; 2009; 325:578–585
27. Bastardie F, Nielsen JR, Ulrich C, Egekvist J, Degel H. Detailed mapping of fishing effort and landings by coupling fishing logbooks with satellite-recorded vessel geo-location. Fisheries Research. Elsevier B.V; 2010;106: 41–53
28. Erisman BE, Paredes GA, Plomozo-Lugo T, Cota-Nieto JJ, Hastings PA, Aburto-Oropeza O. Spatial structure of commercial marine fisheries in Northwest Mexico. ICES Journal of Marine Science. 2011;68: 564–571
29. Aburto-Oropeza O, Ezcurra E, Moxley J, Sánchez-Rodríguez A, Mascareñas-Osorio I, Sánchez-Ortiz C, et al. A framework to assess the health of rocky reefs linking geomorphology, community assemblage, and fish biomass. Ecological Indicators. Elsevier Ltd; 2015;52: 353–361
30. Cisneros-Mata MA. The Importance of Fisheries in the Gulf of California and Ecosystem-Based Sustainable Co-Management for Conservation. In: Brusca RC, editor. The Gulf of California. 2010. pp. 119–135
31. FAO Yearbook—Fishery and Aquaculture Statistics 2012. 2014: 1–105

Index

S

W